Open Questions in Developmental Biology

This book explores the enduring and emerging questions in developmental biology, offering a comprehensive overview that bridges traditional concepts with the innovative inquiries shaping future research. Each chapter delves into a fundamental unresolved question reflecting the insights of the past and the likely vibrancy of future inquiry. These discussions are followed by reviews of novel areas, including mechanobiology, the role of membrane potentials, and the post-transcriptional control mechanisms guiding fate selection and cell differentiation.

Evolutionary Cell Biology

Published Titles

The Notochord: Development, Evolution and contributions to the vertebral column
Eckhard P. Witten and Brian K. Hall

Evolution of Neurosensory Cells and Systems: Gene regulation and cellular networks and processes
Edited by Bernd Fritzsch and Karen L. E. Thompson

The Evolution of Multicellularity
Edited by Matthew D. Herron, Peter Conlin, and William C. Ratcliff

Phenotypic Plasticity & Evolution: Causes, Consequences, Controversies
Edited By David W. Pfennig

Hox Modules in Evolution and Development
Edited By David E. K. Ferrier

Odontodes: The Developmental and Evolutionary Building Blocks of Dentitions
Edited By Donglei Chen

Evolutionary Cell Biology of the Fetal-Maternal Interface: Reflections on Evolutionary Innovation
Günter P. Wagner and Mihaela Pavlicev

Open Questions in Developmental Biology: Molecular, Cellular, and Evolutionary Perspectives
Edited by Igor Adameyko and Gerhard Schlosser

For more information about this series, please visit: www.crcpress.com/Evolutionary-Cell-Biology/book-series/CRCEVOCELBIO

Open Questions in Developmental Biology

Molecular, Cellular, and Evolutionary Perspectives

Edited by
Igor Adameyko
Gerhard Schlosser

CRC Press
Taylor & Francis Group
Boca Raton London New York

CRC Press is an imprint of the
Taylor & Francis Group, an **informa** business

Designed cover image: Igor Adameyko, Gerhard Schlosser, ChatGPT.

First edition published 2027
by CRC Press
2385 NW Executive Center Drive, Suite 320, Boca Raton FL 33431

and by CRC Press
4 Park Square, Milton Park, Abingdon, Oxon, OX14 4RN

CRC Press is an imprint of Taylor & Francis Group, LLC

ISBN: 978-1-032-90971-4 (hbk)
ISBN: 978-1-032-90972-1 (pbk)
ISBN: 978-1-003-56078-4 (ebk)

DOI: 10.1201/9781003560784

Typeset in Times
by SPi Technologies India Pvt Ltd (Straive)

Contents

SECTION IV Shedding New Light on Old Problems

Preface

Developmental biology is currently passing through a pivotal historical moment. On the one hand, it is a field enriched by unprecedented experimental power: single-cell transcriptomics, lineage tracing, high-resolution live imaging, and computational modeling now allow us to observe development at a level of detail unimaginable just two decades ago. These technologies provide access not only to the components of developmental systems but also to their dynamics, heterogeneity, and variability. On the other hand, many of the most fundamental questions remain unresolved. How do cells make decisions? How is multipotency realized, maintained, and progressively constrained? What defines a cell type in a continuously changing system? How do organisms reliably construct themselves despite noise, stochasticity, and perturbation? And how did these processes emerge and diversify during evolution?

These fundamental questions in developmental biology remain unresolved, likely because our ability to measure has outpaced our ability to interpret. Modern approaches provide highly detailed descriptions of molecular states and cell populations, yet development is a dynamic and causal process that cannot be fully understood from static or correlative data alone. At the same time, developmental systems operate across multiple scales – from genes to tissues – and most methods capture only fragments of this complexity, making integration into a coherent framework challenging. The persistence of the same old open questions, thus, mostly does not reflect a lack of data, but a need for new theoretical frameworks capable of organizing and explaining complexity. Developmental biology is thus entering a phase where conceptual innovation is as critical as technological progress.

While many classical questions in developmental biology remain currently unresolved, additional and fundamentally new questions appear. These emerge in part from the rapid development of neighboring scientific and technological fields – such as systems biology, information theory, software engineering, and artificial intelligence – which are increasingly influencing how we think about biological systems. Concepts from these disciplines prompt us to reconsider development not only as a collection of physical and biochemical processes but also as a form of organization capable of self-maintenance and information processing. This raises a series of profound questions: How do cells encode, store, and process information? What constitutes a functional unit of organization or information within a developing organism? How are genomic activity, combinatorial signaling inputs, bioelectrical states, and mechanical forces integrated into coherent decisions that guide developmental trajectories?

Historically, developmental biology has focused primarily on the "hardware" of life – the molecular components, cellular structures, and tissue architectures that underlie embryogenesis. Yet it is becoming increasingly clear that this perspective alone is insufficient. To understand development in a deeper sense, we must also begin to address its "software": the rules, logic, and dynamic processes by which biological systems interpret information and robustly generate coordinated outcomes. This shift does not simply require new tools, but a transformation in how we conceptualize living systems – moving toward frameworks that can accommodate distributed computation, emergent behavior, and multiscale integration.

It is precisely this recognition of a need for conceptual renewal that underlies the present volume. It is built on the conviction that further progress in the field will not come solely from the continued accumulation of data but will require us to confront these emerging and longstanding challenges through a change in scientific perspective. However, the chapters assembled here do not aim to provide a comprehensive or systematic overview. Instead, they are inevitably selective and subjective, focusing on areas where uncertainty, disagreement, or conceptual gaps persist. Each contribution highlights a different dimension of what remains unknown and, crucially, why these questions have proven resistant to resolution. Some chapters take a broad, conceptual, and even philosophical perspective, while others focus on well-defined problems explored through specific biological systems. Together, they form a mosaic of perspectives reflecting the current intellectual landscape of the field.

The chapters of the book are grouped into four thematic parts, each addressing a different level at which open questions in developmental biology emerge or persist.

Section I – "The Developing Organism as a Complex System" – argues for the need for a system-level understanding of development. Here, development is framed not as a linear execution of genetic instructions, but as a self-organizing, evolving system shaped by modularity, functionality, autonomy, and historical contingency. In his chapter, Schlosser revisits François Jacob's metaphor of evolution as a "tinkerer," arguing that developmental systems bear the marks of this evolutionary bricolage. He proposes that understanding development requires attention to how biological systems maintain and reproduce themselves and evolve under constraints, rather than being engineered for optimal design. Levin's contribution extends this perspective by introducing the idea of "biophysical intelligence," proposing that development is not simply gene-driven but emerges from distributed information-processing across cells and tissues. His framework bridges genotype and phenotype through bioelectric, mechanical, and collective properties of living systems, challenging gene-centric views and emphasizing agency and regulation at multiple levels. Together, these chapters highlight that many open questions arise because our current conceptual frameworks struggle to capture development as a complex, adaptive system.

Section II – "Computational Approaches to Development" – focuses on how new analytical and conceptual tools from the computational sciences are reshaping our understanding of developmental processes. The chapters by Adameyko and by Cornwall-Scoones and Briscoe converge on one of the central unresolved issues: how cells make decisions. Adameyko traces the evolution of ideas about cell fate, moving from deterministic hierarchies toward models in which cells operate within probabilistic and context-dependent landscapes shaped by noise, signaling, and epigenetic variability. He introduces the concept of collective multipotency, suggesting that fate decisions are not purely cell-autonomous but emerge from interactions within cell populations. In parallel, Cornwall-Scoones and Briscoe dissect gene regulatory networks as dynamical systems, emphasizing that understanding development requires not only mapping network components but quantifying how regulatory inputs are integrated at the level of cis-regulatory elements. They highlight emerging approaches – from massively parallel reporter assays to deep learning – that aim to uncover the logic by which gene expression patterns are computed. Together, these chapters illustrate a shift toward viewing development as a form of distributed computation embedded in space and time and underscore that many open questions persist not due to a lack of data but due to limitations in how we interpret and formalize it.

Section III – "The Cell as the Unit of Development" – returns to the biological foundation of the field, emphasizing the central role of the cell as both an autonomous unit and a participant in higher-order organization. Wagner argues that evolutionary developmental biology must be grounded in cell biology, proposing that many evolutionary innovations – such as placentation or tissue organization – can only be understood through cellular properties like metabolism, interaction, and functional integration. Mayor, in turn, revisits classical and enduring problems such as regulative development, scaling, competence, and robustness. He highlights that these questions, which were central to early embryology, remain largely unresolved despite molecular advances and calls for renewed integration of genetics, biophysics, and systems-level thinking with a focus on the interactions between cells. Both chapters argue that to fully understand the robustness and regulative capacities of development, we have to abandon the purely gene-centric approaches that have dominated developmental biology for many decades and have to refocus on the cell as the most elementary autonomous unit of the developing organism. This section underscores a key theme of the book: reductionist approaches, while indispensable, are insufficient on their own to explain how development operates as an integrated, self-regulating process.

Section IV – "Shedding New Light on Old Problems" – brings together perspectives that challenge established assumptions and reopen long-standing debates. Tanaka's chapter on regeneration synthesizes a complex body of work showing that cellular potential during regeneration is highly context-dependent and cannot be reduced to simple lineage rules. Her analysis reveals how embryonic heritage, tissue context, and injury-induced signals interact to shape regenerative outcomes. Levine and Treen, in turn, revisit the evolutionary origins of chordates with their "fallen fish" hypothesis, suggesting that tunicates may represent a secondarily simplified descendant of a more complex, vertebrate-like ancestor. Their interpretation challenges conventional evolutionary narratives and raises fundamental questions about how key vertebrate features such as neural crest and placodes emerged. These chapters exemplify how progress in developmental biology often arises not from incremental accumulation of new data but from rethinking foundational assumptions in light of new evidence.

The grouping of chapters thus proceeds from organism-level principles, through computational and regulatory frameworks, to cellular mechanisms, and finally to specific contexts where these ideas are tested, challenged, and reinterpreted. Across these levels, a unifying theme emerges: development is not a fixed program but a dynamic, context-dependent process involving interactions across scales – molecular-, cellular-, tissue-level – and shaped by evolution. But we still struggle to find the best conceptual approaches that allow us to untangle this complex web of interactions. Our book aims to guide the reader through this landscape of current uncertainty in developmental biology. Rather than presenting definitive answers, the chapters collectively map out where the key conceptual boundaries lie and where they are beginning to shift. In doing so, the book invites the reader not only to think about development but to reconsider how we formulate and approach questions about it.

The chapters of the book are, therefore, not intended to function as conventional reviews. We have explicitly encouraged all contributors to present their personal perspectives, interpretations, and, where appropriate, speculative ideas. As a result, the chapters reflect not only the current state of knowledge but also the authors' individual viewpoints on where the field stands and where it might go next. This approach inevitably introduces diversity in style and opinion, but it is precisely this diversity that gives the book its value. By allowing room for informed speculation and critical reflection, we aim to provide a more vivid and forward-looking picture of developmental biology – one that captures the uncertainties, debates, and emerging ideas that define the field today.

About the Editors

Igor Adameyko is Professor and a Department Chair at the Center for Brain Research of the Medical University of Vienna and a group leader at Karolinska Institutet in Stockholm. Prof. Adameyko is known for the discovery of the multipotency of nerve-associated Schwann cell precursors, a population of neural crest-derived cells with exceptional plasticity and distribution in the body. His studies introduced a radically new concept for developmental biology in that it defined precursor pools existing in a highly specialized niche that use nerves as conduits to migrate and differentiate through temporally and spatially delineated nerve–Schwann cell communication.

Gerhard Schlosser is Lecturer Above The Bar in the School of Biological and Chemical Sciences at the National University of Ireland in Galway. He has received two PhDs – the first in Philosophy from the Albert-Ludwigs-Universität Freiburg and the second in Biology from the University of Bremen. After two post-docs – one at the University of Bremen and another at the University of California in San Diego – he joined the faculty at the University of Galway, Ireland. He's the author or co-author of dozens of peer-reviewed journal articles.

Contributors

Igor Adameyko
Department of Neuroimmunology
Center for Brain Research
Medical University of Vienna
Vienna, Austria

and

Department of Physiology and Pharmacology
Karolinska Institutet
Stockholm, Sweden

James Briscoe
The Francis Crick Institute
London, United Kingdom

Jake Cornwall-Scoones
The Francis Crick Institute
London, United Kingdom

Michael Levin
Allen Discovery Center at Tufts University
Medford, Massachusetts, USA

and

Wyss Institute for Biologically Inspired Engineering
Harvard University
Boston, Massachusetts, USA

Michael Levine
Lewis-Sigler Institute for Integrative Genomics
Princeton University
Princeton, New Jersey, USA

and

Department of Molecular Biology
Princeton University
Princeton, New Jersey, USA

Roberto Mayor
Department of Cell and Developmental Biology
University College London
London, United Kingdom

and

Centre for Integrative Biology
Faculty of Sciences
Universidad Mayor
Santiago, Chile

Gerhard Schlosser
School of Biological and Chemical Sciences
University of Galway
Galway, Ireland

Elly M. Tanaka
Institute of Molecular Biotechnology
Austrian Academy of Sciences
Vienna, Austria

Nicholas Treen
Lewis-Sigler Institute for Integrative Genomics
Princeton University
Princeton, New Jersey

Günter P. Wagner
Alison Richard Professor
Emeritus
Department of Ecology and
Evolutionary Biology
Yale University
New Haven, Connecticut, USA

and

Department of Evolutionary Biology
University of Vienna
Vienna, Austria

and

Department of Animal Science and
Hagler Institute for Advanced Studies
Texas A&M University
College Station, Texas, USA

Section I

The Developing Organism as a Complex System

1 In the Tinkerer's Workshop

Structures, Functions and the Decomposition of Development

Gerhard Schlosser

1.1 INTRODUCTION

Since the early days of Experimental Embryology, we have made enormous progress in our understanding of developmental processes. Experimental Embryology identified important tissue interactions during development from the 1880s onwards. Following the discovery of the DNA structure in 1953 and the elucidation of basic mechanisms of transcription and translation in the 1950s and 1960s, developmental genetics began to unravel the molecular basis of developmental processes in the late 1960s. The evolutionary conservation of many developmental processes became increasingly clear in the 1980s, greatly boosting the fledgling discipline of evolutionary developmental biology (EvoDevo). With rapidly accumulating insights into the molecular components of developmental processes and the flood of data created by the new high-throughput sequencing technologies since the 1990s, the need for integrative approaches, for example involving network models and computational simulations and subsumed under the heading of "Systems Biology", has been increasingly recognized (Ideker et al., 2001; Noble, 2002; Kitano, 2002; Brigandt, 2013; Alon, 2020). This is a laudable project, which is currently driven very much by the need to develop tools to make sense of high-throughput data.

However, I will argue in this chapter that the remaining questions and challenges of developmental biology call for more attention to be paid to conceptual issues. Some of these have already been implicitly recognized by the early experimental biologists (see below) and have been explicitly discussed by earlier generations of system thinkers (e.g. Uexküll, 1928; von Bertalanffy, 1950; Bonner, 1974; Riedl, 1975; Varela et al., 1974) but have been buried under a heap of molecular details and a somewhat motley assortment of computer models since. More specifically, I wish to argue that our approaches to systems biology should be guided more by insights into the organization of living systems, in particular their autonomy, modularity and

DOI: 10.1201/9781003560784-2

evolvability. As pointed out by Francois Jacob in his famous essay, evolution has opportunistically re-assembled organisms by playing and tinkering with pre-existing parts while keeping them alive and thriving (Jacob, 1977). This process has left deep marks in the organization of living beings today, and I will argue that these can help us to guide our research.

To set the stage for my argument, I will first provide a very abbreviated historical sketch before outlining what I see as the main remaining challenges in developmental biology. In the following section, I will briefly discuss what to expect from an explanation in developmental biology. I will then highlight some of the major organizational properties of living and developing organisms, before sketching how insights into these properties may inform and benefit our approaches to explaining developmental processes.

1.2 FROM EXPERIMENTAL EMBRYOLOGY TO DEVELOPMENTAL GENETICS

In 1885, Wilhelm Roux, the father of experimental biology (or "Entwicklungsmechanik" as he called it), suggested that the key question of this new discipline was whether an organism or its parts develop autonomously ("Selbstdifferenzierung" – self-differentiation) or require interactions with other parts or the environment ("abhängige Differenzierung" – dependent differentiation) (Roux, 1885). As an adherent of August Weismann's germ plasm theory (Weismann, 1885), Roux believed that most embryos are mosaics of self-differentiating parts (with determinants for different organs already localized to different regions of the fertilized egg) and, therefore, favoured the first idea. To test this, he used a hot needle to destroy one of the first two blastomeres of a frog embryo (Roux, 1888). When the remaining blastomere formed only half an embryo, he took this as evidence in support of his view. Soon, however, this simple picture was shattered by accumulating evidence that many embryos are able to replace missing parts and to compensate for even major perturbations of development. This was first shown in sea urchins by Hans Driesch, where isolated early blastomeres formed complete pluteus larvae (Driesch, 1891).

These observations primed Driesch to leave science, convinced that mysterious and inexplicable vital forces were in operation, incompatible with known physico-chemical interactions. Others, however, in particular in the labs of Ross G. Harrison in the US and Hans Spemann in Germany, began to experimentally dissect the time and location of tissue interactions ("inductions") underlying the regulative capacities of developing embryos (reviewed in Maienschein, 2014; Schlosser, 2024). General arguments about whether development was dominated by self- or dependent differentiation were now replaced by attempts to identify the units of development in space and time.

Between 1900 and the 1930s, these research programs led to two important insights. First, it became evident that the development of most tissues or organs proceeds from dependent to self-differentiation: while initially interactions with surrounding tissues are required, these become dispensable at some stage (e.g. Spemann, 1905). Grafting a tissue primordium to a new location during the first stage prevented the formation of the tissue or organ, while grafting at the second stage did not.

The transition to self-differentiation was termed "determination" and was itself shown to proceed in a stepwise manner, from "labile" to "stable" determination (what we now call "specification" and "commitment", respectively) (Spemann, 1918; Spemann and Geinitz, 1927).

Second, it became clear that units of determination exist at different hierarchical levels and that processes of determination at these different levels are to some extent dissociable from each other. A tissue or organ may be determined at a certain point in time, while the determination of the fates and relative positions of its parts (i.e. the "patterning" of the tissue or organ) may and often do occur later. Moreover, patterning may result from purely intrinsic interactions in some tissues (and such tissues became known as "morphogenetic fields"; Spemann, 1936; Huxley and de Beer, 1934; Weiss, 1939; De Robertis et al., 1991) but may require extrinsic interactions with adjacent tissues in others.

An example of the former is the vertebrate limb bud. Experiments by Ross Harrison and others showed that such limb buds are already stably determined, as evidenced by the formation of relatively normal limbs after grafting salamander limb buds to a new position in the trunk (Harrison, 1907). Nevertheless, they are still able to form a properly patterned limb even after the removal of half a limb bud or the fusion of two limb buds, suggesting the existence of intrinsic patterning mechanisms ensuring the determination of properly arranged parts at a later stage (Harrison, 1918). We now know that intrinsic signalling centres in the apical ectodermal ridge (secreting fibroblast growth factors [FGFs)], zone of polarizing activity (secreting sonic hedgehog [Shh]) and dorsal limb epidermis (secreting wingless-integrated proteins [Wnts]) and their complex interactions are involved in the patterning of the limb bud along the proximo-distal, antero-posterior and dorso-ventral axis, respectively (reviewed in Zuniga, 2015; Delgado and Torres, 2017).

In contrast to the limb buds, other tissues require extrinsic interactions for their proper patterning. This can be illustrated by the vertebrate inner ear. Grafting experiments again demonstrated that the otic placode (the precursor of the otic vesicle, which forms the inner ear) is determined early, at neurula stages, in salamander embryos (Röhlich, 1929; Yntema, 1933, 1939). Like the limb bud, the otic placode is still able to compensate for missing or added parts at this stage (Kaan, 1927; Harrison, 1936). Experiments involving the rotation and re-implantation of the otic placode then showed that patterning of the otic placode/vesicle along the antero-posterior and dorso-ventral axis occurs at slightly later stages (Harrison, 1924, 1936). However, different from the limb bud, proper patterning did not occur in otic placodes positioned at any distance from their normal location. Although ectopic otic vesicles did form sensory areas, these were never arranged in a normal pattern and did not develop into the complex three-dimensional labyrinth of the inner ear with its various vestibular and auditory subdivisions (Kaan, 1927; Röhlich, 1929; Yntema, 1933, 1939). These findings indicated that the inner ear depends on interactions with neighbouring tissues for proper patterning, and we now know that retinoic acid signalling from posterior mesenchyme and FGF signalling from rhombomere 4 are crucial for its antero-posterior patterning, whereas dorso-ventral patterning relies on Sonic hedgehog from the ventral notochord and Wnt signalling from the dorsal hindbrain (reviewed in Groves and Fekete, 2012; Wu and Kelley, 2012). Even tissues that act as morphogenetic fields with intrinsic patterning, like the limb buds, are usually subject to some extrinsic patterning, which may control regional differences in how they and their parts develop (forelimb vs. hindlimb).

Whereas Experimental Embryology gained some important insights into the architecture of the developing embryo and its composition of spatiotemporally distinct units, it revealed very little about the molecular mechanisms underlying the behaviour of these units. The advent of developmental genetics in the wake of the molecular revolution of the 1960s then paved the way for new research questions (Judson, 1979; Crowe et al., 2015; Hopwood, 2019). It can now be determined which genes are expressed when and where in development, and perturbation of gene expression allows to study the role of particular genes for specific developmental processes (Barresi and Gilbert, 2023; Tickle et al., 2025). Loss of function (LOF) experiments involving the perturbation of gene expression via mutation of particular genes or the blockage of their transcription or translation allow to analyse whether expression of a particular gene is *required* for normal development. Increasingly sophisticated techniques to target these perturbations to defined times and locations in the developing embryo (e.g. the Cre-Lox technique of conditional mutagenesis in mice) help to establish the requirement of gene expression in a particular region and a specified time for a particular developmental process. Gain of function (GOF) experiments involving the overexpression of gene products (e.g. by injection or electroporation of DNA or mRNA) complement LOF approaches by showing that supplementation of a particular gene product may be *sufficient* to activate a particular developmental process under certain conditions (e.g. in a subset of tissues expressing all the other genes additionally required for the process during a limited stage of development).

1.3 CURRENT APPROACHES IN DEVELOPMENTAL BIOLOGY AND THEIR LIMITATIONS

Genes with potentially important roles for a developmental process are often identified by what is known as a "candidate gene approach" (Zhu and Zhao, 2007; Lawson and Wolfe, 2011). This involves first choosing promising candidates and then testing them in LOF studies, usually using a limited set of "markers" to test whether the process is compromised or not. Neither the choice of candidates nor the choice of "markers" in analysing the LOF phenotype is typically justified in much detail, but rather taken as a given, but these choices deserve to be looked at more carefully to help us appreciate what may be limiting our current understanding of development. The prioritization of candidates involves first to consider only those genes that are expressed in the right place at the right time. To further narrow down interesting candidates, typically only certain categories of genes (e.g. those encoding membrane proteins or transcription factors [TFs]) will be considered. And finally, a subset of these genes is chosen based on the known roles of related genes in other species or in other developmental processes of the same species. The choice of "markers" to determine the developmental outcome of LOF experiments is necessarily limited to a few morphological features and/or expression patterns of genes or proteins considered to be "representative" for the process.

While the candidate gene approach allows us to identify some of the components (e.g. genes expressed at a certain location and time) that are required for a certain developmental process, it has several weaknesses. The first major weakness is that it relies heavily on the correlation of expression and previous knowledge about some genes in identifying promising candidates, but apart from this depends on trial and error and provides little guidance in how to identify all the relevant components of the process as well as the patterns of interactions between them that are singly necessary and jointly sufficient for the process to proceed normally. The problem is exacerbated if there are genes with redundant functions or alternate developmental pathways that do not rely on the compromised gene and produce a normal developmental outcome in different ways. In such cases, single-gene LOF studies may not have an abnormal phenotype. Thus, essential roles of genes for certain developmental processes can be easily missed unless multiple genes are perturbed simultaneously. Similarly. establishing which interactions between components are important is usually challenging, since it requires careful experiments in which two or more genes are simultaneously perturbed in a combinatorial fashion.

Unbiased research strategies like forward genetic screens, genome-wide association studies (GWAS) or quantitative trait locus (QTL) analyses can help to identify a more complete set of genes required for a developmental process (Members-of-the-complex-trait-consortium, 2003; Cano-Gamez and Trynka, 2020). However, ultimately, these approaches suffer from similar limitations: Like the candidate gene approach, they are normally unable to detect important genes if these are backed up by genes and pathways with redundant functions, and they typically do not provide insights into the required patterns of interactions between essential genes (although correlative evidence for interactions may be obtained for very large sample sizes).

Current genomic approaches using high-throughput sequencing can alleviate some of these problems. In spite of present technical limitations (e.g. "dropouts" of genes with low expression levels), which will not be discussed here (Shapiro et al., 2013; Kharchenko, 2021), these techniques can offer a much more comprehensive picture of gene expression levels during various stages of a developmental process, in particular for single-cell sequencing. The clustering of cells based on similarity of gene expression, TF binding patterns, chromatin marks or genome accessibility, together with the pseudo-temporal ordering of cells, allows to generate complex hypotheses about which genes may play essential roles for a particular developmental process (e.g. a cell fate decision) and how they may interact with other genes (e.g. Briggs et al., 2018; Wagner et al., 2018; Soldatov et al., 2019). Many tools have already been developed to predict gene regulatory networks (GRNs) from high-throughput data, including the prediction of regulons – sets of genes co-regulated by a specific TF during a particular developmental transition – from co-expression patterns and predicted TF binding sites (Hecker et al., 2009; Aibar et al., 2017; Nguyen et al., 2021; Badia et al., 2023). However, since these tools rely on correlations observed in sequencing data, these hypothetical GRNs need to be validated by careful experimental LOF and GOF approaches. The new genomic approaches, therefore, offer an embarrassment of riches: While the prediction of gene regulatory interactions becomes easier and easier, experimental validation of these interactions remains as time- and labour-consuming as ever. This inevitably results in a rapidly growing rift between the number of predicted and validated components and interactions of developmental processes.

The first open question of developmental biology today, thus, is one of *completeness*: How do we close this ever-widening gap and how can we obtain a relatively complete explanation of all and only the relevant interactions between components of a particular development process? To address the problem of explanatory completeness, it has been proposed that analytic research strategies, which aim to decompose the developing organism, have to be complemented by synthetic, integrative approaches (Kitano, 2002; Bechtel, 2011; Villaverde and Banga, 2014). This is the goal of systems biology, which aims to use various empirically informed modelling approaches to reconstitute a given developmental process in silico, allowing quantitative predictions about process outcomes from different inputs. However, such modelling approaches usually require starkly simplifying assumptions and are limited to a few interacting components. The more interactions are to be considered and the more precise the dynamics are to be modelled (e.g. by differential equations rather than Boolean functions), the more computationally demanding and eventually intractable these models will become (Bolouri and Davidson, 2002; Hecker et al., 2009).

Whether a computational model can be considered an adequate representation of a developmental process also depends on our demands for how precisely we aim to predict its developmental outcomes. While simple models may be able to roughly predict the behaviour of the process under a constrained range of common inputs, more complex models may be required to precisely foretell the developmental outcomes under a broader range of input conditions. What is currently known as "Systems Biology" more or less subsumes all kinds of approaches that use computer models. What is missing is a general conceptual perspective on the organization of organisms and how this may help us to distinguish relevant from irrelevant components and interactions to be considered in our modelling approaches. I will argue below that to address this question of completeness, we should take a step back and – like the experimental embryologists – pay more attention to the architecture of a developing organism and how it can be decomposed into semi-autonomous units.

A second weakness of the candidate gene approach, and one that it also shares with the other research strategies discussed so far, is that there are no well-defined criteria for what comprises normal versus abnormal development. How different does the outcome of a developmental process after LOF of some gene have to be from the average outcome in control embryos to count as deficient or disrupted? In other words, given that there will always be some variability in the outcome of a developmental process between individuals, which range of outcomes is normal, and which range is not? This issue has moved centre stage with recent high-throughput genomic approaches, which have alerted us to the high degree of variability of gene expression profiles – or "microheterogeneity" as it has been called – between different individual cells of the same type (Kamenev et al., 2025; Adameyko, this volume). The existence of buffering mechanisms, such as negative feedback regulation and alternate developmental pathways, appears to generate stable (homeostatic) developmental outcomes, nevertheless. But this buffering capacity will have limits, raising the question of what the range of variability is that can be tolerated and still lead to a proper outcome, and when this range is exceeded and leads to abnormal development or disease. A "proper" outcome here should be recognized as one that is functional and supports the survival and reproduction of the developing organism.

The second open question of developmental biology today, thus, is one of *functional adequacy*: How do we distinguish normal from abnormal development and functional from dysfunctional states? Again, I will argue that we need to reconsider the architecture of developing organisms to address this question. To resolve questions on functionality and normality, we particularly need to take the self-maintaining and self-reproducing nature of living systems into account.

A final, third weakness of all current approaches in developmental biology is their exclusive focus on the genetic drivers of developmental processes. Considering only genes and their interactions (as represented in GRNs) ignores the important roles that interactions of proteins with other proteins, membranes or metabolites, posttranslational modifications of proteins and their responses to chemical and mechanical forces play for many developmental processes (e.g. Jenuwein and Allis, 2001; Nelson and Bissell, 2006; Heisenberg and Bellaiche, 2013; Mammoto et al., 2013; Snider et al., 2015; Lobingier et al., 2017; Yadav et al., 2018). Only when such epigenetic interactions (in the widest sense) are included can we hope to obtain a reasonably complete understanding of functional developmental processes. This will be especially important for our understanding of morphogenetic processes that guide the formation of the three-dimensional shape of the embryo (Heisenberg and Bellaiche, 2013; Davies, 2023; Barresi and Gilbert, 2023; Tickle et al., 2025).

Thus, the third major open question of developmental biology today is one that we may call *epigenetic adequacy*: How can we expand our gene-centred explanations to include other, epigenetic interactions? This is an important question that will be addressed in several other chapters of this book. Here, however, I will focus on the first and second questions introduced above, which will be of central importance regardless of whether we consider interactions between genes, proteins or any other components of living systems. Before presenting my argument for the need for a new developmental systems biology that more explicitly considers the organization of living systems, I will illustrate our current approaches and their limitations with one specific example. I will stick with the inner ear and focus on the differentiation of sensory hair cells.

1.4 DISSECTING THE DEVELOPMENTAL PROCESS OF HAIR CELL DIFFERENTIATION

Hair cells are mechanosensory cells residing in several sensory areas of the inner ear of vertebrates (Figure 1.1 A, B), with some of these areas involved in maintaining body orientation and balance (vestibular part of the ear) and others in hearing (auditory part). In aquatic vertebrates, hair cells are also found in the lateral line system on the skin, which is able to detect water movements (reviewed in Coffin et al., 2005; Goodyear et al., 2006; Fritzsch et al., 2007; McPherson, 2018; Schlosser, 2021a). On their apical side (facing the fluid-filled cavity in the inner ear or the environment in the lateral line), hair cells are equipped with one cilium and a bundle of microvilli (Figure 1.1 B). Bending of the microvilli by mechanical stimuli opens ion channels,

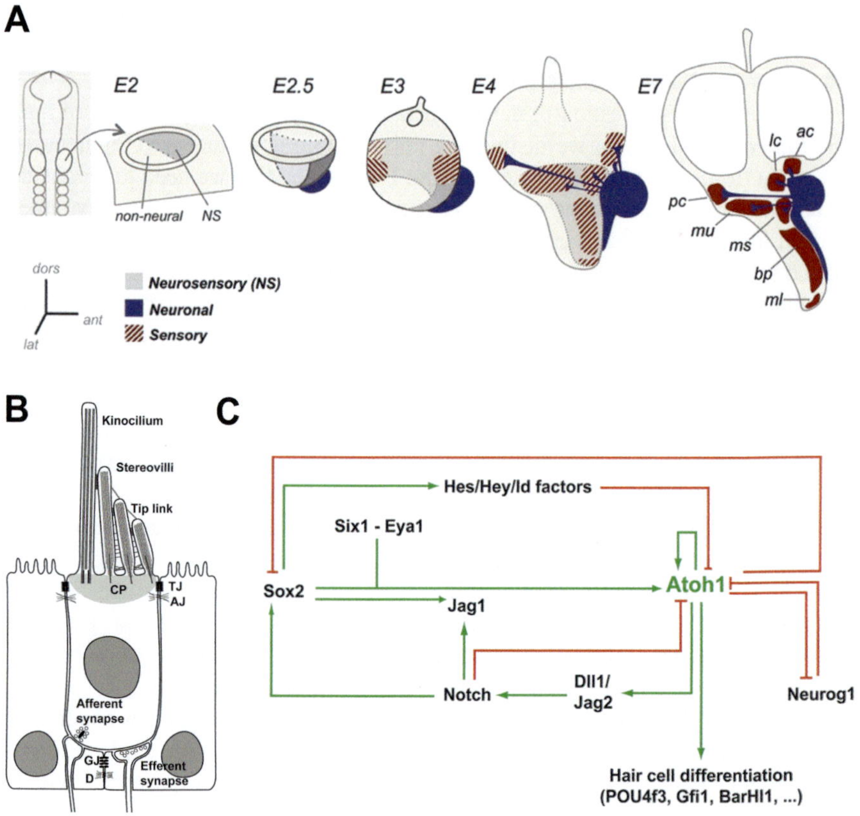

FIGURE 1.1 Differentiation of hair cells in the vertebrate inner ear. A: Development of the inner ear in the chick from embryonic days (E) 2–7. The neurosensory (NS) domain is established early, before invagination. It subsequently gives rise to the hair cells and supporting cells of the sensory patches (red) and the neurons of the vestibulocochlear ganglion (blue) connecting the ear to the brain. **B**: Diagram of a vertebrate hair cell. AJ, adherens junction; CP, cuticular plate; D, desmosome; GJ, gap junction; TJ, tight junction. **C**: Central role of Atoh1 in the core regulatory network regulating differentiation of vertebrate hair cells.

(A: Reprinted with permission from Neves et al., 2013; B: Reprinted from Schlosser, 2021a, after Goodyear et al., 2006; C: Reprinted from Schlosser, 2021a).

leading to the depolarization of the hair cell and the release of synaptic vesicles from the opposite, basal side. This activates sensory neurons, which transmit a signal to the brain. Both the hair cells and the sensory neurons of the inner ear originate from common neurosensory areas of the embryonic otic vesicle (Figure 1.1 A) (Alsina et al., 2009; Wu and Kelley, 2012; Maier et al., 2014; Whitfield, 2015). In mammals, hair cells cannot be replaced when damaged or lost in the adult, leading to irreversible hearing loss. In contrast, many other vertebrates can regenerate hair cells throughout life (Brignull et al., 2009; Groves, 2010; Groves et al., 2013; Atkinson et al., 2015). Understanding the mechanisms of hair cell differentiation and regeneration and how they differ between mammals and other vertebrates has, therefore, attracted a lot of attention.

Several genes with important roles for hair cell differentiation were identified by a candidate gene approach, prioritizing genes that (1) were expressed in neurosensory areas and subsequently in hair cells, (2) encoded TFs and (3) belonged to the basic helix loop helix (bHLH) family of TFs. The latter were considered to be particularly promising because bHLH genes were already known to play a central role in the specification of other cell types, such as MyoD for vertebrate muscle cells (Weintraub et al., 1989; Berkes and Tapscott, 2005). Furthermore, homologs of some vertebrate bHLH genes, such as Atoh1 and Ascl, had previously been shown to play central roles in the specification of sensory neurons and sensory cells in insects and other invertebrates (Ghysen and Dambly-Chaudière, 1989; Jarman et al., 1993; Zhao and Emmons, 1995; Bertrand et al., 2002; Huang et al., 2014). The bHLH gene encoding the TF Atoh1, which is expressed in the neurosensory areas and hair cells of the inner ear, was, consequently, chosen as one of the prime candidates to be tested in LOF studies (Bermingham et al., 1999).

Indeed, an LOF mutation in the mouse Atoh1 gene led to a complete loss of hair cells as judged by the absence of cells with the typical morphology of hair cells in scanning electron micrographs and loss of immunostaining for the hair cell marker proteins myosin VI and calretinin (Bermingham et al., 1999), indicating that Atoh1 expression in the inner ear is *necessary* for hair cell differentiation (Figure 1.1 C). This was later confirmed by specifically knocking out Atoh1 in the developing inner ear in mice, in which the Cre recombinase introducing an LOF mutation in the loxP-flanked Atoh1 locus is expressed only in the inner ear under the control of an ear-specific Pax2 promoter (Pan et al., 2011). LOF experiments in zebrafish, in which translation of Atoh1 mRNA was blocked (using Morpholino antisense oligonucleotides), led to similar hair cell loss, indicating a central role of Atoh1 for hair cell differentiation throughout vertebrates (Millimaki et al., 2007). However, due to gene duplication in teleosts, two Atoh1 genes exist in zebrafish (*atoh1a*, *atoh1b*) with largely redundant functions. Blocking translation of only one of them results in only minor loss of hair cells, and only when both genes are targeted are hair cells completely lost (Millimaki et al., 2007). This illustrates the problem of single-gene LOF studies in recognizing essential genes, if there are genes with redundant functions.

The importance of Atoh1 for hair cell differentiation was confirmed in GOF studies, in which ectopic hair cells were observed after Atoh1 overexpression in other cell types of the inner ear or immediately neighbouring ectoderm (e.g. Zheng and Gao, 2000; Millimaki et al., 2007). These studies established that activation of Atoh1 is *sufficient* to promote hair cell differentiation but only in very specific tissues. Taken together, these LOF and GOF studies confirm that Atoh1 expression is one of the necessary factors for hair cell differentiation, but also suggest that it needs to interact with other (additionally necessary) factors to fulfil that role. The studies also point to the region where these other factors are expressed, but otherwise leave us in the dark about their nature and how they interact with Atoh1. Candidate gene studies in mice have identified a few of these additional factors. LOF of several other genes expressed in the inner ear and encoding TFs, in particular POU4f3 and Gfi1, showed severe hair cell loss, suggesting that these are also required for hair cell differentiation

(Erkman et al., 1996; Wallis et al., 2003). Subsequent studies have begun to elucidate how these TFs interact with Atoh1. They have shown that both POU4f3 and Gfi1 are (directly and/or indirectly) transcriptionally activated by Atoh1 (Figure 1.1 C) and that Atoh1 can efficiently induce hair cell fate in vitro or in juvenile mice only in combination with POU4f3 and Gfi1 TFs (Costa and Henrique, 2015; Costa et al., 2022; Iyer et al., 2022). The latter TFs appear to cooperate with Atoh1 in multiple pathways, including a repression of neuronal genes (Gfi1) and the opening of chromatin (POU4f3) to allow Atoh1 binding (Matern et al., 2020; Yu et al., 2021).

Notably, additional LOF and GOF studies have shown that Atoh1 is also required for the differentiation of other cell types, including the mechanoreceptive Merkel cell in the skin (Van Keymeulen et al., 2009; Maricich et al., 2009), and is able to promote differentiation of supernumerary Merkel cells specifically in epidermal cells (Ostrowski et al., 2015). These findings further highlight the context-dependent role of Atoh1 and indicate that it needs to interact with different factors to promote hair cell differentiation than to promote Merkel cell differentiation. Since POU4f3 and Gfi1 are also co-expressed and presumably cooperate with Atoh1 during the development of Merkel cells in the epidermis (Wallis et al., 2003; Masuda et al., 2011; Jarvis et al., 2025; Yu et al., 2021), they also suggest that additional genes with expression in the inner ear but not the epidermis must be required for hair cell differentiation. One of the promising candidates is the gene encoding the TF Pax2, since it is expressed in the inner ear, Pax2 LOF leads to deficits (although quite mild) in hair cell differentiation, and Pax2 has been shown to synergize with Atoh1 in promoting hair cell formation (Favor et al., 1996; Chen et al., 2013). Several further genes (e.g. Eya1, Six1, Sox2, GATA3, Foxg1, Lmx1a) have been implicated in hair cell differentiation, but whether and how they need to interact with Atoh1 in that process is currently unknown (Schlosser, 2021a; Elliott et al., 2021; Iyer and Groves, 2021).

Taken together, these findings illustrate the first major limitation of the candidate gene approach: its struggle in identifying a complete set of essential components and interactions underlying a particular developmental process. While unbiased approaches such as GWAS studies of hearing deficits have helped to identify a few additional candidates for genes involved in hair cell differentiation and/or function (e.g. Girotto et al., 2011; Trpchevska et al., 2022; Ahmed et al., 2025), they are equally silent regarding the required patterns of interactions between them. In recent years, hair cell differentiation and regeneration have been studied in many vertebrates using high-throughput sequencing approaches (e.g. Kolla et al., 2020; Jan et al., 2021; Baek et al., 2022; Iyer et al., 2022). These studies have confirmed the upregulation of many of the previously known central players like Atoh1, POU4f3 and Gfi1 during these processes and have identified multiple additional co-regulated genes, without, however, revealing any major new factors of central importance for this process that have been validated in functional studies.

Upregulation of Atoh1, POU4f3 and Gfi1 can convert some non-sensory cells of the mouse cochlea into hair cells even in 1-week-old juvenile mice (Iyer et al., 2022). These cells look relatively normal morphologically and cluster with normal hair cells in single-cell RNA-Seq studies, but whether they indeed can function as hair cells that allow sound perception in vivo has not been determined by electrophysiological or audiological studies. This illustrates the second major limitation of our current research strategies: the difficulty in determining normal and functional states and distinguishing them from those that are abnormal and dysfunctional. This deficit has also hampered our ability to identify factors that interfere with hair cell regeneration in the postnatal mammalian ear or factors that facilitate such regenerative processes in other vertebrates.

The current limitations in identifying all relevant but only the relevant factors for explaining a developmental process, such as hair cell differentiation, and in distinguishing normal from abnormal development suggest that our experimental approaches could benefit from more conceptual guidance. Before I elaborate further, let me briefly discuss what we should expect from explanations in developmental biology.

1.5 TOWARDS A NEW DEVELOPMENTAL SYSTEMS BIOLOGY

Explanation in developmental biology (as in other biological sciences) aims to reveal the mechanisms underlying developmental processes (Machamer et al., 2000; Bechtel, 2011; Brigandt, 2013; Bechtel and Richardson, 2010; Craver and Darden, 2013). "Mechanisms" here is used very broadly (and not in a purely mechanical sense) to refer to underlying structures (including pathways or networks of interacting elements) that have the capacity to transform a certain initial state into a certain end state, for example, a pattern of gene expression or cellular interactions at an earlier stage of development into another pattern at a later stage. Elucidating the mechanisms for a developmental process usually involves the decomposition of the process into lower-level components. A complete mechanistic explanation specifies the structure and interactions (connectivity) of all relevant but only the relevant components for the process to proceed (Craver and Kaplan, 2020), that is, all components that are singly necessary and jointly sufficient for the process to occur. This does not imply that all necessary components must be broken down to the molecular level (Craver and Kaplan, 2020). Often, only some of these factors can be specified in molecular detail, while others may be acknowledged as higher-level background conditions, potentially awaiting further dissection into molecular components and interactions.

In our example of hair cell differentiation, the transcriptional regulatory relationship of Atoh1, POU4f3 and Gfi1 and some of their target genes can be specified in some detail, while the genetic background of inner ear cells (its gene expression and genomic accessibility status), which is also required for allowing hair cell differentiation in response to Atoh1, POU4f3 and Gfi1 activation, has to be acknowledged as background condition. Because of the prevalence of feedback circuits and non-linear interactions in developmental and other biological processes, their dynamic behaviour is often counter-intuitive and may be predictable only with computational models of the hypothesized mechanisms (Bechtel, 2011; Brigandt, 2013).

Whereas the centrality of this mechanistic type of explanation in developmental biology is widely recognized by both researchers and philosophers of biology, it is often overlooked that such explanations make best sense if our decomposition strategies follow the natural fault lines of biological organization and "carve nature at its joints" (Plato). Following the suggestion of Willard van Orman Quine and others, I maintain here that such "natural kinds" have to be defined in relation to a process, in which they act as a unit (Quine, 1969; Wagner, 1996).

Our research strategies into the decomposition of developing organisms should therefore be guided by insights into the peculiarities of the organization of living systems. Only for developmental processes that proceed normally by the interaction of a limited set of components and are relatively independent of other developmental processes and thus act as quasi-independent units (modules) of development, can we strive to provide a reasonably complete mechanistic explanation. Only developmental processes that contribute to the self-maintenance and self-reproduction of the organism can be considered to be functional and proceed normally. And only developmental processes that proceed relatively independently of other developmental processes and are functionally separable from them in the face of accumulating heritable mutations persist as units in an evolutionary lineage and, thus, comprise evolutionarily stable units of development (modules or units of evolution) (Schlosser, 2002, 2004, 2023; Wagner, 1995, 2014). The latter not only can be traced as quasi-independent units through phylogeny, each with its own evolutionary history, decoupled from the history of other units, but also serve as building blocks to be redeployed and recombined into new units (novelties) during evolution.

To see this more clearly, I will now present a very abbreviated sketch of what I consider the core features of the architecture of living and developing organisms. Explicit recognition of these features should help us to cast systems biology in a new mould that allows us to address the open questions of developmental biology in a more rigorous conceptual framework. I have developed these ideas in much more detail elsewhere and refer the interested reader to these other publications for more information (Schlosser, 1993, 1998, 2002, 2007, 2023).

1.6 THE ARCHITECTURE OF DEVELOPING ORGANISMS

It has long been recognized that organisms are processes that exhibit a peculiar dynamic stability that is maintained by complex patterns of recurrent interactions among their components. Only those parts of an organism that are necessary under at least some conditions to ensure the maintenance (recurrence) of the process that we call "organism" will be considered as its components here (I have termed such constitutive elements "constituents" elsewhere: Schlosser, 1993, 2002, 2007). Such components can be identified at different hierarchical levels (e.g. genes, cells, organs). As parts of the process "organism", its components have to be conceptualized as processes themselves, that is, as a set of input–output transformations or types of activities at certain points in space and time. The components of developmental processes are not genes (e.g. the gene *Atoh1*), proteins, cells or tissues, but their activities at specific locations during particular stages (e.g. *Atoh1* activity in the embryonic inner ear) of development.

However, these activities depend on structures (component structures) with particular capacities to process inputs into outputs and to interact (Figure 1.2). A gene with binding sites for certain TFs in its enhancers has the capacity to be transcriptionally regulated by the binding of these TFs, and the structure of the protein encoded by the gene in turn determines its capacity to transcriptionally regulate other genes.

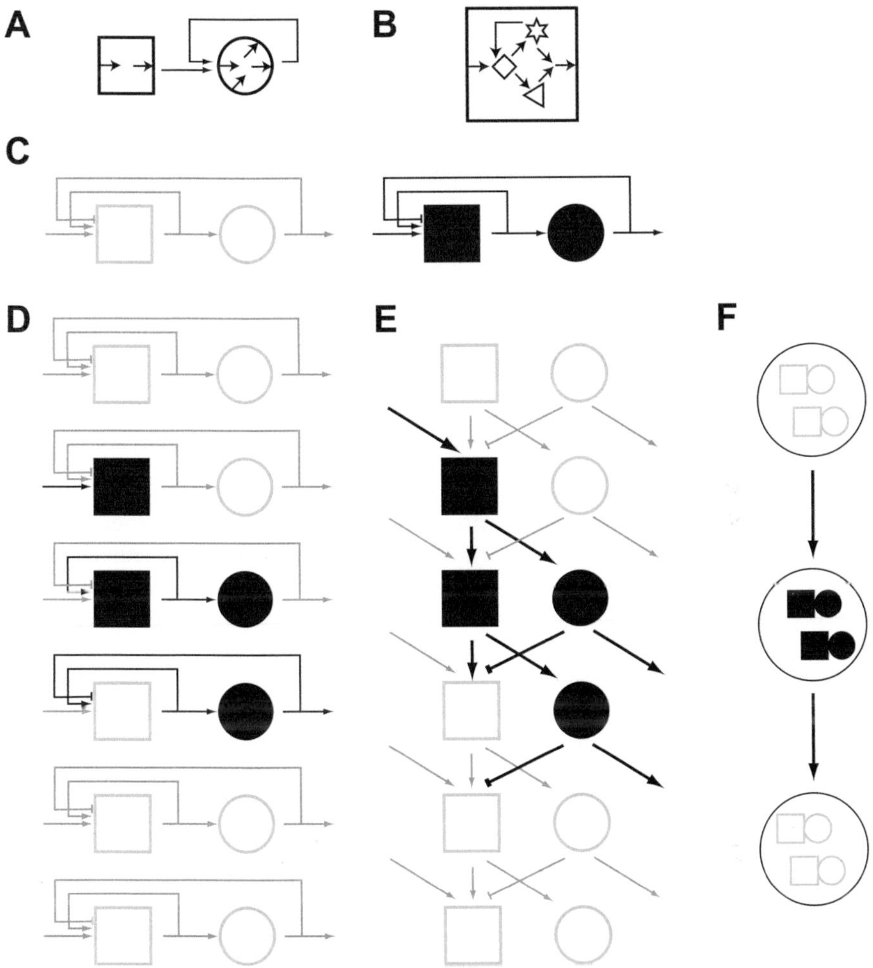

FIGURE 1.2 Developmental processes as activities of networks of structures with certain capacities. A: Schematic representation of structures (geometric symbols) with particular capacities (arrows inside symbols). Such structures underlie the activities of components of a system. The capacities of a structure embody a rule of transformation, which determines the future state of activity and output of the structure as a function (in the mathematical sense) of its present state and input. These structures can be linked with each other (connecting inputs and outputs of structures; arrows between symbols) to form networks. **B**: Structures may form a hierarchy so that the capacity of a higher-level structure (square) can be decomposed into a network of interactions between lower-level structures (small symbols). **C**: Example of a simple network with activating (arrows) and inhibitory (lines with bars) interactions shown in an inactive (left) and active (right) state. **D**: Diagram illustrating changes in network activity at discrete time steps (time axis from top to bottom). An activating input to the square at the second time step is processed based on the interaction capacities of the network structures and their connectivity. **E**: A simplified representation of the same network behaviour (grey arrows and symbols: inactive network links and structures; black arrows: active network links and structures). **F**: The network behaviour shown in D and E can also be represented as a transition between network states, where each state is a set (circle) of network activity patterns (shown as groups of symbols).

The interaction capacity of the gene, which is embodied in its structure, therefore determines its future state of activity and its output (contribution to the regulation of other genes) as a function (in the mathematical sense) of its present state and input (Figure 1.2 A, B). The interaction capacities of structures allow them to be linked together in interaction networks (e.g. GRNs), whose behaviour is determined by the interaction capacity of the individual structures, their connectivity and the initial state of activity of the network and any of its inputs (Figure 1.2 C–E). Elucidating the mechanism of a developmental process is to determine the network of all singly necessary and jointly sufficient component structures and their interactions that will explain its proper behaviour (the functional subset of input–output activity transformations). Acknowledging the organization of living systems can help us to focus on the relevant components and interactions in this quest.

1.6.1 Self-Maintenance, Self-Reproduction and Functionality

Living systems can only live in a range of supporting environments because they require recurring interactions with environmental factors (uptake of nutrients, avoidance of predators, etc.) to survive. The components of a living system including both its internal components and essential environmental interactions are generated in a complex network of collectively autocatalytic interactions involving each of these components (Varela et al., 1974: Oyama, 1985; Schlosser, 1993, 1998, 2002, 1993, 2002; Griffiths and Gray, 1994; Fontana and Buss, 1994a; Szathmary, 1995; Szathmary and Maynard Smith, 1997). Metaphorically speaking, living systems are working factories engaged in manufacturing their own machines and collecting all necessary supplies. Importantly, such autocatalytic networks involve not only the recurrence of patterns of interactions but also the multiplicative construction of components (e.g. expression patterns of the same gene in different tissues; the differentiation of cells of the same type, like hair cells, in different tissues, such as the inner ear and lateral line organs). Each of those components will depend on a structure with the same capacities, but may potentially engage in different interactions (Figure 1.3 A, B). Systems with this kind of organization have been called autopoietic (Varela et al., 1974; Maturana, 1975), self-maintaining (Fontana and Buss, 1994a; Fontana and Buss, 1994b) or complex self-reproducing systems (Schlosser, 1993, 1998).

"Re-production" with "-" here refers to the general phenomenon of recurrent production of some state (i.e. self-maintenance) rather than to the multiplicative generation of progeny ("reproduction" without "-"). The ability to reproduce, however, is another feature of living systems. In addition to the collectively autocatalytic reproduction of the components of an organism, it involves the construction of more organisms of the same kind (Szathmary, 1995; Maynard Smith and Szathmary, 1995; Szathmary and Maynard Smith, 1995, 1997; Griesemer, 2000; Schlosser, 2002), that is, a reproduction (and possibly multiplication) of the entire "life cycle" (Bonner, 1974) (Figure 1.3 A, C). In multicellular organisms like animals or plants, reproduction involves development, that is, the reconstruction of a complex adult with many different cell types arranged in well-patterned and well-shaped tissues and organs from a much simpler and often unicellular propagule (e.g. the zygote formed by fusion of male and female gametes).

Self-maintenance and self-reproduction ensure the survival of living systems via a quasi-cyclical generation of states and processes, including the re-enactment of developmental processes in subsequent generations (Figure 1.3 A–C). Only organismal processes (including developmental processes) that proceed in a manner that supports self-maintenance and reproduction, and consequently their own recurrence, can be considered components of the organism. We call such processes "functional" and their outcomes "functions" (Schlosser, 1998). Developmental (and other) biologists often refer to functions whenever they talk about the effect of some process, losing sight of the fact that not all effects are functions and that functions are firmly grounded in the quasi-cyclical organization of living systems. Only when this is recognized can we productively use functionality as a benchmark to distinguish normal from abnormal developmental processes.

When we talk about a particular developmental process, we refer to a set of activities or input–output transformations that take place relatively (but not absolutely!) invariably in different individuals of the same species or of larger taxonomic groups. Hair cell differentiation is an example of such a process. It takes place in all individual vertebrates but will not proceed in exactly the same fashion in all of them.

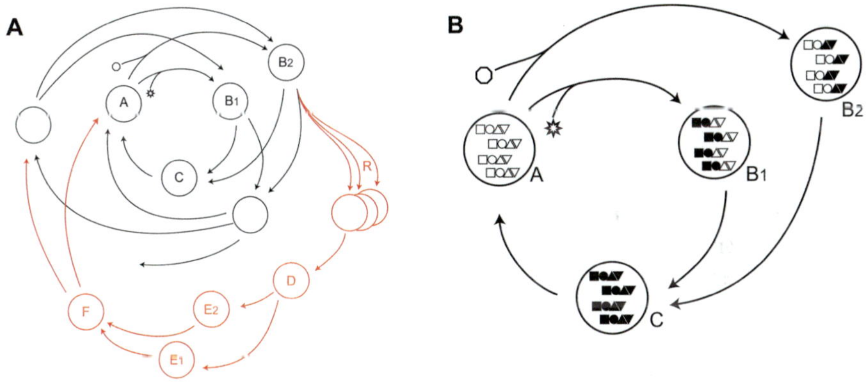

FIGURE 1.3 Self-maintenance, self-reproduction and functionality. A: Schematic depiction of a self-maintaining system as a quasi-cyclical system of state transitions. Each state is depicted as a circle representing a set of individual system states (defined by specific activity patterns of component structures and their interactions). Some state transitions depend on interaction with external factors (e.g. star and polygon). Functions are outcomes of states (subsequent states) that are part of a self-reproducing cycle of state transitions (e.g. B_1 and B_2 are functions of A, being part of the cycle A-B_1/B_2-C-A). Multiplicative reproduction (R) of the system and re-establishment of a mature (adult) organization by development is indicated by the red circle and arrows. Plasticity is indicated by alternative state transitions, for example, A-B_1-C in the presence of one type of conditions (star) or A-B_2-C in the presence of another type (polygon) (similarly D-E_1-F or D-E_2-F). **B**: State transition A-B_1/B_2-C shown in more detail with network activity patterns shown schematically for each set as groups of symbols (here considered to be invariant for each set; see D for a more realistic picture). It is assumed that state transition A-B_1-C is mediated by one modular unit of components (square and circle) and A-B_2-C by another unit (triangles).

(Continued)

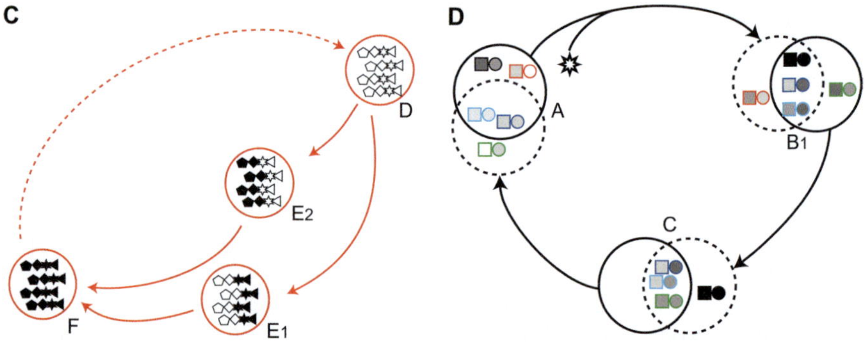

FIGURE 1.3 (Continued) **C**: State transition D-E_1/E_2-F shown in more detail with network activity patterns shown schematically for each set as groups of symbols assuming that state transition D-E_1-F is mediated by one unit of components (star and triangle) and D-E_2-F by another unit (polygon and rhombus). **D**: State transition A-B_1-C shown in more detail, taking into account the variability of network activity patterns for each state. To simplify, activity patterns (shades of grey) for only two components (square and circle) are shown and coloured outlines are used to trace changes of particular activity patterns through state transitions. Each state transition is a transformation of a set of initial activity patterns (circle with solid outline) into a set of final activity patterns (circle with broken outline. However, only a subset of those state transitions are functional, viz., those that allow re-production of a state (i.e. survival of the system and re-establishment of the state in this or following generations). Such state transitions have to be between activity patterns that occupy the intersections between solid and broken circles for each transition of the functional cycle (blue outlines). Other activity patterns may be able to be generated during state transition A-B_1 but fail to support subsequent state transitions of the functional cycle (red outlines) or may support a subsequent state transition but cannot be generated during state transition A-B_1 (green outlines).

Ignoring interspecific differences here (hair cell differentiation in mice will be somewhat different from hair cell differentiation in zebrafish), there will even be some differences between individuals of the same species, in particular between healthy mice and mutant mice that are born deaf (i.e. there will be subsets of the process for different genotypes with different fitness). To gain insight into normal development, we need to elucidate the mechanism of hair cell development for the functional subset of the process and, therefore, should be able to distinguish functional (proper) from non-functional development (Figure 1.3 D). Because the functionality of one process depends on the functionality of all the processes concatenated to each other in the self-maintaining and self-reproducing organism, this may be difficult in practice. However, to begin to determine whether a developmental process is functioning properly, it is already useful to consider if the outcome of a developmental process would support an immediate follow up functional process (e.g. if hair cell differentiation supports synapse formation to sensory neurons projecting axons to the auditory centres of the brainstem; proper outgrowth of these axons in turn would be the precondition for proper auditory information processing and hearing, which allows appropriate behavioural responses to environmental stimuli supporting survival and reproduction).

1.6.2 Modularity

The developing organism is not an irreducibly complex system in which all components intricately interact with each other all the time. Instead, it is composed of quasi-independent building blocks or modules. Modules are processes that once initiated show relatively stable behaviour due to the integrated interaction of their components, while acting relatively independently of other components of the containing system, such as the developing organism (Raff, 1996; Wagner, 1996; Wagner and Altenberg, 1996; Gerhart and Kirschner, 1997; Schlosser, 2002, 2004).

I have previously argued that whether some process acts as a module can only be determined relative to some containing process (e.g. development of an individual organism or evolution of a genealogical lineage) and class of perturbations (e.g. environmental changes or mutations) (Schlosser, 2002, 2004). Modules or units of development are those developmental processes that develop relatively independently of each other when challenged by developmental perturbations (due to environmental factors or mutations in somatic cells) (Figure 1.4). Modules or units of evolution are those developmental processes that evolve relatively independently of each other when challenged by heritable variations (mutations in germ line cells) (Figure 1.4). Only processes that have at least a partly different genetic basis and make relatively independent fitness contributions can act as modules of evolution (Schlosser, 2002, 2004; Wagner, 1995). Modules of evolution, therefore, serve as evolutionarily stable individualized characters that evolve in a relatively integrated manner but independently of other modules of evolution, resulting in "mosaic evolution" (see also Wagner, 1989, 1996, 2014; Müller, 2003, 2010; Schlosser, 2002, 2023).

It has been proposed that animals and other multicellular organisms have a character architecture comprising a core of components connected by mutually stabilizing interactions and a periphery of components linked to the core but dispensable for core activity. Such an architecture facilitates the evolutionary stability of characters (conserved core activity) while permitting changes of character states (altered links to the periphery; Figure 1.5) (Davidson and Erwin, 2006; Peter and Davidson, 2015; Wagner, 2007, 2014; DiFrisco et al., 2020; McKenna et al., 2021). The core regulatory networks (CoRNs) underlying the stability of characters (developmental processes) were initially proposed to be formed by core networks of regulatory genes, termed "kernels" (Davidson and Erwin, 2006) or "character identity networks" (ChINs) (Wagner, 2007, 2014), with the lability of character states reflecting the malleable links of this core to "differentiation gene batteries" or "realizer genes". This model has been recently extended by acknowledging that intercellular signalling, cell adhesion and other pattern formation mechanisms may cooperate with cross-regulating TFs as part of CoRNs at different levels of the biological hierarchy, thereby forming "character identity mechanisms" (ChIMs) or "character identity modules" (ChIMos) (DiFrisco et al., 2020; McKenna et al., 2021). Here, I will use the generic term "Core Regulatory Network" (see also Schlosser, 2021a, b, 2023) to refer to such networks underlying character stability (Figure 1.5).

While modules of development in the sense introduced here, such as the developing limb bud or the differentiating hair cell, typically also serve as modules of

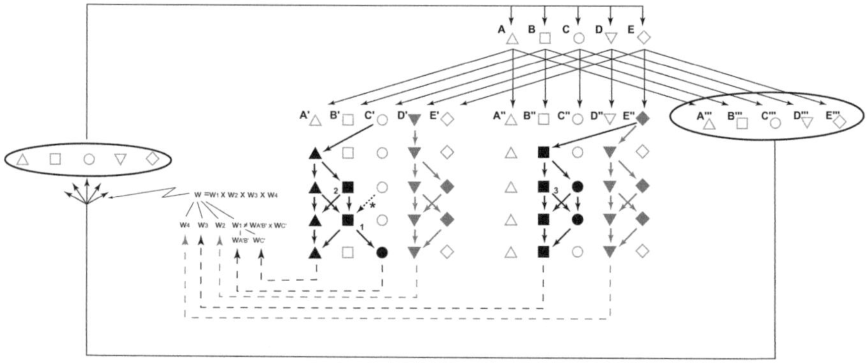

FIGURE 1.4 Modules of development and modules of evolution in a self-reproducing system. The figure schematically depicts the life cycle of a reproducing system (only active interactions depicted; environmental interactions not shown for simplicity). A–E (symbols) indicate various types of components of the life cycle that are multiplicated (thin arrows) during the life cycle (A'–E'; A"–E", A'''–E'''). For example, A–E may represent different genes, which after cell division are present in multiple cell types of a multicellular organism, one of which may represent a germ line cell (ellipse incorporating A'''–E''') and the others somatic cells (incorporating A'–E' and A"–E", respectively). Gene activity patterns in different cell types may differ, as indicated. The initial differences in gene activity patterns among various cells may either be established via some stochastic or environmentally induced symmetry break or via asymmetries in the activity of non-genetic components reproduced via germ cells (not shown). Effects of system activities on fitness are indicated by hatched lines. The fitness w measures the rate of reproduction and is composed of several independent (multiplicative) fitness contributions (w_1 to w_4). Components that behave in an integrated but quasi-independent manner during development (produce integrated and context-insensitive input–output transformations) form developmental modules. Three types of such developmental modules are illustrated: the interactions between D and E (instantiated in two domains D'–E' and D"–E"), the interactions between A' and B' and the interactions between B" and C". Note that A'–B' will only qualify as a module if it is relatively context-insensitive, that is, when the input from C' (asterisk) is weak. However, it may have strong outputs (1) on C' that may be indispensable for the activity of the latter. Developmental modules with such indispensable effects on other components of the system may not act as modules of evolutionary transformations, because they are functionally linked to these other components and there is likely to be epistasis between the fitness w of A'B' and C' ($w_4 \neq w_{A'B'} \times w_{C'}$). In addition, different developmental modules may be linked evolutionarily due to developmental (genetic) linkage: A'–B' and B"–C" may act as developmental modules, but not as modules of evolutionary transformations, since they may be independently perturbable during development (e.g. interactions 2 and 3 may be separately affected by perturbation of B' and B", respectively), but not by heritable variations (mutations) in germ line cells (e.g. 2 and 3 may tend to be jointly affected by perturbation of B'''). For the same reason, developmental modules D'–E' and D"–E" will form a single module of evolution, even if they make independent fitness contributions.

Modified from Schlosser, 2007.

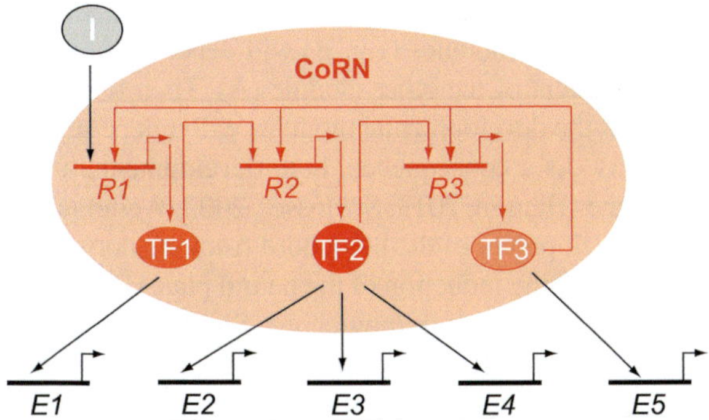

FIGURE 1.5 Core regulatory network (CoRN). In this schematic example, a CoRN is formed by a small set of regulator genes (R1 to R3) encoding transcription factors (TF1 to TF3). After activation by an inducer (I), the CoRN members maintain their own expression by cross-regulation and regulate the expression of downstream effector genes (E1 to E5). Although illustrated here as a gene regulatory network involving primarily transcriptional cross-regulation of genes, CoRNs may also involve other interactions (e.g. signalling or adhesive interactions between cells), in particular for higher-level modules.

Reprinted from Schlosser, 2023.

evolution, this is not always the case. Sometimes, independent modules of development may be tied together in a single module of evolution. There are two main reasons for this: (1) developmental (genetic) linkage or (2) functional linkage between the two modules. First, units that are independently perturbable during development (e.g. by experimental manipulation of their component somatic cells or by mutations of genes in these cells) and, thus, qualify as distinct modules of development may have a partly shared genetic basis and are therefore genetically linked. Mutations of shared genes in the germ line may therefore pleiotropically affect both units in the chain of generations linking the fate of these units evolutionarily (Figure 1.4). Consequently, only units of the so-called "genotype-phenotype map", that is, units that respond coherently and relatively independently to heritable variations, serve as individualized modules of evolution (Cheverud, 1984; Riedl, 1975; Wagner, 1996, 2014; Wagner and Altenberg, 1996; Wagner et al., 2007). This is easy to see for units instantiated multiple times in development, such as the many cells belonging to the same cell type (e.g. hair cells, neurons, muscle cells) or tissues and organs developing in more than one copy (e.g. hairs, feathers, vertebrae, limbs). Left and right limb buds or the hair cells of the left and right inner ear (or of different sensory areas in the same ear), for example, develop independently of each other, but they are regulated by the activity of the same genes.

Second, if a developmental process A goes ahead independently of another developmental process B, but makes an indispensable contribution to B, A will form a distinct developmental module but will be linked with B evolutionarily, because A and B make an interdependent contribution to fitness (i.e. to the probability of reproduction in the next generation) (Figure 1.4). Whenever components of A and B interact in the generation of a common function, the functionality of any change in the component of one of the modules (e.g. B) will depend on its ability to properly interact with the component of the other module (A). Therefore, at least some heritable changes in one of the developmental modules will induce selection on the other due to fitness epistasis. As a consequence, both developmental units will co-adapt and co-evolve over time (Hansen, 2013; Schlosser, 2002; Weinreich et al., 2005). For example, while the development of the notochord from the dorsal blastopore lip (A) proceeds independently of the induction of the neural plate (B), the latter depends on signals from the dorsal blastopore lip/notochord. Similarly, once initiated hair cell differentiation (A) proceeds cell-autonomously and appears not to require interactions with supporting cells. Conversely, however, the behaviour of supporting cells (B; including their ability to produce more hair cells) is dependent on interactions with hair cells (with Notch signals from hair cells preventing reprogramming of supporting cells; Iyer et al., 2022). Only units that have a certain degree of both genetic and functional separability from other such units, therefore, serve as evolutionarily stable individualized units or modules of evolution (reflected in a high "co-evolution probability" among their components; see Schlosser, 2002, 2004).

Recognition of developmental processes that serve as quasi-independent developmental modules is important because only such units show a unitary behaviour that is determined by all and only their components and their interactions. Therefore, only for such units can we hope to provide a relatively complete explanation elucidating the mechanism underlying the mesh of singly necessary and jointly sufficient interactions between its components. For a reasonably complete understanding of the constraints governing the evolutionary fate of developmental processes – the conservation of some of its components and their interactions, and the modifications of others in evolution – we have to broaden our analysis further and consider evolutionary (rather than only developmental) modules.

1.6.3 Hierarchies and Cross-Cutting Organization of Modules

Developmental processes and their underlying structures (e.g. regulatory networks of interacting genes, cells or tissues) can often be decomposed in a hierarchical fashion into interacting components, which in turn may be further decomposed into lower-level components. For example, the process of inner ear development can be understood by considering the differentiation (and morphogenetic shape changes, which will be ignored here) of its different cell types and how these are coordinated with each other by the exchange of signalling molecules or mechanical stimuli. The differentiation of hair cells and other inner ear cell types, in turn, can be understood by considering the regulatory relationships between some of the genes expressed in hair cells, including Atoh1, POU4f3 and Gfi1.

Importantly, such decomposition only makes sense if the components of the process themselves act as relatively integrated and quasi-independent units (modules). The behaviour of the upper-level developmental process (e.g. hair cell differentiation)| can then be mechanistically understood from the interactions of its components (regulatory activity of Atoh1, POU4f3 and Gfi1 in differentiating hair cells), whose behaviour is determined by the regular interaction capacities of underlying structures (determined by TF binding sites in the enhancers of *Atoh1, POU4f3* and *Gfi1* genes as well as the capacity of Atoh1, POU4f3 and Gfi1 proteins to bind to specific sequences in the enhancers of other genes), the way they are connected to each other in a network and their initial states of activity.

As a corollary of the hierarchical embedding of lower-level units as components of higher-level units, the identity of higher-level units is sometimes evolutionarily preserved in spite of the replacement of some of its components by others that fulfil similar regulatory roles. This is known as "genetic piracy" or "developmental systems drift" (DSD) (Roth, 1988; True and Haag, 2001). For instance, although the antero-posterior patterning system involving Gap and Hox genes is relatively well conserved in insects, in flies, the *bicoid* gene, which only evolved in dipterans, has taken up the early role of *hunchback* in initiating the regionalized expression pattern of these genes (Lynch and Desplan, 2003; McGregor, 2005). Similarly, in nematodes, a very similar structured vulva is set up by different cell–cell interactions and signalling pathways (EGF vs. Wnt) in different species (Kiontke et al., 2007; Sommer, 2012). DSD has also been described for hair cell differentiation, where the role of two core TFs, Atoh1 and Sox2, appears to have been swapped in one of the branches after the divergence of teleosts and amniotes with respect to both the temporal sequence of expression and the phenotypic effects of overexpression (with Atoh1 acting as the more upstream regulator in zebrafish and presumably other teleosts but as the more downstream regulator in amniotes; Millimaki et al., 2010; Raft and Groves, 2015).

Although a hierarchical organization of developing organisms is widely recognized, there is an important caveat that deserves more attention. While hierarchical, the organization of developing organisms does not lend itself to a decomposition into a single, simple hierarchy of components. Instead, we can recognize cross-cutting units in a Quasi-Cartesian system of organization (Figure 1.6; Wagner, 2014). This is mostly due to the fact that many developmental processes/units develop in multiple copies and the regulatory interactions that underlie the development of a unit with a certain identity (e.g. limb development; differentiation of hair cells) can be distinguished from regulatory interactions involved in pattern formation, that is, those that determine its region-specific properties (forelimbs vs. hindlimbs; vestibular vs. auditory hair cells). Mechanisms of pattern formation, such as antero-posterior or dorso-ventral patterning of the entire body axis or of a developing organ, serve to coordinate the development of multiple developmental units (forelimbs as well as thoracic vertebrae; vestibular hair cells as well as supporting cells), ensuring the development of each of them in a regionally and temporally appropriate way. The regulatory networks underlying antero-posterior and dorso-ventral patterning mechanisms, therefore, operate largely independently of each other along different body axes (with each of them embedded in its own hierarchy) and also independent of networks that regulate the general differentiation of a developmental unit (again embedded into a separate hierarchy).

The regulatory networks involved in pattern formation and those involved in differentiation of various developmental units intersect in that the former help to specify region- and time-specific subtypes of the latter. This can be illustrated by considering the role of the Hox patterning system in regulating the segment-specific subtype identity of different organs. Complex cross-regulatory interactions among Hox genes

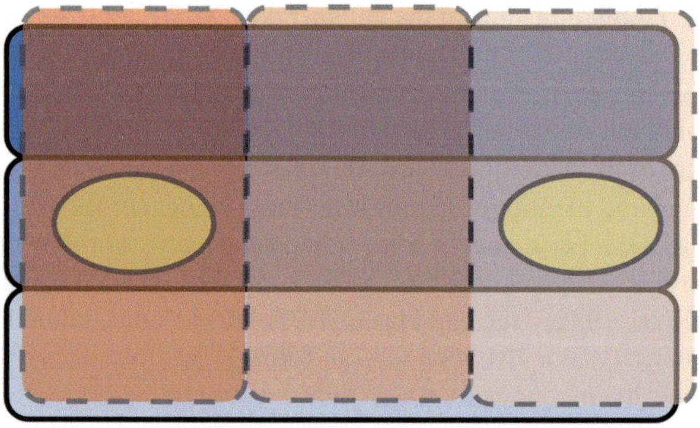

FIGURE 1.6 Cross-cutting organization of modules. Cross-cutting modules may be organized in a Quasi-Cartesian system (Wagner, 2014). For example, the development of some repeated character (yellow ellipse) may be regionally modulated by orthogonal patterning mechanisms such as antero-posterior (red, hatched outlines) and dorso-ventral patterning mechanisms (blue, solid outlines).

Reprinted from Schlosser, 2023.

and some additional cofactors, which are only partly understood, underlie the collinear expression of these genes in development (Mallo and Alonso, 2013; Deschamps and Duboule, 2017), with Hox genes located towards the 3' end of the Hox cluster being more anteriorly expressed than Hox genes located towards the 5' end. There have been profound changes in the genomic organization of Hox genes and some of their regulatory interactions throughout evolution (Duboule, 2007). Nevertheless, the outcome of this regulatory network resulting in a spatiotemporally ordered, collinear expression of Hox genes has been evolutionarily well preserved in almost all branches of the metazoan phylogenetic tree, as has the central role of Hox genes for antero-posterior body patterning during development (Akam, 1995; Garcia-Fernandez, 2005; Duboule, 2007). Hox genes expressed in a certain defined region along the antero-posterior body axis typically affect the region-specific subtype identity of most cell types or organs developing in this region.

For example, in arthropods, the Hox gene Antennapedia (Antp) specifies second thoracic (mesothorax) segment identity for dorsal appendages (forewing), ventral appendages (second thoracic leg) and many other structures (e.g. sensory organs and tracheal pits) developing from the mesothorax (Hughes and Kaufman, 2002). Another Hox gene, Ultrabithorax (Ubx), will specify a different subtype identity for the same structures in the adjacent third thoracic segment. Although Antp and Ubx each modulate the development of multiple units in a region-specific way, the regulatory networks underlying the development of each of these units will respond differently. While Antp will repress antennal genes and promote the expression of forewing genes during the development of the dorsal appendage of the mesothorax, it will activate different target genes in the sensory organ primordia of the mesothorax, presumably due to cooperation with different cofactors and the accessibility of different cis-regulatory regions in each of the two units (Hughes and Kaufman, 2002).

1.6.4 Plasticity

Apart from modularity, another central organizational property of living systems is their plasticity. I will use this term broadly here to refer to the capacity of living systems for self-maintenance and self-reproduction under a variety of conditions (Gerhart and Kirschner, 1997; West-Eberhard, 2003; Beldade et al., 2011). This involves regulatory capacities, which often involve complex feedback interactions. These allow the modulation of developmental processes to accommodate stochastic or environmentally induced variations in other developmental units. Size, distribution or numbers of a particular developmental unit may, for example, be modulated by signals from other developmental units (in particular those with which they subsequently interact functionally), ensuring a match in the size, distribution or number of interacting developmental units. Such "exploratory mechanisms" have been described at multiple levels of organization, such as in the formation of microtubules, blood vessels, muscles, synapses and immune cells (Gerhart and Kirschner, 1997). In the inner ear, survival of the sensory neurons that innervate the hair cells depends on their exposure to neurotrophins released by the hair cells, ensuring a proper match of these two cell populations (Fritzsch et al., 2004).

Regulatory capacities also guarantee the canalization of many developmental processes, that is, their relatively invariant (homeostatic) completion by several alternative trajectories depending on stochastic or environmentally induced variation in the initial conditions (Waddington, 1957; Gibson and Wagner, 2000; Rutherford, 2000; Beldade et al., 2011). Recent single-cell sequencing studies have shown that while many progenitor cells are surprisingly heterogeneous in their gene expression patterns, they nevertheless undergo relatively invariant binary cell fate decisions during normal development, highlighting this capacity for canalization (Laslo et al., 2006; Briggs et al., 2018; Soldatov et al., 2019). Although the mechanisms allowing invariable cell fate decisions in the face of extensive progenitor heterogeneity are only beginning to be understood, they often appear to involve mutual regulatory repression paired with self-reinforcement of alternative differentiation states (Laslo et al., 2006; Soldatov et al., 2019).

1.6.5 Evolvability

While life as we know it originated only once, it never became extinct in all of the four billion years of its glorious existence on Earth. While some of the branches of the tree of life eventually died off, others sprouted more profusely. The success story of living systems and their persistence over many aeons is linked to their evolvability, that is, their ability to change, adapt and diversify, while maintaining a functional living organization in each of its branches (Wagner and Altenberg, 1996; Gerhart and Kirschner, 1997; Kirschner and Gerhart, 1998). This requires, first and foremost, the ability for overall faithful and potentially multiplicative reproduction, which in multicellular organisms involves development, that is, the faithful reconstruction of an organism with the same organization (inherited from its ancestor) in the next generation. Complete faithfulness, however, would not support evolution, so in addition, there must be opportunities for relatively rare changes that are then themselves passed on to the next generation (heritable variation). Taken together, the capacities for heritable variation and for multiplicative reproduction support the generation of different variants, which permit both evolutionary change and diversification. Because different variants may differ in fitness (i.e. in their ability to survive and reproduce) and there are environmental limitations to the number of organisms that can survive and reproduce, heritable variation also provides the substrate for natural selection and, thus, adaptation.

The highly (but not completely) faithful replication of DNA provides the central mechanism for both faithful reproduction (inheritance) and the introduction of variation, which is itself faithfully reproduced and, thus, is the basis for heredity. Following Maynard Smith and Szathmáry, I distinguish inheritance (their "heredity" in the wide sense) and heredity (their "unlimited heredity") and consider heredity as a property of only those entities, which will reliably reproduce any variation out of a more or less extensive set of variations resulting from the direct perturbation of these entities or their components (Maynard Smith and Szathmary, 1995; Szathmary, 1995; Szathmary and Maynard Smith, 1995, 1997). According to this definition, both DNA and proteins are inherited (reproduced) by progeny, but only DNA has heredity. In contrast to other inherited entities, those with heredity are *directly* selectable (causally independent of other selectable entities) and *sustainably* selectable (open for the repeated introduction of variants) (Schlosser, 2002). Therefore, in spite of increasing support for the transgenerational inheritance of some epigenetic modifications, there is little evidence for the long-term persistence or the direct and sustainable selectability of such modifications (Jablonka and Raz, 2009; Boskovic and Rando, 2018). They are, therefore, unlikely to make important contributions to persistent evolutionary changes, in contrast to genetic mutations. The recognition of the privileged role of genes for heredity distinguishes the approach advocated here from an otherwise kindred position in the philosophy of science known as "Developmental Systems Theory" (Oyama, 1985; Gray, 1992; Griffiths and Gray, 1994).

The capacity for sexual reproduction contributes greatly to evolvability because it involves the recombination of genes and, therefore, presumably, helps to break the linkage between unfavourable and favourable variants of different components of the self-maintaining and reproducing organism (Maynard Smith, 1978; Barton and Charlesworth, 1998). But most importantly in the context of the present argument, evolvability is also greatly facilitated by the two organizational hallmarks of living systems discussed above, viz., their modularity and plasticity. Modularity allows the "mosaic evolution" (e.g. Futuyma, 1986) of different characters, because selection for variations in one (evolutionary) module can proceed relatively independently of selection for variations in another module. Modularity also facilitates the recombination and redeployment of units, which underlie the formation of evolutionary novelties and the increase of organismal complexity in a combinatorial manner (Schlosser, 2002, 2023). Plasticity allows the buffering not only of environmental but also genetic perturbations, thereby promoting robust maintenance of the different quasi-independent modules and ultimately of the functional self-maintaining and reproducing system in the face of heritable variation.

Together, modularity and plasticity contribute to both "robustness and flexibility" (Gerhart and Kirschner, 1997; Kirschner and Gerhart, 1998) of evolutionary transformations: they help to preserve the integrity and coherence of intramodular interactions, while playing with and accommodating intermodular connections (Raff, 1996; Wagner and Altenberg, 1996; Gerhart and Kirschner, 1997; Schlosser, 2002; Beldade et al., 2011; Moczek et al., 2011). Organisms are able to evolve because they are composed of building blocks that operate quite independently of each other, each fulfilling a largely separable function often under a variety of circumstances. These building blocks tend to be evolutionarily stable but may accumulate changes over time independently of other building blocks. They may also be newly rejigged and recombined to form novel units. In fact, they are what's available in the workshop of evolution to reconfigure the self-maintaining and self-reproducing contraption that we know as the organism, while keeping it running all the time and never shutting it down.

It is, therefore, evolution, the great tinkerer that can point us the way to a new developmental systems biology and will enable us to address the two main open questions of developmental biology introduced above: first, in shaping evolutionarily stable units of development over time, it presents us with natural units of development that display relatively integrated and coherent behaviour and for which we can hope to give relatively complete mechanistic explanations. And second, in safeguarding the self-maintenance and self-reproduction of the developing organism and the functionality of its components, it provides us with criteria to distinguish normal (functional) from abnormal (dysfunctional) behaviour of these units. Bringing a new developmental systems biology to life based on these principles is very much a project for the future. But more conceptual clarity along the lines I have merely sketched here will be essential to develop novel approaches to make sense of the avalanche of data that are currently generated. A historic example for the importance of conceptual clarity in spawning a fruitful new research program is Willi Henning's foundational work on phylogenetic systematics (Hennig, 1966), which forms the basis of current phylogenetics. A similar framework for systems biology is sorely needed. In the last section of this chapter, I will briefly indicate how greater awareness of this conceptual background may, in fact, help us to design new research strategies for developmental biology.

1.7 A TINKERER'S GUIDE TO DEVELOPMENTAL SYSTEMS BIOLOGY

Considering that evolving organisms perpetually maintain their functional organization, while their architecture is every so often reconstructed by evolutionary tinkering, can inform our quest for both completeness and functional adequacy of developmental explanations. First, because development is shaped by evolutionary tinkering, comparisons between different species can be exploited to recognize evolutionarily stable units of development and guide our research towards the relevant components of developmental processes, thereby enabling an explanation of the process that is reasonably complete. Whereas currently EvoDevo is mostly dedicated to unravelling the mechanisms of evolutionary change, it has a larger role to play in helping us to elucidate the mechanisms of (evolutionarily conserved) developmental processes.

In the current age of high-throughput sequencing, comparative approaches should adopt a more central role to identify core regulators of developmental processes and distinguish relevant from irrelevant factors. Among the vast number of gene expression changes observed between subsequent stages of a developmental process or between control and experimental conditions in LOF or GOF studies, changes that are conserved between different phylogenetic lineages are most likely to be in core regulatory components of the process (belonging to the set of its singly necessary and jointly sufficient components) (Figure 1.7). However, because of the possibility of DSD, a subset of the non-conserved changes also deserves closer scrutiny and may reveal additional core regulators. Because DSD is expected to be a relatively

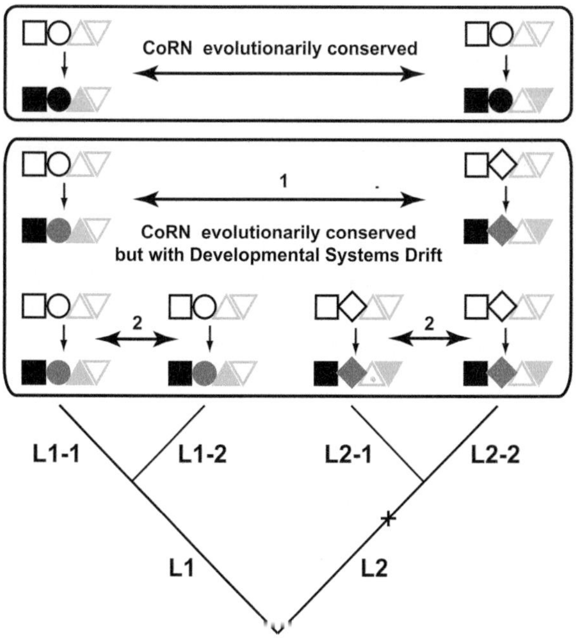

FIGURE 1.7 Comparative approach to identify core regulators of evolutionarily stable developmental processes. The activity patterns of genes (open or filled symbols) and their changes (arrows from top to bottom) during development or in an LOF or a GOF experiment are depicted here for different evolutionary lineages (L1 and L2) and their sublineages (L1-1/L1-2 and L2-1/L2-2). Upper box: Comparison of correlated gene expression changes (open to filled symbols) between two lineages (double headed arrow) reveals evolutionarily conserved putative members of the core regulatory network (CoRN) (black symbols). Lower box: A refined strategy takes into account the possibility of developmental systems drift (DSD). In step 1, putative members of the CoRN are identified as above based on evolutionary conservation between L1 and L2 (black square). In step 2, the analysis is extended by comparison with lineages more closely related to the ones compared in step 1. This enables the identification of additional components that are not conserved between L1 and L2 but are conserved between L1-1 and L1-2 (dark grey circle) and L2-1 and 2-2 (dark grey rhombus), possibly reflecting one or several DSD events in L1 and/or L2 (e.g. at x).

rare event compared to the general pattern of evolutionary conservation of important regulatory interactions underlying a developmental process, it should be possible to detect it by screening for gene expression changes that are not conserved between distantly related species but are conserved between more closely related species (Figure 1.7). There are clearly multiple operational challenges to be overcome for such a comparative approach to succeed, and cross-species comparisons between single-cell genomic data sets have been notoriously challenging. However, several tools have already been developed to improve the mapping of such data between species, and further progress in this field is to be expected (e.g. Sebe-Pedros et al., 2018a; Sebe-Pedros et al., 2018b; Tarashansky et al., 2019, 2021).

Second, because functionality is preserved during evolutionary transformations, interspecific comparisons of developmental processes may also aid our understanding of normal versus abnormal developmental processes. The range of interspecific variation between gene expression changes during the same developmental process in different species will at least provide insight into a minimum range of variation that is compatible with functional completion of the developmental process. Such studies, therefore, may be a valuable complement to studies that aim to define conditions for normal and abnormal development by comparison of intraspecific variation (e.g. between experimental LOF or GOF individuals and controls).

It would be beyond the scope of this chapter to delve further into specific operational strategies, so this shall suffice here to indicate some of the opportunities arising from a more conceptually mature systems biology. Apart from suggesting new ways to extract valuable insights for our understanding of developmental processes from comparative approaches, it promises to re-vitalize developmental biology by helping to re-focus our research strategies to proper and natural units of development. In a way, this will require us to revisit some of the questions that motivated the early experimental embryologists, but paired with a closer attention to evolutionary processes and equipped with a vastly richer knowledge of molecular interactions. Once we become such attentive and knowledgeable apprentices in the workshop of evolution, the great tinkerer, the future of developmental biology will be bright.

ACKNOWLEDGEMENTS

I thank Gunter Wagner and Igor Adameyko for helpful comments on the manuscript.

REFERENCES

Ahmed, S., Vaden, K.I., Jr., Leitao, D., Dubno, J.R., Drogemoller, B.I., 2025. Large-scale audiometric phenotyping identifies distinct genes and pathways involved in hearing loss subtypes. *Commun. Biol.* 8, 1515.

Aibar, S., Gonzalez-Blas, C.B., Moerman, T., Huynh-Thu, V.A., Imrichova, H., Hulselmans, G., Rambow, F., Marine, J.C., Geurts, P., Aerts, J., van den Oord, J., Atak, Z.K., Wouters, J., Aerts, S., 2017. SCENIC: single-cell regulatory network inference and clustering. *Nat. Methods* 14, 1083–1086.

Akam, M., 1995. Hox genes and the evolution of diverse body plans. *Philos. Trans. R. Soc. Lond. B Biol. Sci.* 349, 313–319.

Alon, U., 2020. *An introduction to systems biology*, 2nd ed. CRC Press, Boca Raton.

Alsina, B., Giráldez, F., Pujades, C., 2009. Patterning and cell fate in ear development. *Int. J. Dev. Biol.* 53 1503–1513.

Atkinson, P.J., Huarcaya Najarro, E., Sayyid, Z.N., Cheng, A.G., 2015. Sensory hair cell development and regeneration: similarities and differences. *Development* 142, 1561–1571.

Badia, I.M.P., Wessels, L., Muller-Dott, S., Trimbour, R., Ramirez Flores, R.O., Argelaguet, R., Saez-Rodriguez, J., 2023. Gene regulatory network inference in the era of single-cell multi-omics. *Nat. Rev. Genet.* 24, 739–754.

Baek, S., Tran, N.T.T., Diaz, D.C., Tsai, Y.Y., Acedo, J.N., Lush, M.E., Piotrowski, T., 2022. Single-cell transcriptome analysis reveals three sequential phases of gene expression during zebrafish sensory hair cell regeneration. *Dev. Cell* 57, 799–819 e796.

Barresi, M.J., Gilbert, S.F., 2023. *Developmental biology*, 13th ed. Oxford University Press, New York.

Barton, N.H., Charlesworth, B., 1998. Why sex and recombination? *Science* 281, 1986–1990.

Bechtel, W., 2011. Mechanism and biological explanation. *Philos. Sci.* 78, 533–557.

Bechtel, W., Richardson, R.C., 2010. *Discovering complexity: Decomposition and localization as strategies in scientific research*, 2nd ed. MIT Press, Cambridge.

Beldade, P., Mateus, A.R., Keller, R.A., 2011. Evolution and molecular mechanisms of adaptive developmental plasticity. *Mol. Ecol.* 20, 1347–1363.

Berkes, C.A., Tapscott, S.J., 2005. MyoD and the transcriptional control of myogenesis. *Semin. Cell Dev. Biol.* 16, 585–595.

Bermingham, N.A., Hassan, B.A., Price, S.D., Vollrath, M.A., Ben Arie, N., Eatock, R.A., Bellen, H.J., Lysakowski, A., Zoghbi, H.Y., 1999. Math1: an essential gene for the generation of inner ear hair cells. *Science* 284, 1837–1841.

Bertrand, N., Castro, D.S., Guillemot, F., 2002. Proneural genes and the specification of neural cell types. *Nat. Rev. Neurosci.* 3, 517–530.

Bolouri, H., Davidson, E.H., 2002. Modeling transcriptional regulatory networks. *Bioessays* 24, 1118–1129.

Bonner, J.T., 1974. *On development.* Harvard University Press, Cambridge.

Boskovic, A., Rando, O.J., 2018. Transgenerational epigenetic inheritance. *Annu. Rev. Genet.* 52, 21–41.

Brigandt, I., 2013. Systems biology and the integration of mechanistic explanation and mathematical explanation. *Stud. Hist. Philos. Biol. Biomed. Sci.* 44, 477–492.

Briggs, J.A., Weinreb, C., Wagner, D.E., Megason, S., Peshkin, L., Kirschner, M.W., Klein, A.M., 2018. The dynamics of gene expression in vertebrate embryogenesis at single-cell resolution. *Science* 360, eaar5780.

Brignull, H.R., Raible, D.W., Stone, J.S., 2009. Feathers and fins: non-mammalian models for hair cell regeneration. *Brain Res.* 1277, 12–23.

Cano-Gamez, E., Trynka, G., 2020. From GWAS to function: using functional genomics to identify the mechanisms underlying complex diseases. *Front. Genet.* 11, 424.

Chen, Y., Yu, H., Zhang, Y., Li, W., Lu, N., Ni, W., He, Y., Li, J., Sun, S., Wang, Z., Li, H., 2013. Cotransfection of Pax2 and Math1 promote in situ cochlear hair cell regeneration after neomycin insult. *Sci. Rep.* 3, 2996.

Cheverud, J.M., 1984. Quantitative genetics and developmental constraints on evolution by selection. *J. Theor. Biol.* 110, 155–171.

Coffin, A., Kelley, M., Manley, G.A., Popper, A.N., 2005. Evolution of sensory hair cells, in: Manley, G.A., Popper, A.N., Fay, R.R. (Eds.), *Evolution of the vertebrate auditory system*. Springer, New York.

Costa, A., Henrique, D., 2015. Transcriptome profiling of induced hair cells (iHCs) generated by combined expression of Gfi1, Pou4f3 and Atoh1 during embryonic stem cell differentiation. *Genom. Data* 6, 77–80.

Costa, A., Powell, L.M., Malaguti, M., Soufi, A., Lowell, S., Jarman, A.P., 2022. Repurposing the lineage-determining transcription factor Atoh1 without redistributing its genomic binding sites. *Front. Cell Dev. Biol.* 10, 1016367.

Craver, C.F., Darden, L., 2013. *In search of mechanisms: Discoveries across the life sciences*. University of Chicago Press, Chicago.
Craver, C.F., Kaplan, D.M., 2020. Are more details better? On the norms of completeness for mechanistic explanations. *Brit. J. Phil. Sci.* 71, 287–319.
Crowe, N., Dietrich, M.R., Alomepe, B.S., Antrim, A.F., ByrneSim, B.L., He, Y., 2015. The diversification of developmental biology. *Stud. Hist. Philos. Biol. Biomed. Sci.* 53, 1–15.
Davidson, E.H., Erwin, D.H., 2006. Gene regulatory networks and the evolution of animal body plans. *Science* 311, 796–800.
Davies, J.A., 2023. *Mechanisms of morphogenesis*, 3rd ed. Academic Press, London.
De Robertis, E.M., Morita, E.A., Cho, K.W.Y., 1991. Gradient fields and homeobox genes. *Development* 112, 669–678.
Delgado, I., Torres, M., 2017. Coordination of limb development by crosstalk among axial patterning pathways. *Dev. Biol.* 429, 382–386.
Deschamps, J., Duboule, D., 2017. Embryonic timing, axial stem cells, chromatin dynamics, and the Hox clock. *Genes Dev.* 31, 1406–1416.
DiFrisco, J., Love, A.C., Wagner, G.P., 2020. Character identity mechanisms: a conceptual model for comparative-mechanistic biology. *Biol. Philos.* 35, 44.
Driesch, H., 1891. Entwicklungsmechanische Studien. I. Der Werth der beiden ersten Furchungszellen in der Echinodermenentwicklung. *Experimentelle Erzeugen von Theil- und Doppelbildungen. Z. Zool.* 53, 160–178.
Duboule, D., 2007. The rise and fall of Hox gene clusters. *Development* 134, 2549–2560.
Elliott, K.L., Pavlinkova, G., Chizhikov, V.V., Yamoah, E.N., Fritzsch, B., 2021. Development in the mammalian auditory system depends on transcription factors. *Int. J. Mol. Sci.* 22.
Erkman, L., McEvilly, R.J., Luo, L., Ryan, A.K., Hooshmand, F., O'Connell, S.M., Keithley, E.M., Rapaport, D.H., Ryan, A.F., Rosenfeld, M.G., 1996. Role of transcription factors Brn-3.1 and Brn-3.2 in auditory and visual system development. *Nature* 381, 603–606.
Favor, J., Sandulache, R., Neuhäuser-Klaus, A., Pretsch, W., Chatterjee, B., Senft, E., Wurst, W., Blanquet, V., Grimes, P., Spörle, R., Schughart, K., 1996. The mouse Pax21Neu mutation is identical to a human Pax2 mutation in a family with renal-coloboma syndrome and results in developmental defects of the brain, ear, eye and kidney. *Proc. Natl. Acad. Sci. USA* 93, 13870–13875.
Fontana, W., Buss, L., 1994a. The arrival of the fittest: toward a theory of biological organization. *Bull. Math. Biol.* 56, 1–64.
Fontana, W., Buss, L.W., 1994b. What would be conserved if "the tape were played twice"? *Proc. Natl. Acad. Sci. USA* 91, 757–761.
Fritzsch, B., Beisel, K.W., Pauley, S., Soukup, G., 2007. Molecular evolution of the vertebrate mechanosensory cell and ear. *Int. J. Dev. Biol.* 51, 663–678.
Fritzsch, B., Tessarollo, L., Coppola, E., Reichardt, L.F., 2004. Neurotrophins in the ear: their roles in sensory neuron survival and fiber guidance. *Prog. Brain Res.* 146, 265–278.
Futuyma, D.J., 1986. *Evolutionary biology*. Sinauer, Sunderland.
Garcia-Fernandez, J., 2005. The genesis and evolution of homeobox gene clusters. *Nat. Rev. Genet.* 6, 881–892.
Gerhart, J., Kirschner, J., 1997. *Cells, embryos, and evolution*. Blackwell Science, Malden.
Ghysen, A., Dambly-Chaudière, C., 1989. Genesis of the *Drosophila* peripheral nervous system. *Trends Genet.* 5, 251–255.
Gibson, G., Wagner, G., 2000. Canalization in evolutionary genetics: a stabilizing theory? *Bioessays* 22, 372–380.
Girotto, G., Pirastu, N., Sorice, R., Biino, G., Campbell, H., d'Adamo, A.P., Hastie, N.D., Nutile, T., Polasek, O., Portas, L., Rudan, I., Ulivi, S., Zemunik, T., Wright, A.F., Ciullo, M., Hayward, C., Pirastu, M., Gasparini, P., 2011. Hearing function and thresholds: a genome-wide association study in European isolated populations identifies new loci and pathways. *J. Med. Genet.* 48, 369–374.

Goodyear, R.J., Kros, C.J., Richardson, G.P., 2006. The development of hair cells in the inner ear, in: Eatock, R.A., Fay, R.R., Popper, A.N. (Eds.), *Vertebrate hair cells*. Springer, New York, pp. 20–94.

Gray, R., 1992. Death of the gene: developmental systems strike back, in: Griffiths, P. (Ed.), *Trees of life*. Kluwer, Dordrecht, pp. 165–209.

Griesemer, J., 2000. Development, culture, and the units of inheritance. *Phil. Sci.(Suppl.)* 67, S348–S368.

Griffiths, P.E., Gray, R.D., 1994. Developmental systems and evolutionary explanation. *J. Philos.* 91, 277–304.

Groves, A.K., 2010. The challenge of hair cell regeneration. *Exp. Biol. Med. (Maywood.)* 235, 434–446.

Groves, A.K., Fekete, D.M., 2012. Shaping sound in space: the regulation of inner ear patterning. *Development* 139, 245–257.

Groves, A.K., Zhang, K.D., Fekete, D.M., 2013. The genetics of hair cell development and regeneration. *Annu. Rev. Neurosci.* 36, 361–381.

Hansen, T.F., 2013. Why epistasis is important for selection and adaptation. *Evolution* 67, 3501–3511.

Harrison, R.G., 1907. Experiments in transplanting limbs and their bearing upon the problems of the development of nerves. *J. Exp. Zool.* 4, 239–281.

Harrison, R.G., 1918. Experiments on the development of the fore limb of Amblystoma, a self-differentiating, equipotential system. *J. Exp. Zool.* 25, 413–461.

Harrison, R.G., 1924. Experiments on the development of the internal ear. *Science* 59, 448.

Harrison, R.G., 1936. Relations of symmetry in the developing ear of Amblystoma punctatum. *Proc. Natl. Acad. Sci. USA* 22, 238–247.

Hecker, M., Lambeck, S., Toepfer, S., van Someren, E., Guthke, R., 2009. Gene regulatory network inference: data integration in dynamic models-a review. *Biosystems* 96, 86–103.

Heisenberg, C.P., Bellaiche, Y., 2013. Forces in tissue morphogenesis and patterning. *Cell* 153, 948–962.

Hennig, W., 1966. *Phylogenetic systematics*. University of Illinois Press, Urbana.

Hopwood, N., 2019. Inclusion and exclusion in the history of developmental biology. *Development* 146, dev175448.

Huang, C., Chan, J.A., Schuurmans, C., 2014. Proneural bHLH genes in development and disease. *Curr. Top. Dev. Biol.* 110, 75–127.

Hughes, C.L., Kaufman, T.C., 2002. Hox genes and the evolution of the arthropod body plan. *Evol. Dev.* 4, 459–499.

Huxley, J.S., de Beer, G.R., 1934. *The elements of experimental embryology*. Cambridge University Press, Cambridge.

Ideker, T., Galitski, T., Hood, L., 2001. A new approach to decoding life: systems biology. *Annu. Rev. Genomics Hum. Genet.* 2, 343–372.

Iyer, A.A., Groves, A.K., 2021. Transcription factor reprogramming in the inner ear: turning on cell fate switches to regenerate sensory hair cells. *Front. Cell. Neurosci.* 15, 660748.

Iyer, A.A., Hosamani, I., Nguyen, J.D., Cai, T., Singh, S., McGovern, M.M., Beyer, L., Zhang, H., Jen, H.I., Yousaf, R., Birol, O., Sun, J.J., Ray, R.S., Raphael, Y., Segil, N., Groves, A.K., 2022. Cellular reprogramming with ATOH1, GFI1, and POU4F3 implicate epigenetic changes and cell-cell signaling as obstacles to hair cell regeneration in mature mammals. *Elife* 11, e79712.

Jablonka, E., Raz, G., 2009. Transgenerational epigenetic inheritance: prevalence, mechanisms, and implications for the study of heredity and evolution. *Q. Rev. Biol.* 84, 131–176.

Jacob, F., 1977. Evolution and tinkering. *Science* 196, 1161–1166.

Jan, T.A., Eltawil, Y., Ling, A.H., Chen, L., Ellwanger, D.C., Heller, S., Cheng, A.G., 2021. Spatiotemporal dynamics of inner ear sensory and non-sensory cells revealed by single-cell transcriptomics. *Cell Rep.* 36, 109358.

Jarman, A.P., Grau, Y., Jan, L.Y., Jan, Y.N., 1993. atonal is a proneural gene that directs chordotonal organ formation in the *Drosophila* peripheral nervous system. *Cell* 73, 1307–1321.
Jarvis, J.E., Miao, L., Hallaert, P., Martin, M.R., Collado, L., Hill, N.T., Brownell, I., 2025. POU4F3 is necessary for normal Merkel cell formation, whereas POU4F1 is dispensable. *J. Invest. Dermatol.* 145, 193–196 e195.
Jenuwein, T., Allis, C.D., 2001. Translating the histone code. *Science* 293, 1074–1080.
Judson, H.F., 1979. *The eighth day of creation*. Simon and Schuster.
Kaan, H.W., 1927. Experiments on the development of the ear of Amblystoma punctatum *J. Exp. Zool.* 46, 13–61.
Kamenev, D., Kameneva, P., Adameyko, I., 2025. The role of microheterogeneity in cell fate decisions in neural progenitors and neural crest. *Curr. Opin. Neurobiol.* 92, 103031.
Kharchenko, P.V., 2021. The triumphs and limitations of computational methods for scRNA-seq. *Nat Methods* 18, 723–732.
Kiontke, K., Barriere, A., Kolotuev, I., Podbilewicz, B., Sommer, R., Fitch, D.H., Felix, M.A., 2007. Trends, stasis, and drift in the evolution of nematode vulva development. *Curr. Biol.* 17, 1925–1937.
Kirschner, M., Gerhart, J., 1998. Evolvability. *Proc. Natl. Acad. Sci. USA* 95, 8420–8427.
Kitano, H., 2002. Systems biology: a brief overview. *Science* 295, 1662–1664.
Kolla, L., Kelly, M.C., Mann, Z.F., Anaya-Rocha, A., Ellis, K., Lemons, A., Palermo, A.T., So, K.S., Mays, J.C., Orvis, J., Burns, J.C., Hertzano, R., Driver, E.C., Kelley, M.W., 2020. Characterization of the development of the mouse cochlear epithelium at the single cell level. *Nat. Commun.* 11, 2389.
Laslo, P., Spooner, C.J., Warmflash, A., Lancki, D.W., Lee, H.J., Sciammas, R., Gantner, B.N., Dinner, A.R., Singh, H., 2006. Multilineage transcriptional priming and determination of alternate hematopoietic cell fates. *Cell* 126, 755–766.
Lawson, N.D., Wolfe, S.A., 2011. Forward and reverse genetic approaches for the analysis of vertebrate development in the zebrafish. *Dev. Cell* 21, 48–64.
Lobingier, B.T., Huttenhain, R., Eichel, K., Miller, K.B., Ting, A.Y., von Zastrow, M., Krogan, N.J., 2017. An approach to spatiotemporally resolve protein interaction networks in living cells. *Cell* 169, 350–360 e312.
Lynch, J., Desplan, C., 2003. Evolution of development: beyond bicoid. *Curr. Biol.* 13, R557–559.
Machamer, P., Darden, L., Craver, C.F., 2000. Thinking about mechanisms. *Philos. Sci.* 67, 1–25.
Maienschein, J., 2014. *Embryos under the microscope*. Harvard University Press, Cambridge, MA.
Maier, E.C., Saxena, A., Alsina, B., Bronner, M.E., Whitfield, T.T., 2014. Sensational placodes: neurogenesis in the otic and olfactory systems. *Dev Biol.* 389, 50–67.
Mallo, M., Alonso, C.R., 2013. The regulation of Hox gene expression during animal development. *Development* 140, 3951–3963.
Mammoto, T., Mammoto, A., Ingber, D.E., 2013. Mechanobiology and developmental control. *Annu. Rev. Cell Dev. Biol.* 29, 27–61.
Maricich, S.M., Wellnitz, S.A., Nelson, A.M., Lesniak, D.R., Gerling, G.J., Lumpkin, E.A., Zoghbi, H.Y., 2009. Merkel cells are essential for light-touch responses. *Science* 324, 1580–1582.
Masuda, M., Dulon, D., Pak, K., Mullen, L.M., Li, Y., Erkman, L., Ryan, A.F., 2011. Regulation of POU4F3 gene expression in hair cells by 5' DNA in mice. *Neuroscience* 197, 48–64.
Matern, M.S., Milon, B., Lipford, E.L., McMurray, M., Ogawa, Y., Tkaczuk, A., Song, Y., Elkon, R., Hertzano, R., 2020. GFI1 functions to repress neuronal gene expression in the developing inner ear hair cells. *Development* 147, dev186015.
Maturana, H.R., 1975. The organization of the living: a theory of the living organization. *Int. J. Man-Machine Stud.* 7, 313–332.
Maynard Smith, J., 1978. *The evolution of sex*. Cambridge University Press, Cambridge.

Maynard Smith, J., Szathmary, E., 1995. *The major transitions in evolution*. Freeman, Oxford.
McGregor, A.P., 2005. How to get ahead: the origin, evolution and function of bicoid. *Bioessays* 27, 904–913.
McKenna, K.Z., Wagner, G.P., Cooper, K.L., 2021. A developmental perspective of homology and evolutionary novelty. *Curr. Top. Dev. Biol.* 141, 1–38.
McPherson, D.R., 2018. Sensory hair cells: an introduction to structure and physiology. *Integr. Comp. Biol.* 58, 282–300.
Members-of-the-complex-trait-consortium, 2003. The nature and identification of quantitative trait loci: A community's view. *Nat. Rev. Genet.* 4, 911–915.
Millimaki, B.B., Sweet, E.M., Dhason, M.S., Riley, B.B., 2007. Zebrafish atoh1 genes: classic proneural activity in the inner ear and regulation by FGF and Notch. *Development* 134, 295–305.
Millimaki, B.B., Sweet, E.M., Riley, B.B., 2010. Sox2 is required for maintenance and regeneration, but not initial development, of hair cells in the zebrafish inner ear. *Dev. Biol.* 338, 262–269.
Moczek, A.P., Sultan, S., Foster, S., Ledon-Rettig, C., Dworkin, I., Nijhout, H.F., Abouheif, E., Pfennig, D.W., 2011. The role of developmental plasticity in evolutionary innovation. *Proc. Biol. Sci.* 278, 2705–2713.
Müller, G.B., 2003. Homology: the evolution of morphological organization, in: Müller, G.B., Newman, S.A. (Eds.), *Origination of organismal form*. MIT Press, Cambridge, MA, pp. 51–69.
Müller, G.B., 2010. Epigenetic innovation, in: Pigliucci, M., Müller, G.B. (Eds.), *Evolution - The extended synthesis*. MIT Press, Boston, MA, pp. 307–332.
Nelson, C.M., Bissell, M.J., 2006. Of extracellular matrix, scaffolds, and signaling: tissue architecture regulates development, homeostasis, and cancer. *Annu. Rev. Cell Dev. Biol.* 22, 287–309.
Neves, J., Abello, G., Petrovic, J., Giraldez, F., 2013. Patterning and cell fate in the inner ear: a case for Notch in the chicken embryo. *Dev. Growth Differ.* 55, 96–112.
Nguyen, H., Tran, D., Tran, B., Pehlivan, B., Nguyen, T., 2021. A comprehensive survey of regulatory network inference methods using single cell RNA sequencing data. *Brief. Bioinform.* 22, bbaa190.
Noble, D., 2002. Modelling the heart: insights, failures and progress. *Bioessays* 24, 1155–1163.
Ostrowski, S.M., Wright, M.C., Bolock, A.M., Geng, X., Maricich, S.M. 2015. Ectopic Atoh1 expression drives Merkel cell production in embryonic, postnatal and adult mouse epidermis. *Development* 142, 2533–2544.
Oyama, S., 1985. *The ontogeny of information*. Cambridge University Press, Cambridge.
Pan, N., Jahan, I., Kersigo, J., Kopecky, B., Santi, P., Johnson, S., Schmitz, H., Fritzsch, B., 2011. Conditional deletion of Atoh1 using Pax2-Cre results in viable mice without differentiated cochlear hair cells that have lost most of the organ of Corti. *Hear. Res.* 275, 66–80.
Peter, I.S., Davidson, E.H., 2015. *Genomic control process: Development and evolution*. Elsevier, Amsterdam.
Quine, W.V., 1969. Natural kinds, in: Quine, W.V. (Ed.), *Ontological relativity and other essays*. Columbia University Press, New York, pp. 114–138.
Raff, R.A., 1996. *The shape of life*. University of Chicago Press, Chicago.
Raft, S., Groves, A.K., 2015. Segregating neural and mechanosensory fates in the developing ear: patterning, signaling, and transcriptional control. *Cell Tissue Res.* 359, 315–332.
Riedl, R., 1975. *Die Ordnung des Lebendigen*. Parey, Hamburg.
Röhlich, K., 1929. Experimentelle Untersuchungen über den Zeitpunkt der Determination der Gehörblase bei Amblystoma-Embryonen. *Wilhelm Roux' Archiv für Entwicklungsmechanik der Organismen* 118, 164–199.

Roth, V.L., 1988. The biological basis of homology, in: Humphries, C.J. (Ed.), *Ontogeny and systematics*. Columbia University Press, New York, pp. 1–26.

Roux, W., 1885. Beiträge zur Entwickelungsmechanik des Embryos. I. Zur Orientirung über einige Probleme der organischen Entwickelung. *Z. Biol.* 21, 411–524.

Roux, W., 1888. Beiträge zur Entwickelungsmechanik des Embryo. V. Ueber die künstliche Hervorbringung „halber" Embryonen durch Zerstörung einer der beiden ersten Furchungszellen, sowie über die Nachentwickelung (Post generation) der fehlenden Körperhälfte. *VIRCHOWS Archiv* 114, 419–521.

Rutherford, S.L., 2000. From genotype to phenotype: buffering mechanisms and the storage of genetic information. *Bioessays* 22, 1095–1105.

Schlosser, G., 1993. *Einheit der Welt und Einheitswissenschaft. Grundlegung einer Allgemeinen Systemtheorie*. Vieweg, Braunschweig.

Schlosser, G., 1998. Self-re-production and functionality: a systems-theoretical approach to teleological explanation. *Synthese* 116, 303–354.

Schlosser, G., 2002. Modularity and the units of evolution. *Theory. Biosci.* 121, 1–80.

Schlosser, G., 2004. The role of modules in development and evolution, in: Schlosser, G., Wagner, G.P. (Eds.), *Modularity in development and evolution*. University of Chicago Press, Chicago, pp. 519–582.

Schlosser, G., 2007. Functional and developmental constraints on life cycle evolution: an attempt on the architecture of constraints, in: Brandon, R., Sansom, R. (Eds.), *Integrating evolution and development: From theory to practice*. MIT Press, Cambridge, pp. 113–172.

Schlosser, G., 2021a. *Development of sensory and neurosecretory cell types. Vertebrate cranial placodes, Vol.1*. CRC Press, Boca Raton.

Schlosser, G., 2021b. *Evolutionary origin of sensory and neurosecretory cell types. Vertebrate cranial placodes, Vol. 2*. CRC Press, Boca Raton.

Schlosser, G., 2023. Rebuilding ships while at sea-Character individuality, homology, and evolutionary innovation. *J. Morphol.* 284, e21522.

Schlosser, G., 2024. From "self-differentiation" to organoids-the quest for the units of development. *Dev. Genes Evol.* 234, 57–64.

Sebe-Pedros, A., Chomsky, E., Pang, K., Lara-Astiaso, D., Gaiti, F., Mukamel, Z., Amit, I., Hejnol, A., Degnan, B.M., Tanay, A., 2018a. Early metazoan cell type diversity and the evolution of multicellular gene regulation. *Nat. Ecol. Evol.* 2, 1176–1188.

Sebe-Pedros, A., Saudemont, B., Chomsky, E., Plessier, F., Mailhe, M.P., Renno, J., Loe-Mie, Y., Lifshitz, A., Mukamel, Z., Schmutz, S., Novault, S., Steinmetz, P.R.H., Spitz, F., Tanay, A., Marlow, H., 2018b. Cnidarian cell type diversity and regulation revealed by whole-organism single-cell RNA-Seq. *Cell* 173, 1520–1534 e1520.

Shapiro, E., Biezuner, T., Linnarsson, S., 2013. Single-cell sequencing-based technologies will revolutionize whole-organism science. *Nat. Rev. Genet.* 14, 618–630.

Snider, J., Kotlyar, M., Saraon, P., Yao, Z., Jurisica, I., Stagljar, I., 2015. Fundamentals of protein interaction network mapping. *Mol. Syst. Biol.* 11, 848.

Soldatov, R., Kaucka, M., Kastriti, M.E., Petersen, J., Chontorotzea, T., Englmaier, L., Akkuratova, N., Yang, Y., Haring, M., Dyachuk, V., Bock, C., Farlik, M., Piacentino, M.L., Boismoreau, F., Hilscher, M.M., Yokota, C., Qian, X., Nilsson, M., Bronner, M.E., Croci, L., Hsiao, W.Y., Guertin, D.A., Brunet, J.F., Consalez, G.G., Ernfors, P., Fried, K., Kharchenko, P.V., Adameyko, I., 2019. Spatiotemporal structure of cell fate decisions in murine neural crest. *Science* 364, eaas9536.

Sommer, R.J., 2012. Evolution of regulatory networks: nematode vulva induction as an example of developmental systems drift. *Adv. Exp. Med. Biol.* 751, 79–91.

Spemann, H., 1905. Über Linsenbildung nach experimenteller Entfernung der primären Linsenbildungszellen. *Zool. Anzeiger* 28, 419–432.

Spemann, H., 1918. Über die Determination der ersten Organanlagen des Amphibienembryo. *Archiv für Entwicklungsmechanik* 43, 448–555.
Spemann, H., 1936. *Experimentelle Beiträge zu einer Theorie der Entwicklung*. Springer, Berlin.
Spemann, H., Geinitz, B., 1927. Über Weckung organisatorischer Fähigkeiten durch Verpflanzung in organisatorische Umgebung. *Archiv für Entwicklungsmechanik* 109, 129–175.
Szathmary, E., 1995. A classification of replicators and lambda-calculus models of biological organization. *Proc. Biol. Sci.* 260, 279–286.
Szathmary, E., Maynard Smith, J., 1995. The major evolutionary transitions. *Nature* 374, 227–232.
Szathmary, E., Maynard Smith, J., 1997. From replicators to reproducers: the first major transitions leading to life. *J. Theor. Biol.* 187, 555–571.
Tarashansky, A.J., Musser, J.M., Khariton, M., Li, P., Arendt, D., Quake, S.R., Wang, B., 2021. Mapping single-cell atlases throughout Metazoa unravels cell type evolution. *Elife* 10, 66747.
Tarashansky, A.J., Xue, Y., Li, P., Quake, S.R., Wang, B., 2019. Self-assembling manifolds in single-cell RNA sequencing data. *Elife* 8, 48994.
Tickle, C., Martinez Arias, A., Placzek, M., Wolpert, L., 2025. *Wolpert's principles of development*, 7th ed. Oxford University Press, Oxford.
Trpchevska, N., Freidin, M.B., Broer, L., Oosterloo, B.C., Yao, S., Zhou, Y., Vona, B., Bishop, C., Bizaki-Vallaskangas, A., Canlon, B., Castellana, F., Chasman, D.I., Cherny, S., Christensen, K., Concas, M.P., Correa, A., Elkon, R., Estonian Biobank Research, T., Mengel-From, J., Gao, Y., Giersch, A.B.S., Girotto, G., Gudjonsson, A., Gudnason,V., Heard-Costa, N.L., Hertzano, R., Hjelmborg, J.V.B., Hjerling-Leffler, J., Hoffman, H.J., Kaprio, J., Kettunen, J., Krebs, K., Kahler, A.K., Lallemend, F., Launer, L.J., Lee, I.M., Leonard, H., Li, C.M., Lowenheim, H., Magnusson, P.K.E., van Meurs, J., Milani, L., Morton, C.C., Makitie, A., Nalls, M.A., Nardone, G.G., Nygaard, M., Palviainen, T., Pratt, S., Quaranta, N., Ramo, J., Saarentaus, E., Sardone, R., Satizabal,C.L., Schweinfurth, J.M., Seshadri, S., Shiroma, E., Shulman,E., Simonsick,E., Spankovich,C., Tropitzsch, A., Lauschke, V.M., Sullivan, P.F., Goedegebure, A., Cederroth, C.R., Williams, F.M.K., Nagtegaal, A.P., 2022. Genome-wide association meta-analysis identifies 48 risk variants and highlights the role of the stria vascularis in hearing loss. *Am. J. Hum. Genet.* 109, 1077–1091.
True, J.R., Haag, E.S., 2001. Developmental system drift and flexibility in evolutionary trajectories. *Evol. Dev.* 3, 109–119.
Uexküll, J.V., 1928. *Theoretische Biologie*. Springer, Berlin.
Van Keymeulen, A., Mascre, G., Youseff, K.K., Harel, I., Michaux, C., De Geest, N., Szpalski, C., Achouri, Y., Bloch, W., Hassan, B.A., Blanpain, C., 2009. Epidermal progenitors give rise to Merkel cells during embryonic development and adult homeostasis. *J. Cell Biol.* 187, 91–100.
Varela, F., Maturana, H.R., Uribe, R.B., 1974. Autopoiesis: the organization of living systems, its characterization and a model. *Biosystems* 5, 187–196.
Villaverde, A.F., Banga, J.R., 2014. Reverse engineering and identification in systems biology: strategies, perspectives and challenges. *J. R. Soc. Interface* 11, 20130505.
von Bertalanffy, L., 1950. An outline of general system theory. *Br. J. Philos Sci.* 1, 134–165.
Waddington, C.H., 1957. *The strategy of the genes*. George Allen and Unwin, London.
Wagner, D.E., Weinreb, C., Collins, Z.M., Briggs, J.A., Megason, S.G., Klein, A.M., 2018. Single-cell mapping of gene expression landscapes and lineage in the zebrafish embryo. *Science* 360, 981–987.
Wagner, G.P., 1989. The biological homology concept. *Annu. Rev. Ecol. Syst.* 20, 51–69.
Wagner, G.P., 1995. The biological role of homologues: a building block hypothesis. *N. Jb.Geol.Palä ont.Abh.* 195, 36–43.

Wagner, G.P., 1996. Homologues, natural kinds and the evolution of modularity. *Am. Zool.* 36, 36–43.
Wagner, G.P., 2007. The developmental genetics of homology. *Nat. Rev. Genet.* 8, 473–479.
Wagner, G.P., 2014. *Homology, genes, and evolutionary innovation*. Princeton University Press, Princeton.
Wagner, G.P., Altenberg, L., 1996. Complex adaptations and the evolution of evolvability. *Evolution* 50, 967–976.
Wagner, G.P., Pavlicev, M., Cheverud, J.M., 2007. The road to modularity. *Nat. Rev. Genet.* 8, 921–931.
Wallis, D., Hamblen, M., Zhou, Y., Venken, K.J., Schumacher, A., Grimes, H.L., Zoghbi, H.Y., Orkin, S.H., Bellen, H.J., 2003. The zinc finger transcription factor Gfi1, implicated in lymphomagenesis, is required for inner ear hair cell differentiation and survival. *Development* 130, 221–232.
Weinreich, D.M., Watson, R.A., Chao, L., 2005. Perspective: sign epistasis and genetic constraint on evolutionary trajectories. *Evolution* 59, 1165–1174.
Weintraub, H., Tapscott, S.J., Davis, R.L., Thayer, M.J., Adam, M.A., Lassar, A.B., Miller, A.D., 1989. Activation of muscle-specific genes in pigment, nerve, fat, liver, and fibroblast cell lines by forced expression of MyoD. *Proc. Natl. Acad. Sci. U. S. A.* 86, 5434–5438.
Weismann, A., 1885. *Die Continuität des Keimplasma's als Grundlage einer Theorie der Vererbung*. Gustav Fischer, Jena.
Weiss, P., 1939. *Principles of development*. Henry Holt and Co., New York.
West-Eberhard, M.J., 2003. *Developmental plasticity and evolution*. Oxford University Press, Oxford.
Whitfield, T.T., 2015. Development of the inner ear. *Curr. Opin. Genet. Dev.* 32, 112–118.
Wu, D.K., Kelley, M.W., 2012. Molecular mechanisms of inner ear development. *Cold Spring Harb. Perspect. Biol.* 4, a008409.
Yadav, T., Quivy, J.P., Almouzni, G., 2018. Chromatin plasticity: a versatile landscape that underlies cell fate and identity. *Science* 361, 1332–1336.
Yntema, C.L., 1933. Experiments on the determination of the ear ectoderm in the embryo of *Amblystoma punctatum*. *J. Exp. Zool.* 65, 317–357.
Yntema, C.L., 1939. Self-differentiation of heterotopic ear ectoderm in the embryo of *Amblystoma punctatum*. *J. Exp. Zool.* 80, 1–17.
Yu, H.V., Tao, L., Llamas, J., Wang, X., Nguyen, J.D., Trecek, T., Segil, N., 2021. POU4F3 pioneer activity enables ATOH1 to drive diverse mechanoreceptor differentiation through a feed-forward epigenetic mechanism. *Proc. Natl. Acad. Sci. U. S. A.* 118, e2105137118.
Zhao, C., Emmons, S.W., 1995. A transcription factor controlling development of peripheral sense organs in *C. elegans*. *Nature* 373, 74–78.
Zheng, J.L., Gao, W.Q., 2000. Overexpression of Math1 induces robust production of extra hair cells in postnatal rat inner ears. *Nat. Neurosci.* 3, 580–586.
Zhu, M., Zhao, S., 2007. Candidate gene identification approach: progress and challenges. *Int. J. Biol. Sci.* 3, 420–427.
Zuniga, A., 2015. Next generation limb development and evolution: old questions, new perspectives. *Development* 142, 3810–3820.

2 Biophysical Intelligence Between Genotype and Phenotype

The Agential Material of Life From EvoDevo to Regenerative Medicine

Michael Levin

2.1 INTRODUCTION

Each of us has made the remarkable journey from a single cell to a cognitive being capable of understanding our own origin story. Of course, that same kind of journey also takes place on an evolutionary time scale, during the eons in which we moved from our microbial ancestors to complex multicellular beings working in laboratories to understand the mechanisms and the umwelt of their ancestors. For the purposes of this chapter, I will use "morphogenesis" to refer to the entirety of this process – the ability of a single egg cell to eventually self-assemble into a complex form of anatomy and behavior.

2.1.1 Morphogenesis at the Center of the Big Questions of Life and Mind

In a sense, developmental biology is the queen of all the sciences. Within the morphogenetic process are hidden answers to all of life's big questions. First, the increase of spatial complexity – self-organizing order on many scales, which connects to questions of emergence, entropy, and collective intelligence. Second, the path to definitive regenerative medicine: here in front of our eyes are processes that regenerate an entire body from one cell (development), repair against drastic injury (split embryos become normal twins), bind individual cell behaviors toward a larger agenda (cancer suppression and reversion), and produce organisms of age 0, not starting at the age of their parent cells or their entire lineage (antiaging). Third, our path across many sciences: starting off as an oocyte – the province of chemistry and physics – we eventually became subjects of biology and physiology, then of behavioral science, and

 DOI: 10.1201/9781003560784-3

eventually of psychoanalysis (not to mention friendship, love, etc.). Thus, morphogenesis challenges us not only to understand the mechanisms that enable it but also to define the lenses through which we see life – the various disciplines appropriate for the study of our different stages of transformation.

Perhaps the most important lessons morphogenesis teaches are those of gradualism. Yes, we've given convenient names to individual stages of development, and it is clear that adults have different capabilities than embryos or egg cells. The standard story, with crisp boundaries between sciences, claims that the final product of morphogenesis is a being with an inner world – having a first-person perspective – which the quiescent oocyte does not have (being a "mere blob of chemicals"). But, developmental biology also tells us that there is no bright red line – no magic point in which we tick over from being chemico-physical machines to cognitive agents with minds; what it reveals is a gradual process of metamorphosis. Whatever hopes, dreams, preferences, beliefs, and consciousness we may have develop slowly and gradually, as does our body. The process can be slowed down, temporally zoomed in as far as one wants, to see clearly that "and then the brain develops" and similar stages are extremely lengthy, involved continuous processes. The sharp categories, names, and disciplines we paint onto this process are a feature of our attempts to simplify and understand, not necessarily intrinsic features.

This aspect is rarely appreciated outside of our field. For example, many books and papers in the philosophy of mind talk about "the human" with its mind, moral sense, responsibility, etc. They do not generally ask about ontogenetic and phylogenetic ancestors, to discuss when and how single cells became something that has these properties. A lack of appreciation of our origins (and of the plasticity of our embodiment) obscures the many unknowns that lie just beneath the convenience of expedient and discrete terms such as "organism", "machine", "brain", "mind", "intelligence", etc., all of which give a false sense of understanding of the deepest unanswered questions at their edges. Thus, developmental biology and its sister disciplines, bioengineering, synthetic biology, artificial life, etc., are essential backgrounds for progress in many aspects of law, philosophy, and other humanities. All of those communities assume that we have answers that underlie their frameworks; we need to either provide them or communicate the needed changes in those frameworks.

It is crucial that morphogenesis is the study not just of the origin of the bodily anatomy but also of the mind. Alan Turing understood this. His seminal paper (Turing, 1952) addressed the question of where patterns come from *ab initio*, from a chemical soup into structures of embryogenesis. Why would someone who was so interested in intelligence in unfamiliar implementations, in reprogrammability (plasticity) (Turing, 1950), in the power (and limits) of algorithms, and in the foundations of mathematics write one of the first papers quantitatively modeling the origin of order from homogeneous media? Because Turing realized the profound symmetries between the origins of the body and those of the mind, he saw that questions of intelligence, of patterns arising (and being detected) from chaos, go beyond specific embodiments, problem spaces, or origin stories. In our quest to understand what it means to be an embodied mind in our universe, we must be prepared to recognize it in unfamiliar guises. Transitional forms, such as those of embryogenesis, provide an ideal training ground in which to improve our conceptual frameworks and our technologies.

2.1.2 An Idiosyncratic Perspective on Morphogenesis: A Preview

I argue that one thing that morphogenesis makes clear is that what we seek are models of *transformation.* We are not looking for sharp definitions of "life" and of various cognitive terms. The search for clear definitions has held back the field because of a focus on pseudoproblems of classification and because it prevents the transfer of tools across disciplines. Instead, what we must search for are principles of *scaling* – mechanisms and patterns by which capabilities increase and project into different problem spaces as oocytes become complex organisms (Levin, 2022a). What forces are driving beings across the spectrum from chemical machine to mature mind? What mechanisms scale the tiny setpoints of physiological loops in single cells into the grandiose goal states of cell groups as they implement complex organs or the planetary-scale goals of human minds?

In this chapter, I discuss what I think are some of the most interesting aspects of morphogenesis, from an admittedly unconventional and interdisciplinary perspective that has guided my group's work in this field for several decades. Specifically, I propose that morphogenesis, as the layer between the genotype and phenotype, is not merely a complex mapping but a fundamentally cognitive one. Living matter is an *agential material*, representing a next challenge for engineers past the passive matter of past millennia and the active and computational matter of the last few decades. I also review an emerging field of developmental bioelectricity, which shows that the cognitive glue (Levin, 2023) – the mechanisms that align cells toward large-scale anatomical goals – is highly conserved between the behavior of brains and of bodies. To understand this perspective, a few things must be kept in mind.

First, what needs to be explained about morphogenesis? It is not merely the reliability (consistency) or the rise in complexity of the outcome of developmental processes. I focus here on examples typically omitted from modern developmental biology classes: phenomena that reveal the context-sensitive problem-solving aspects of the process (a.k.a., intelligence). Second, what is the role of the genome? I discuss recent advances that help understand its role, the software modes of the genetically specified hardware of life, and what the genome does and doesn't encode. Third, what are the implications of all this for evolution? I argue that evolving an agential material, vs. a passive one, has many interesting consequences (not the least of which is an agency ratchet that kicks in prior to self-replication and selection).

What is at stake, in moving from a bottom up view of molecular biology as a privileged layer of explanation (Noble, 2012) to a multiscale competency architecture model more consistent with recent advances in causal information theory (Albantakis et al., 2019; Hoel and Levin, 2020; Hoel, 2017; Hoel, 2018; Hoel et al., 2016a; Hoel et al., 2013a; Klein and Hoel, 2019)? It's hard to overemphasize the importance of getting this right – moving beyond constraining mechanist and organicist models to understand the deep symmetry at the heart of life and mind. The first impacts will concern the biomedicine of birth defects, traumatic injury, cancer, and aging (Levin, 2024b). Additional impacts will affect the basic understanding of evolution, enabling tools for bioengineering and synthetic morphology (the next generation past synthetic biology) (Davies and Levin, 2023), transhumanist efforts to modify our form and cognition beyond what random mutation and selection have given us so far, and the design of artificial intelligence (AI) and robotics built on the deep lessons from the agential material of life.

2.2 THE UNREASONABLE EFFECTIVENESS OF THE BEHAVIORAL SCIENCES IN DEVELOPMENTAL BIOLOGY

A famous essay (Wigner, 1960) by Eugene Wigner asks why mathematics is so surprisingly effective in explaining and controlling the physical world. But any estimate of whether a formalism is unreasonably (surprisingly) effective requires that we have good priors with respect to which we can form expectations. Progress is inhibited when assumptions about what tools are appropriate for phenomena like morphogenesis are treated as axioms. All such assumptions, such as the ubiquitous but often unspoken claim that formalisms suitable for low-level machines ("mechanisms") are more facilitative of progress than those suitable for richer agents, are currently being tested.

2.2.1 What Formalism Is Appropriate?

An important tool for understanding processes like morphogenesis is computational modeling. Computer models require making assumptions explicit and effectively highlight gaps in knowledge (as exist, e.g., when trying to flesh out the "arrow models", popular in developmental biology papers, into something that will actually run). Computational models also force a clarity of perspective of how one thinks of the system being modeled: creating an abstraction of cells, signals, or shapes, one has to specify their properties and capabilities.

Cellular automata (CA) models (Akberdin et al., 2007; Hatzikirou and Deutsch, 2008; Lehotzky and Zupanc, 2019; Markus et al., 1999) have been popular in developmental biology because they seem to capture the fundamental dynamic: a set of local rules by which cells interact. This seems reasonable at first, because development is so reliable – builds the same thing (nearly) every time – and because CAs don't require a central controller – the cells work in parallel. Also, they embody a fundamental assumption of the modern age: there is no goal, no sense in which the future outcome is represented, just local actions. This is considered a beneficial feature because of Morgan's Canon (Morgan, 1903) – the idea that all systems should be explained without recourse to cognitive concepts wherever possible. It also matches the pervasive intuition that chemical processes in morphogenetic systems simply *can't* know anything – they have to be automata.

This view has three major problems. First, it is so widely prevalent that workers in the field (especially students raised on this idea) forget that it is only an assumption, not a result, and that estimates of the level of intelligence of any system are an empirical question, not a linguistic or philosophical one, which must be settled by experiment. Second, it neglects neuroscience and cybernetics – two modern sciences that have made great progress on natural and artificial systems, which literally encode representations of goal states and exert effort to meet those goals (with variable degrees of ingenuity). We now know that it doesn't take magic to implement goal-seeking systems, and thus such models in developmental biology are not an appeal to mysterianism but rather a natural attempt at extension of rigorous tools used with great profit elsewhere. Third, it appeals to "emergence" to explain the actual complex forms that result. What does "emergence" mean? One thing it signifies is surprise: if the appearance of outcomes from simple rules were not foreseen by a typical scientist presented with the rules, it's emergent. If a typical modern human can look at the rules and immediately see what will happen, it's not emergent. The notion of emergence as observer–relative surprise is not particularly useful. What it obscures is the serious problem faced by workers who are certain that chemical material can't know anything, but (usually) believe that they, and their chemical brain, really do know things. The highly rewarded ability to keep those mentally separate, and to believe that developmental biology doesn't need to concern itself with cognition, is a major barrier to progress.

This framework is very limiting for biomedicine and bioengineering, because morphogenesis, as viewed by CA models, is not reversible – it only goes in one direction. Given the rules, one simply runs them forward to see what emerges. But what if, as in engineering, synthetic morphology, or regenerative medicine, one wants to go the other way: given what we want to happen (system-level, large-scale outcomes), what should the rules and starting conditions be? That is, in general, an intractable problem (Lobo et al., 2014). CA models simply offer no tractable way to design outcomes because the gulf between the parameter knobs (rules) and outcomes is too large and convoluted. There are, of course, systems (such as the Game of Life classic automaton) where that is the best we can do. But biology offers alternatives. For example, in homeostatic systems (which are ubiquitous in living material), one can shift the encoded setpoint and the system will do the rest. More advanced versions, such as those in brainy species, have goal states that can be reset by training or communication, and offer many highly useful competencies for achieving those goals without having to be micromanaged to do so. The question is, what does it take to be a goal-directed system, and do morphogenetic systems have it?

I have coined the idea of a "Spectrum of Persuadability" (Levin, 2022a) to adopt a practical, engineering perspective focused on interventions and interactions with a system. It is critical to note that *where* any given system fits on the Spectrum of Persuadability (Figure 2.1) is not knowable from a philosophical armchair – we cannot simply skew low (as reductionism and Morgan's Cannon would have us do), nor can we always skew high (as animist and mysterian approaches often do). We start with plausible estimates, and then we do experiments to see how that works out, being ready to revise our estimate in either direction. This works because cognitive claims are *interaction protocol* claims. When one says that a system, such as an embryo, has a certain level of intelligence, they are making a claim as to what kind of operational toolkits – physical rewiring, cybernetics/control theory, behavior science, psychoanalysis, etc. – will offer the best prediction and control (or more broadly, the most beneficial interaction). It is an iterative, empirical process because,

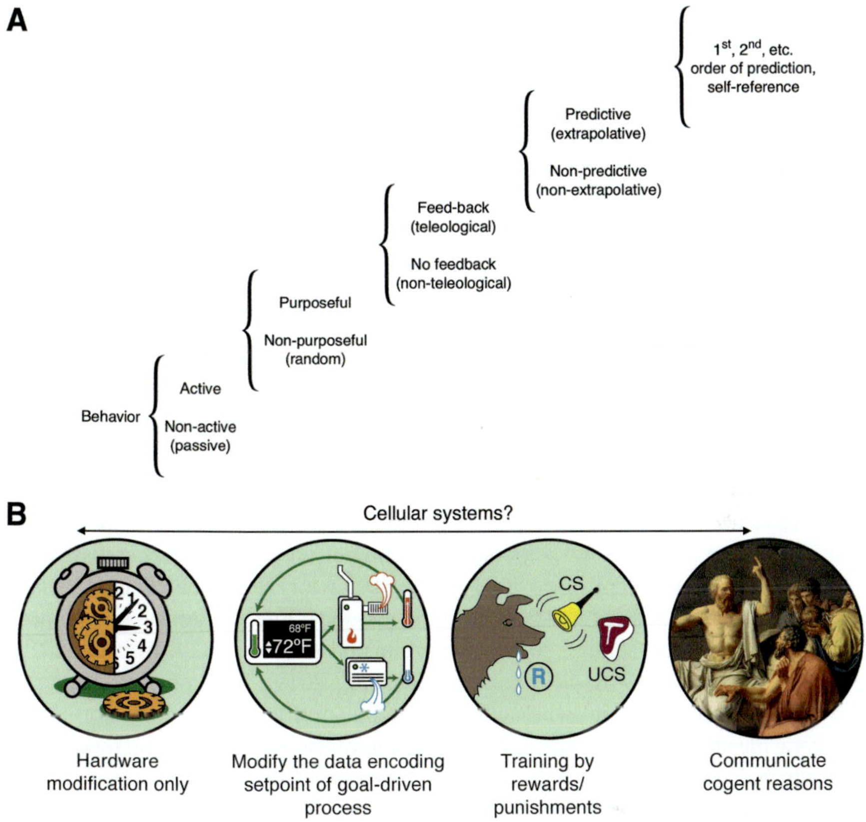

FIGURE 2.1 Agentic models are not mysterian. A common assumption holds that only first-order (feedforward) models, in which outcomes are "emergent" from the action of subunits where nothing knows the destination of the process, can be rigorous. In contrast, cybernetics, control theory, and other fields have long produced tools to understand and manage systems with feedback, including a range of progressively more sophisticated architectures of machines that have goals without magic, ranging from passive matter to human level metacognition (A) (Rosenblueth et al., 1943). A developing system moving from an oocyte to a human being with a brain traverses the types of dynamical control shown in panel A. The Spectrum of Persuadability (B) illustrates four waypoints on this continuum and is organized from a pragmatic, empirically-focused perspective (Levin, 2022a). On this view, all systems (including cellular collectives) must be experimentally probed, not philosophically decreed, to determine what kind of interaction is optimal: hardware rewiring (e.g., genetic engineering and molecular biology), setpoint resetting (editing biophysically encoded homeostatic setpoints), training (positive and negative reinforcement stimuli), or other forms of high-level communication.

Panel B by Jeremy Guay of Peregrine Creative.

having made the claim, everyone can see how it worked out. One can't simply know the intelligence level of any system by observations of its behavior alone. What one must do, especially to gauge the level of sophistication in goal-directed activity, is perform *perturbative experiments*: for example, place a barrier between the system and its hypothesized goals (in whatever space it hypothesizes it to be navigating), and see how much it can implement James' definition of intelligence: "same goal by variable means" (James, 1890). Feedforward, cellular automaton models do not capture these kinds of dynamics. Fortunately, if we accept that favoring open-loop (feedforward, goal-less, know-nothing) mechanisms is only an assumption that we are not bound to, with a whole spectrum of rigorous tools available (Figure 2.1A), a number of other options suggest themselves as interesting research agendas that have already begun paying off.

2.2.2 On the Application of Mentalistic Terminology in Developmental Biology

The presence of homeostatic loops (basal version of goal-seeking mechanisms) in biology is well-known. It's seen in physiological regulation (Cannon, 1929, 1932), problem-solving behavior, and the ability of cancer to evade various therapeutics over time. But homeostasis is the first rung on the ladder of intelligence, implementing a basic ability to store a setpoint (memory), determine current state in order (sensing), and implement that setpoint by exerting effort and minimizing distance to it (navigation of a problem space, goal-seeking behavior) (Rosenblueth et al., 1943). Subsequent elaborations of this basic scheme include homeodynamics, homeorhesis (Chuang et al., 2019; Colditz, 2020; Matsushita and Kaneko, 2020; Webster, 2019), allostasis (McEwen, 1998; Schulkin and Sterling, 2019), and more sophisticated mechanisms recognized in cybernetics and behavioral science, which correspond to increasing degrees of competency and ingenuity in which living materials have to reach their goals despite unforeseen challenges and perturbations. This enables an empirical research program that asks, what set of tools along the spectrum of intelligence is appropriate for any given biological system? Crucially, this is not a philosophical or linguistic project seeking to redefine intelligence as mere complexity. It is an experimental research program that seeks to *rigorously determine what problem-solving capacities can be found, and exploited (by bioengineers, other biological systems, and evolution itself), in living materials* and beyond. To see why, it's important to dig into how agentic terminology is normally used.

The use of "intelligence" and other cognitive terms applied outside of their familiar context of brainy animals immediately raises questions: might not these terms be misused? Are not morphogenetic systems simply following the rules of chemistry – why anthropomorphize them? In the modern age, we must accept that *all* cognitive systems – we included – exhibit chemistry, not magic, when one drills down to examine the lower levels. Thus, there simply is no special human category of which one can correctly anthropomorphize as somehow being beyond the laws of physics at its base. That word is an anachronism and needs to be retired in favor of an empirically grounded view, updated with the latest findings in causal information theory (Albantakis et al., 2017; Hoel and Levin, 2020; Hoel, 2017; Hoel, 2018; Hoel et al., 2016b; Hoel et al., 2013b; Moore et al., 2018), in which it is perfectly possible for a system to be subject to chemistry and also have additional levels that have causal roles. Indeed, the standard view of modern developmental biology is not really reductionist (in which case it would seek explanations of events in terms of quantum foam), but rather chemo-centric, having picked the level of "chemistry" as an arbitrary gold standard for explanations in the field. Denis Noble (Noble, 2010, 2011, 2012) and others have sufficiently pointed out that there is no privileged level of causation in biology.

Fortunately, we have many tools available to us besides the formalisms of chemistry, including those of cybernetics and behavioral science. I focus on two main points in clarifying the use of cognitive terminology (developed in detail in Levin, 2022b). First, that almost any paradigm can be rescued by enough epicycles; indeed, after one has produced a new effect or reached a new capability, it is easy to say "that's just chemistry doing what chemistry does". The criterion should not be whether some dynamic can be described at the lowest level (chemistry? Quantum mechanics?). Instead, the emphasis should be on novel capabilities and new research programs facilitated (or suppressed) by a given way to understand a system. I propose that attempts to mine the rich toolbox of behavioral science to exploit capabilities of morphogenetic systems will continue to pay off in many (but likely not all) cases. The existing gains facilitated by this view, and the promise for regenerative medicine, are detailed elsewhere (Lagasse and Levin, 2023; Mathews et al., 2023).

A sometimes uncomfortable aspect of my position is that the empirical utility of framings needs to be applied fearlessly and followed wherever it may lead: its empirical consequences must be taken seriously even when they contradict long-cherished a priori commitments to how non-intelligent a given system "must be". In other words, if a specific framing, which uses tools normally reserved for brains, results in fruitful new research programs on bacterial biofilms (Martinez-Corral et al., 2019; Prindle et al., 2015; Yang et al., 2020), plant roots (Bassel, 2018; Calvo et al., 2016; Ciszak et al., 2016; Davis et al., 2023; Johnston and Bassel, 2018; Topham et al., 2017; Yokawa et al., 2018), the training of gene-regulatory networks (Biswas et al., 2023; Biswas et al., 2021; Csermely et al., 2020; Gyurkó et al., 2013), or developmental/regenerative biology (Levin, 2021), then the scientific approach requires that we consider those systems to be bona fide subjects of that corner of the natural world that is supposed to be described by the behavioral science of a spectrum of minds.

To cut through philosophical impasses and hew close to testable hypotheses, empirical progress, and research agendas at the bench, I take an operationalist stance with respect to questions like "but is it really cognitive or only metaphorically so?" In keeping with a fairly standard metaphysics of science, I eschew any use of "really". All we have in science is metaphors, and an unflinching commitment to testing them and seeing which ones offer the best purchase on the world. Thus, consistent with an engineering approach (formalized in the Technological Approach to Mind Everywhere (TAME) framework; Levin, 2022a), I define the perspective of this chapter as follows: what tools offer the most insight, in terms of not only predictive capacity but in terms of generatively driving new research questions and new capabilities? Specifically, might some of those tools be found within the cognitive and behavioral sciences? It should not be a surprise, because of course, the endpoint of development is the construction of a system to which those tools very effectively apply. Do they only kick in after the brain is mature, or might basal versions be found earlier on in the process? We next look at a few specific examples that motivate the use of such models in morphogenetic contexts.

2.2.3 Specific Examples: Progress Made by Porting Tools From Cognitive Science

The perspective I propose is of the following symmetry: much like animal behavior in 3D space is the result of a collective intelligence composed of neurons aligned in networks, navigation of the space of possible anatomies (anatomical morphospace) is also a kind of behavior of the cellular collective. What kind of navigational competencies does morphogenesis offer?

The first is the ability to reach the same goal from different starting positions. A salamander's limb amputated at any position along the proximodistal axis will regenerate the correct amount of tissue, in the correct structure, and *then stop* (Figure 2.2A). This is the most remarkable aspect of regeneration – it stops when a specific outcome is achieved (the "target morphology"). The system reliably navigates from different starting locations in morphospace and continues moving until the delta between target morphology and current position is within acceptable levels of error. Many species' embryos, when cut into pieces, give rise to monozygotic twins, triplets, etc., not half-bodies. Conversely, multiple separate embryos coalesced into one also form normal embryos (even if carcinoma cells are mixed in) (Mintz, 1971; Mintz and Illmensee, 1975). Scrambled tadpole faces give rise to normal frog faces (Pinet et al., 2019; Vandenberg et al., 2012) because craniofacial organs don't simply move in pre-determined paths during metamorphosis, but rather move *as needed* (thus accommodating unexpectedly incorrect starting positions) to achieve a normal frog face (Figure 2.2B). Given that development is basically regeneration of an entire body from one cell, they both share the ability to attempt to find and maintain a specific region of anatomic morphospace. In effect, morphogenesis of development, regeneration, wound healing, and resistance to aging and cancer are all kinds of error minimization driven by feedback loops.

A closely related concept is that of memory: systems that react to the same inputs differently, after they've experienced them. Homeostatic processes have an explicit setpoint to which they minimize distance (or responsively, or even proactively, as in the concept of allostasis (Adar et al., 2025; Harrison et al., 2025; McEwen, 1998; Schulkin and Sterling, 2019; Sennesh et al., 2022)). Storing that setpoint is a kind of memory; interestingly, that memory is often not hardwired – there is a default, but it's rewritable. One example is trophic memory in deer (Bubenik, 1966; Bubenik and Pavlansky, 1965): a wound made at a specific location in a deer antler rack will result in an ectopic branch in *subsequent years* as the antler rack is shed and regenerated (Figure 2.2C). Imagine trying to write a typical arrow diagram for a molecular biology model as is standard in developmental biology papers and textbooks: the structure must ascertain the position of the damage, store this information somewhere else in the body for months (while the whole antler rack falls off), and act on it via specific changes to bone, nerve, vasculature, and skin growth at a later timepoint. This is a kind of physiological memory of damage and a response of tissue growth whose future is different from its past (or from its default) because of experiences it has had.

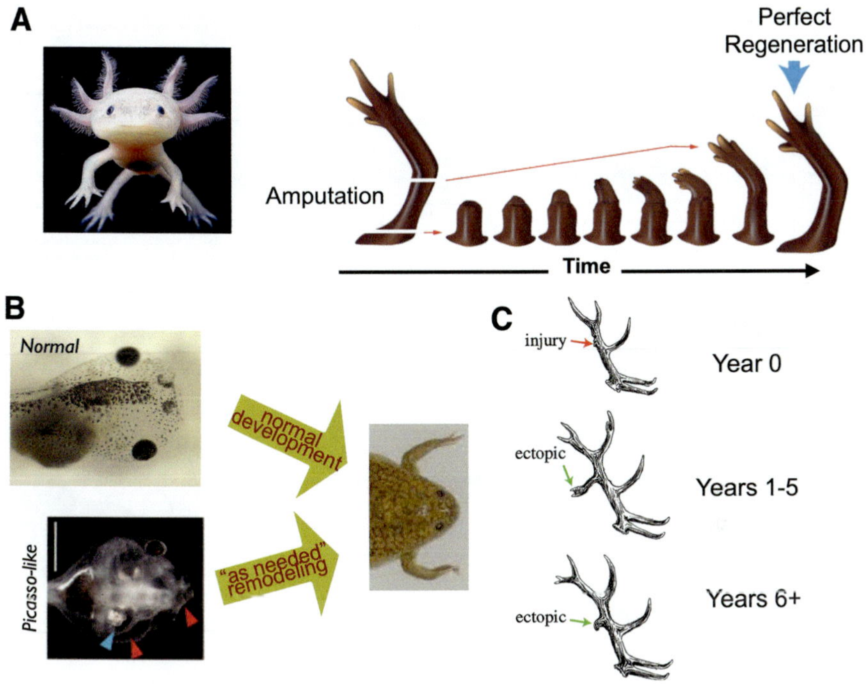

FIGURE 2.2 Feedforward models are a poor fit for fascinating aspects of developmental biology. Axolotls and other amphibians regenerate their limbs, showing how a complex system reliably returns to a specific morphology after it's been deviated and stops when the anatomical goal is achieved (A). Like the limb, which achieves its endpoint regardless of where it was amputated, *Xenopus laevis* tadpoles, which have had their craniofacial organs scrambled (eye on top of the head, aqua arrowhead, jaw structures off to the side, red arrowheads), still become normal frogs (B) because morphogenesis is not a prescribed series of steps, but a context-sensitive process that performs novel actions as needed to reach a specific endpoint state. Image in panel A by Jeremy Guay of Peregrine Creative. Images in panel B used with permission from (Vandenberg et al., 2012). In deer antlers, damage at a specific point, in antlers that are soon shed, causes ectopic tines to grow at that location in subsequent years, illustrating the long-term, nonlocal storage of information that guides future morphogenetic outcomes (C). Dynamic morphogenesis is not well described by open-loop feedforward models but offers many fascinating phenomena that require homeostatic, homeodynamic, or allostatic kinds of models (i.e., systems with setpoints, memory, and ability to reach goals from different starting positions by diverse means).

Image in panel C used with permission from Lobo et al. (2014).

Other examples of rewritable pattern memory will be discussed below in the context of bioelectricity in planaria.

Crucially, living matter implements a multiscale competency architecture, with each level of organization having its own ability to meet goals and implement target states. Life operates in many spaces beyond the 3D space of conventional behavior (Fields and Levin, 2022). Anatomical morphospace is one, but others include

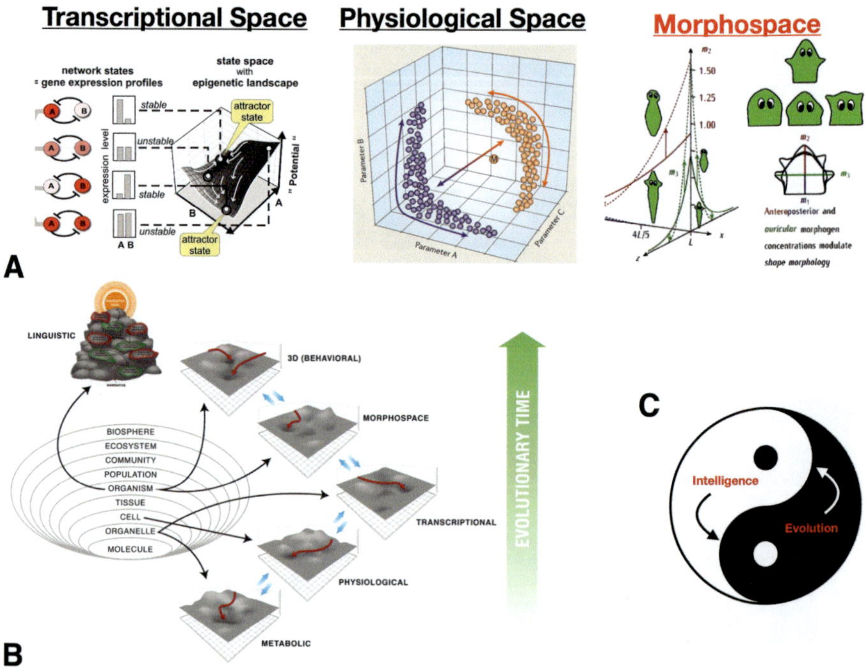

FIGURE 2.3 Living systems exhibit behavior in many spaces besides conventional 3D space. Evolution produces systems that adaptively navigate the space of gene expression, physiological states, and anatomical possibilities (A). Evolution reused some of the same mechanisms and algorithms, pivoting living systems across metabolic, physiological, transcriptional, morphological, behavioral, and linguistic systems over time (B). Intelligence and evolution form an ongoing positive feedback spiral, as autonomous competencies of the living material potentiate evolvability and force evolution of problem-solving agents instead of fixed solutions, which in turn produce systems with higher cognitive capacity (C).

Panels in A used with permission from Huang et al. (2009), Marder and Goaillard (2006), and Cervera et al. (2021), respectively. Panel B by Jeremy Guay of Peregrine Creative.

metabolic space, physiological state space, transcriptional space, etc.; it is plausible that evolution has reused some of the same mechanisms and algorithms across spaces as it expanded and complexity increased (Figure 2.3).

Morphogenesis is the result of cooperative (and competitive, see Gawne et al., 2020) actions of individual cells. But the collective's ability to navigate anatomical morphospace is the result of coordinated behaviors of individual cells, which are themselves made of competent parts. Gene regulatory networks and chemical pathways can, on their own, exhibit six different kinds of learning, including habituation, sensitization, and even Pavlovian conditioning; they can even count to small numbers (Biswas et al., 2023; Biswas et al., 2021). These were discovered when standard assays of behavioral science were applied to the functionality of GRNs; prior models of GRNs as dumb dynamical systems implied that rewiring (e.g., gene therapy) was the only way to change their behavior. Taking seriously the fact that a need for hardware rewiring was only an assumption, testing behaviorist protocols revealed that even the minimal mathematics of GRNs and pathways (coupled ordinary differential equations) were sufficient to result in a learning system whose future behavior could be altered by prior experience without any changes in the hardware or its parameters (thus opening the door to a myriad applications in drug conditioning and cell training).

The scale-free competency also applies *above* the level of individual embryos and the cell–tissue–organ–system levels operating within them. It was recently found that not only do waves of ion flux propagate across groups of embryos, but their communication actually enables large collectives of embryos to resist teratogens better than small collectives or singletons (Tung et al., 2024). This CEMA (Cross-Embryo Morphogenetic Assistance) effect shows that groups of embryos solve physiological problems better than singletons. Indeed, such hyperembryos have their own transcriptomes and express different genes than do single embryos on their own.

2.2.4 Flow of Control in Systems Made of Competent Subsystems

What is happening in living materials is that each level of organization operates as a virtual governor, which deforms the option space for its parts (Dewan, 1976; Pezzulo and Levin, 2016). The components don't know that (or why) their problem-solving actions are actually contributing to the navigation of a space of which they are unaware. What is "an embryo", really? What is there one of, and what are we counting when we look at 10^6 cells in a blastoderm and call it "one embryo" (Figure 2.4)? What we are observing is alignment: both physical (Gray et al., 2011) and teleonomic (Levin, 2019) alignment of cells toward one specific target in anatomical space. Wholes must constantly enforce the alignment of their parts toward specific goals; defects in this process manifest as aging, degeneration, and cancer (Rubin, 1985, 2006, 2007; Rubin et al., 1996).

Now, the question of whether these higher levels are real (in the sense of having their own causal power, as in downward causation; Auletta et al., 2008; Craver and Bechtel, 2007; Ellis, 2009; Ellis and Bloch, 2011; Juarrero, 2009; Okasha, 2012), or are just convenient ways of speaking, used to be a philosophical debate between reductionists and those convinced of a more holistic picture of the world. It is important for developmental biologists to be aware of recent developments in causal information theory, which, remarkably, have moved this debate from the field of philosophy to that of tractable mathematics: there are now tools (Biswas et al., 2021; Cliff et al., 2017; Hoel and Levin, 2020; Lizier, 2014; Lizier et al., 2011; Varley and Hoel, 2022; Wang et al., 2012; Wibral et al., 2014) that can be applied to data from living systems to rigorously answer the question of whether higher levels of organization have their own causal power over and above whatever their mechanistic parts are doing. The definitive answer is that sometimes they can (some systems have maximum causation at levels above their smallest components), and biological systems are being quantified with respect to this, which means that for the first time we may soon have good quantitative definitions of what an "organ" and an "embryo" are.

One of the most fascinating aspects of this multiscale architecture is how control is propagated across scales. Consider the example of what happens when a tail is

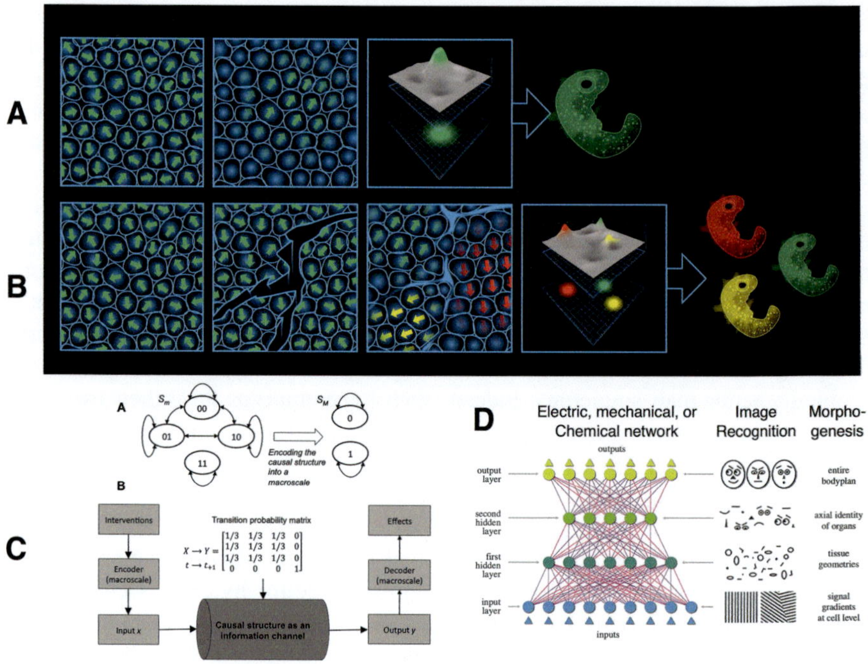

FIGURE 2.4 Collective systems are multiscale agents with causal emergence. An "embryo" is detected when a group of individual cells shares a joint homeostatic goal in anatomical space, cooperating to reach a specific outcome (A). Embryonic blastoderms can be cut into pieces (after Lutz, 1949), where each island will make its own embryo, showing how an excitable medium (the cell field) can give rise to some number (not fixed genetically) of individuals based on the dynamic ability of cells to join into collectives with plastic boundaries (B). Crucially, new tools in information theory show how to quantitatively determine when a whole is more than the sum of its parts – the "embryo" is a macroscopic whole that can exert effects on its cells and chemical pathways, and causality does not always flow up from microscopic molecular events. In panel C (taken with permission from Hoel, 2017) is shown a simple example: a Markov chain with four microscale states that can be encoded into a macroscale chain with two states; the causal structure transforms interventions (functional signaling) into effects. A macro causal model is a form of encoding for the interventions (inputs) and effects (outputs) that can use a greater amount of the capacity of the communication channel, showing how top-down causation is not a philosophical whimsy but a measurable feature of certain systems (biological ones being superb examples, Hoel and Levin, 2020). Panel D shows how a network (electrical or chemical) can coarse-grain information to encode higher-level states: patterned on machine learning architectures (McClelland et al., 2010; McClelland and Rogers, 2003; Watson et al., 2022) and retinal processing (Berry et al., 1999; Grossberg, 1978; Marr, 1982; Melanitis and Nikita, 2019), such as is used in complex image recognition, intermediate layers can encode progressively more generalized information, such as organ-level shape decisions that are transduced by downstream pathways into biochemical states below the single cell level.

Panel taken with permission from Pezzulo and Levin, 2016.

grafted surgically to the middle flank of an amphibian body: it slowly remodels into a limb (Farinella-Ferruzza, 1953, 1956; Holtfreter, 1955). What this illustrates is, first, that the collective has a body-scale target morphology toward which it will try to reduce error in contexts beyond repair of injury or embryogenesis (i.e., context-sensitive remodeling to a specific setpoint) (Pinet and McLaughlin, 2019). Second, consider the perspective of the tail tip cells: they are sitting in their correct environment – no injury, no damage. Why are they turning into fingers? They don't know what a finger is, how many there should be, or why the whole thing is remodeling. What happens is that the morphogenetic goals at the organism level are being transduced down into the cell- and molecular-level signals needed to get the tissue to remodel into a new structure (see Figure 2.4D, Pezzulo and Levin, 2016). The effect is radically non-local, both in lateral space and in vertical levels of organization.

This ability of lower-level mechanisms to respond to higher-level goals is a critical part of cognition and highlights one of the most important symmetries between the embodiment of minds and morphogenetic competencies. This is seen in the everyday magic of the mind-body interface during voluntary motion. Upon awakening, a human mind may have very abstract goals – financial, social, scientific, spiritual, etc. In order for them to execute on those goals via behaviors, these abstract goals must eventually *make calcium and potassium ions cross muscle cell membranes* in ways they otherwise would not do. In other words, our bodies are a remarkable set of transducers that literally make chemistry obey abstract mental structures. This is implemented by bioelectricity (see Section 2.4 below), but could in theory be driven by many mechanisms. What is important is that just as in cognition, morphogenesis reveals how the behavior at the molecular level is driven by the goal states of higher levels. One simply cannot stay at a single level (e.g., the beloved molecular biology) and have full insight into what is happening or how to control it. Doing so is like giving an accounting of air molecules' elastic interactions when documenting the results of two mathematicians working on a wonderful new proof: it misses the whole point of what was happening and closes off the steps of discovery of the next big advance.

2.3 BETWEEN GENOTYPE AND PHENOTYPE: MORPHOGENESIS AS TARGET AND DRIVER OF EVOLUTION

Mutations affect the genotype, but selection happens on the phenotype. Evolution cannot be understood without knowing how they are linked. This seems obvious on the one hand, and yet, claims about the genetic underpinning of anatomical form and function are often made without a serious consideration of their relationship. A fundamental understanding of evolution, and the role of morphogenesis as its target and its driver, requires a focus on what lies between genotype and phenotype.

2.3.1 Morphogenesis as a Problem-Solving Process That Interprets the Genome

The old view of genes encoding specific characters has been replaced by a much more nuanced picture, such as the omnigenic model (Boyle et al., 2017), focusing on emergence and complexity science. And yet, one fundamental thing has remained: the mapping between genes and outcomes may have redundancy, pleiotropy, degeneracy, and other ways to implement a hairball of causality, but it's always seen as passive. In other words, there are mechanisms linking them, but the unstated, near-universal assumption is that these mechanisms are low on the spectrum of intelligence. Genetic material is thus an input into chemical mechanisms, and even when DNA is described as information (fodder for a cybernetic, or even cognitive, system!), the interpreters of that information are thought to be polymerases and ribosomes, a very low level of description. I argue that it is only a hypothesis that large-scale morphogenesis can be explained as a purely feedforward emergent consequence of mechanical events. Morphogenesis is actually a problem-solving process, using many strategies well-recognized by behavioral scientists. A potential consideration of more sophisticated dynamics by the morphogenetic machinery opens new ways of thinking about the kind of plasticity that has massive implications for evolvability and for the course of evolution.

The first thing about morphogenesis is that it offers fascinating reprogrammability. The genetically specified hardware does the expected work reliably under default outcomes (dogs have puppies and cats have kittens, making it seem like we understand heredity). However, there are a variety of ways in which the default setpoints can be rewritten without manipulating the DNA. Much has been said elsewhere about epigenetic inheritance, and this now often refers to chromatin modification mechanisms, but there are other important memory media in cells (Fields and Levin, 2017), including cytoskeletal and membrane components that, like DNA, offer an unbroken link going all the way back to the last universal common ancestor.

What is critical here is the plasticity, or reprogrammability, of living matter and the nano-scale hardware components provided to cells by the DNA. One example is that of cortical inheritance in ciliates (Beisson and Sonneborn, 1965). These unicellular organisms have motile hairs (cilia) on their cell surface, which can funnel food particles into their mouth opening. If a small square of surface was excised, rotated 180 degrees, and grafted back in place, all of that cell's offspring would have such a rotated square in their surface: a permanent line of organisms with a novel morphogenetic feature that is totally invisible to DNA profiling, because that is not where the information is. Moreover, these creatures live on the verge of starvation (because some of the surface is flowing food out of their mouth!) and their perfectly normal genome cannot help them.

Another example (Figure 2.5) is found in planarian flatworms (Oviedo et al., 2010) (discussed more below in the context of bioelectricity). Planaria, when cut into pieces, reliably regenerate worms that have exactly one head and one tail. This is a phenomenon that cannot be explained by typical morphogen gradient models because single cuts result in opposite anatomical fates from cells that were adjacent neighbors prior to the cut (having the same positional information). Reasoning that the head vs. tail decision required long-range coordination of wound cells with the rest of the body to know the position and context of the fragment, we inhibited physiological signaling through gap junctions. The result was two-headed worms. Two-headed planaria were described well over 100 years ago, but no one, to our knowledge, *recut* such animals to see what would happen, until we did it in 2008, presumably because it was obvious: having normal genetics, removal of the ectopic second head should cause repair to the default target morphology (one head, one tail). We did recut them, because we hypothesized that the tissue had a physiological memory that could override the default, and found a remarkable multicellular version of the ciliate example above: two-headed animals, when cut, continue to produce two-headed offspring in perpetuity, with no genetic changes being necessary to create a line of animals with a totally different body plan. Note that what is happening in this example is not just modifying emergent outcomes, but actually modifying the target toward which the cellular processes continue to work in cases of deviation by amputation injury.

These examples, along with the deer antler trophic memory described above and others, reveal that the genome facilitates a very powerful kind of substrate: one whose capabilities are not directly determined by the hardware specification but by the physiological software that it implements. But the wonder of morphogenesis does not stop at being amenable to changes via external stimuli or perturbations: its most

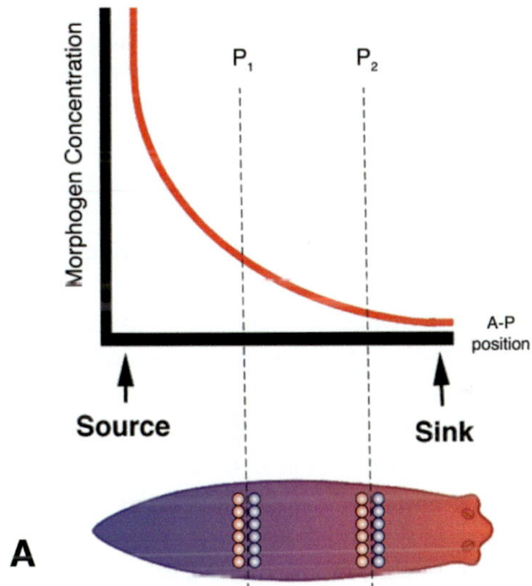

FIGURE 2.5 Resetting non-local pattern memory in vivo. Planarian flatworms reliably regenerate their body upon amputation. Typical models explain the ability of each fragment to determine which end should form a head and which end should form a tail using a gradient of some morphogen mapped onto a middle fragment resulting from two cuts (A), which is high at one end and lower at the other (red curve). However, the more interesting comparison is what happens at *one* cut. Consider the fragments made at the level indicated with P1. The cells on either side of that cut have the same positional information because they are adjacent neighbors – there is no significant difference in the morphogen level, and yet, they have radically different anatomical fates: one side makes a head and the other makes a tail. This shows that local positional information is not sufficient to set (or explain) anatomical fate. Indeed, the same events happen at position P2, and there is no absolute rule about the morphogen level that can correctly explain the fate of each side of the two cuts. Instead, it is the *relative* positional information of the cuts that can tell each one which way it's facing and what it should do. Reasoning that this is mediated by long-range physiological signaling through gap junctions (Oviedo et al., 2010),

(*Continued*)

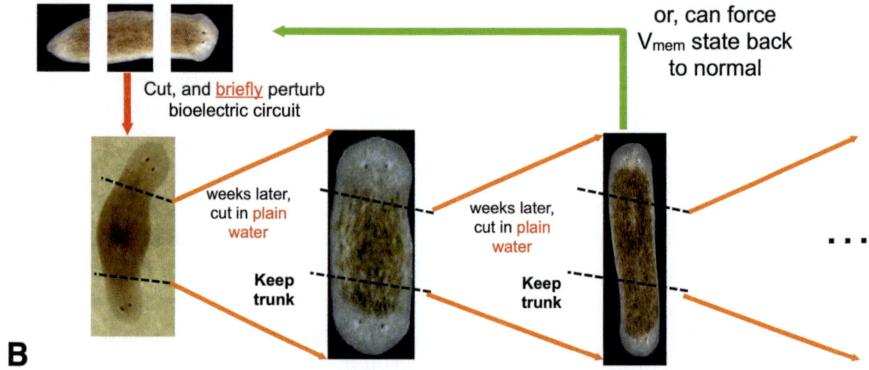

FIGURE 2.5 (Continued) we inhibited this communication (B), finding not only that two-headed animals result, but that this change to the target morphology information is permanent, and persists indefinitely through future rounds of amputation (red arrow indicates that a middle fragment can form a two-headed worm if bioelectric signaling is perturbed; green arrow indicates that a two-headed worm can be induced back to the one-head state by inhibition of a proton pump; Beane et al., 2011).

amazing property is intrinsic context-sensitive responses aimed at specific outcomes (problem-solving capacity able to handle unexpected internal and external circumstances).

One example is illustrated by planarian responses to barium, a non-specific potassium channel blocker. Exposure to barium causes planarian heads to rapidly degenerate, as cells (especially neurons) degrade when unable to participate in potassium flux. Remarkably, when kept in barium, planaria eventually regenerate new heads that are barium-insensitive (Emmons-Bell et al., 2019). Transcriptomic analysis showed that they rapidly identify just a handful of differentially-expressed genes that dealt with their novel stressor – one that they (and likely, their lineage) had never seen before. How did the planarian cells that don't turn over rapidly enough for clonal selection and don't have time for a generate-and-test search identify a small number of transcriptional actions in a ~20,000-dimensional space of possibilities that solved the new problem facing them? We don't know, but the answer to how cells find appropriate responses to novel stressors would do much to advance medical practice. This example takes place in transcriptional and physiological spaces.

The next example of the use of available affordances to solve a new problem (a standard component of IQ tests) is chosen from morphogenetic space. In newts, kidney tubules are created by numerous cells working together (Figure 2.6A). Polyploid newts can be created, which have extra copies of the genetic material, illustrating the remarkable multiscale intelligence of morphogenesis (Fankhauser, 1945a, 1945b). First, the animals are normal and viable despite massive polyploidy. Second, the cells adjust their size accordingly. Third, the animals remain normal size because *fewer cells* participate in tubule morphogenesis. Finally, and most remarkably, if the cells are made so large that more than one cannot fit in a tubule diameter, then just one cell will *bend around itself*, leaving a hole in the middle, to produce the same structure.

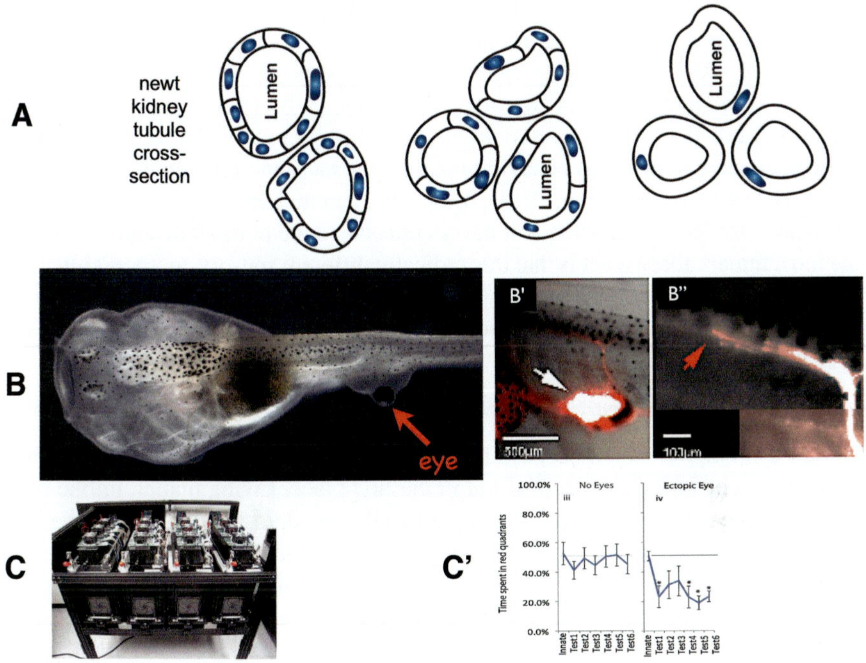

FIGURE 2.6 Adaptive response to novel scenario: plasticity of form and function. (A) Newt kidney tubules usually consist of approximately eight cells working together. When cells are made much larger, the same-sized tubule is constructed from fewer cells. When the cells are made enormous, one single cell wraps around itself to make the same large-scale structure, illustrating how, in the service of the same final anatomical outcome, different molecular mechanisms (cytoskeletal bending vs. cell–cell communication) can be called up to achieve the goal. (B) Tadpoles with eyes induced on their tail (red arrowhead), but missing primary eyes in the head, have ectopic nerves that do not connect to the brain and stop at the spinal cord (white arrow indicates the ectopic eye; red arrows indicate optic nerve emerging from the ectopic eye); despite this, automated devices that train and test them on visual ques (C) show that they can learn behaviors requiring light sensing and context-sensitive responses (C') – they can see, despite a completely novel sensory-motor architecture.

Panel A by Jeremy Guay of Peregrine Creative after Fankhauser (1945a, 1945b); panels B–C' taken with permission from Blackiston et al. (2010), Blackiston and Levin (2013), and Blackiston et al. (2017).

This is, first of all, an example of top-down causation as discussed above: cells will adjust as needed in order to achieve a specific large-scale target morphology. Moreover, it's an example of "same goal by different means" (i.e., problem-solving) – using different molecular mechanisms (cytoskeletal bending instead of cell-to-cell communication) to achieve the same outcome. Finally, it's an example of *creative* problem-solving – using available tools to solve a problem in a new way. Imagine being a newt coming into the world: not only do you face massive uncertainty about the external environment, you also can't even count on your own parts. You don't know how many copies of your genome you will have, how big your cells will be, how many cells you will have (Cooke, 1979, 1981), etc. You must reach the target morphology regardless of those circumstances.

The final example concerns behavioral endpoints of morphogenesis (Figure 2.6B–C). When eye primordia are transplanted to the tail of a *Xenopus* embryo, normal eyes result (i.e., the eye cells are able to build an eye in abnormal locations). These eyes, as seen in visual behavioral learning assays, are *functional* – the animals can see (Blackiston and Levin, 2013), despite the fact that the optic nerve connects to the spinal cord, gut, or nowhere – it never connects to the normal location, the brain. The most remarkable aspect is that this radically different sensory-motor architecture does not require new rounds of mutation, selection, and adaptation – it works "right out of the box" in its new configuration. Why?

2.3.2 Beginner's Mind: Problem-Solving, Not Fixed Solutions

The tail-eye tadpole works because the standard *Xenopus* embryo could never rely on the eye to be in the correct position in the first place. Living matter, unlike most of our computer technology, is a highly unreliable medium: you can never be sure of how many copies of any molecule you have, how long things will last, or what will happen to you. The one thing of which you can definitely be certain is that you will be altered or mutated (depending on the time scale) in ways you can't predict. The only way to overcome the paradox of change (Levin, 2024c; Van Valen, 1973) is by committing to plasticity. Evolution mostly doesn't make solutions to specific problems; it makes problem-solving systems, which under default conditions do the same thing (thus lulling observers like us into models of hardwired characteristics "encoded" by DNA), but due to the intelligence baked in at multiple levels of organization, it can also do many other things in a context-sensitive manner. This is most obvious, to us, in the case of brainy animals navigating the 3D space via motion, but is seen throughout anatomical, physiological, and transcriptional spaces when we know how to look. On this view (Hartl and Levin, 2025), the genome is a boundary condition on the cellular hardware and also a prompt to the intelligence of cell collectives, which can interpret and use it as a resource to address their stressors and setpoints (Figure 2.7). Moreover, the bottleneck or bow-tie node of development (Csete and Doyle, 2004; Doyle and Csete, 2011; Frank, 2007, 2009), which squeezes down to an egg between generations, forces creative interpretation of the memories of past instances – an architecture that is essential to cognitive function, as follows.

Each of us has no access to the past – only to the memory engrams in our brain (or body, Blackiston et al., 2015) that were left as a consequence of past experiences. The interpretation of those engrams, as part of the neurally *constructed* (Davis et al., 2024; Gallo, 2010; Otgaar et al., 2023; Roediger and McDermott, 1995; Shao et al., 2022) (not just read out) memories, is a continuously active process of self-(re)creation. At each moment, we exist at the bow-tie node between the compressed, generalized lessons of the past and the opportunity to interpret our current state for maximum adaptive benefit in the future. This is not necessarily the way past instances of us interpreted the data (although it often is); the collective has competencies to do new things with the ancient prompts. Our self-constructing cognitive Self is able to

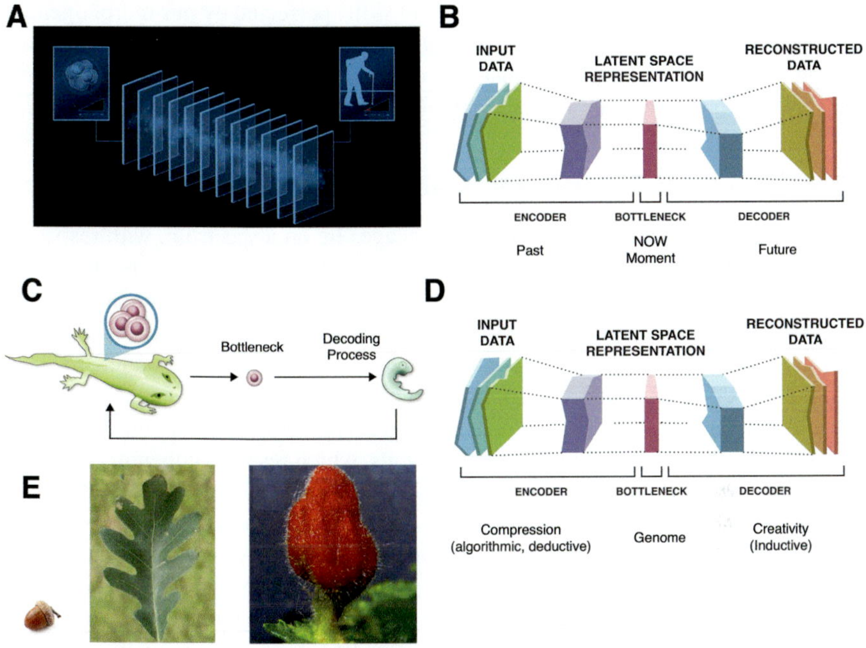

FIGURE 2.7 The paradox of change requires creative interpretation of information. At any moment in time, a cognitive being does not have access to the past – what they have access to as memories are biophysical engrams left for them as messages from their past self (panel A shows this stack of "Selflets"). This means that a cognitive system uses a bow-tie architecture where individual experiences in the past are compressed by learning into sparse engrams that discard a lot of details to generalize from specifics. At each moment, a mind must decode those molecular/biophysical states to give them meaning – interpret them (B). It is well known in neuroscience that this process is not passive recall but active reconstruction. The very nature of the compression of learning means that we never automatically know what molecular engrams mean and must actively improvise memories on-the-fly. A time-extended mind must continuously change to remain adaptively engaged. The same paradox of change affects the body; the one thing that a biological system can count on is that nothing will stay the same; not only will the environment change but also the internal parts will eventually be mutated. This means that the system cannot take for granted (in a fixed manner) how its genomically encoded affordances should be used: the experience of past generations is compressed within the genome but must be decoded by morphogenesis to produce an embryo or some other outcome that fits current circumstances (panel C). Thus, the right side of the bow-tie (D) features a non-algorithmic process that must creatively interpret the genome and the environment in the face of missing information. This is what accounts for the remarkable plasticity (Figure 2.6) and reprogrammability (Figure 2.5) of living tissue not needing genetic change to produce novel outcomes from the same hardware (such as the exquisitely patterned galls that arise from plant cells normally making flat green leaves, E).

Images A–D by Jeremy Guay of Peregrine Creative. Panel E used with permission from Hartl and Levin (2025).

do this, a functional pivot of far more ancient skills perfected by our morphogenetic Self (Maturana and Varela, 1980; Rosen, 1985). The cellular collective, solving problems in the context of unreliable materials and environment, must also interpret (not just slavishly obey) the memory engram left for it by our past Selves: the genome.

2.3.3 Evolution Operating on an Agential Material

Morphogenetic strategies across different lineages lie on a spectrum, with respect to how literally they take their genome. On one side exist organisms like the nematode *Caenorhabditis elegans* – eutelic organisms that all have the same number of cells; development is guided by strict lineage relationships, and each cell can be numbered – each instance of the organism is identical at the level of cells. Such organisms have minimal plasticity/regenerative capacity and a highly prescribed, stereotyped order. In the middle are amphibians – significantly regenerative and plastic, amenable to all sorts of grafting, cutting, and other perturbations, which result in coherent outcomes, not to mention the examples of newt tubules above. On the far end lies a strategy that maximally leans in to uncertainty and the need to reach specific outcomes despite unpredictably changing hardware. A fascinating example is provided by asexual planarian flatworms, which can regenerate every part of the body from even a tiny fragment (Saló et al., 2009). Their genome is a mess (they can be mixoploid, like a tumor) (Guo et al., 2018; Nishimura et al., 2015) because, unlike sexually reproducing animals, they don't purge somatic mutations at each generation. To reproduce, they split and regenerate, meaning that every mutation that doesn't kill a neoblast stem cell propagates itself and its altered genome into the two regenerative offspring.

Is it not shocking, given the standard views of the genome and its role, that the animal with the highest regenerative capacity (morphogenetic stability), cancer resistance, and lack of aging (immortality) is the animal with the messiest, not the cleanest, genome? Moreover, planaria are perhaps the only model system in which no mutant lines exist (except our two-headed lines, and those are not genetic – see below), nor has transgenic technology been shown to be possible in these animals. Why? Our simulations (Shreesha and Levin, 2023) have suggested an explanation for these observations, showing how evolution works differently on a competent but unreliable material. In this context, "competent" means (as in the examples above) the ability to reach specific goal states despite changes in internal composition and external microenvironment – some degree of context-sensitive navigation of possibilities, not only hardwired stereotypical activity. Unreliable means that on multiple time scales, the system can make few assumptions as the genetic complement of its cells (and thus the hardware it affords) varies widely. This means that success requires more effort to be put into the creative improvisation of navigating anatomical and physiological spaces with equipment of which one cannot be certain.

The impact of behavioral intelligence on evolution has been studied extensively (Baldwin, 1896; Dopazo et al., 2001; Hinton and Nowlan, 1987; Odling-Smee, 1995; Piaget, 1976; Thorpe, 1945). But in developmental biology, the consequences for evolution of the role of context-specific problem-solving *at all scales* can be seen most clearly (Baluska et al., 2022a; 2022b; 2025; Miller et al., 2023; Reber et al., 2024a; Reber et al., 2024b). In our simulations of evolution acting on morphogenetic systems (Hartl et al., 2024; Shreesha and Levin, 2023), we observed some fascinating dynamics (Frank, 2009, 2019a, b). Imagine a mutation that has multiple consequences in an amphibian, one of which is that the mouth is moved to an aberrant location. In a mosaic (low-competency) organism, starvation and death would result, preventing evolution from exploring the other consequences of that mutation. Evolution would have to wait until they occurred alone, which, given the pleiotropic nature of development, would take a long time. However, in a plastic, high-competency system, the mouth may move to find its correct location (Pinet et al., 2019; Pinet and McLaughlin, 2019; Vandenberg et al., 2012), rescuing the animal and thus enabling the other consequences of the mutation to have an impact on fitness. Competency of the agential material of life turns many deleterious mutations into neutral ones, greatly facilitating the speed of evolution.

But it has one other important consequence. When the cellular collective repairs defects (anatomical, physiological, or transcriptional), selection does not see the hardware as clearly: a tadpole with its mouth in the right location could have had great structural genetics, or poor ones that have been made up for by the flexibility of morphogenesis. Selection never really sees genomes (since it always acts on phenotypes), but it's critical to note that the *degree* to which information about a genome can be acted upon by selection depends on the degree of active intelligence of the layer mapping genetic information (hardware) into active patterns of form, physiology, and behavior. What we observed is a positive feedback loop: the more competency tissues had, the less time evolution spent optimizing the hardware and the more effort went into creating an algorithm that was effective *despite* faulty hardware, which in turn hid more information about the genomes from selection. Of course, different species (for different reasons of ecology) will achieve different balance points between making sure the hardware is correct and policies that can get to an adaptive endpoint despite suboptimal hardware. Planaria are a great example of pushing all the way to the right of that continuum: because of their reproductive mode and unreliable genetics, evolution was forced to find a set of competency algorithms that allow planarian cells to make a successful worm body no matter what it starts from, effectively ignoring its natural mutations just as it ignores scientists' attempts to introduce transgenes or make mutant lines.

One final aspect to note is the impact of the intelligence of the materials on the course of evolution itself (Baluska and Miller, 2018; Buckley et al., 2024; Kouvaris et al., 2017; Miller et al., 2020; 2018; Noble, 2022; Power et al., 2015; Rago et al., 2019; Shapiro, 2011; Shapiro, 2022; Watson et al., 2022; Watson et al., 2016; Watson and Szathmary, 2016). Even mutations random with respect to the fitness of the resulting organism end up working differently due to the competency of the substrate, which can bend the option space so that mutations' effects are better in the future. In a sense, the ability of cells to reach specific outcomes despite novel challenges implements a kind of "chance favors the prepared mind" scenario. It's not the mutations that have to be smart, nor does there need to be any large-scale plan to the course of its meanderings; nevertheless, the competencies of the material can lend dynamics to the evolutionary process that have simple cybernetic properties (Fields and Levin, 2020) themselves, in effect instantiating a process that is neither completely blind and stupid nor requires high-level forward-planning intelligence.

The spectrum of the field of diverse intelligence (Figures 2.1 and 2.3) is not limited to the cellular and subcellular levels, but may be applied to the largest scales of space (ecosystems) and time (the evolutionary process itself) – a fascinating research program for the future.

2.4 BIOELECTRICITY: A HIGHLY CONSERVED COGNITIVE GLUE MECHANISM

The previous sections focused heavily on the ability of cell groups to achieve specific goal states in anatomical morphospace, even contrasting that capability with the hardware properties encoded in the genome, as targets that evolution can tune independently. All hardware (proteins available to cells) has to be encoded genetically; what mechanisms underlie the morphogenetic policies that are in some way distinct? There are probably many, but there is one that is currently the best understood: bioelectrically-mediated cellular networks that implement computational capabilities that do not themselves have to be coded in the genetic hardware because they operate as physiological software and make use of universal information-processing patterns (e.g., as shown in Figure 2.5B).

2.4.1 Evolutionary Context for Bioelectricity: A Common Medium Underlying Bodies and Minds

The previous strange-sounding statement is just a different way to describe brains. It's a useful way to do because it allows us to generalize beyond the familiar substrate of neuroscience to better understand its evolutionary history and its implications. Brains consist of networks of electrically active cells working in networks to process information, perform active inference, store memories, etc. In particular, electrophysiological events in the brain act as a *cognitive glue* – a mechanism that binds individual neurons into a larger system that has goals, memories, and preferences in problem spaces that no neuron knows anything about (Figure 2.8). All of us have mental lives pursuing target states in highly abstract social, financial, and other spaces, which none of our neurons can access because of this set of policies that binds cells into a network that can operate on larger sets of states. There are other examples of cognitive glue (e.g., stress sharing, Shreesha and Levin, 2024) that are not bioelectric in nature, and bioelectricity is not unique in its ability to scale competent subunits into collectives that access new problem spaces and have a larger cognitive light cone (size of the setpoints it can work toward). But evolution certainly exploited bioelectric dynamics because they are ideal for implementing feedback (memory) systems that integrate information across both space and time.

Remarkably, most of the brain's tricks, including inference of patterns, and mental time travel to the past (memory), to the future (prediction), and to alternate worlds (counterfactual what-ifs), derive from an ancient, well-conserved system of bioelectric networks. These networks consist of cells with ion channels, electrical synapses (gap junctions), and neurotransmitter transducers – many of which go back to the time of bacterial biofilms. Ion channels have long been studied in bacteria (Koshland,

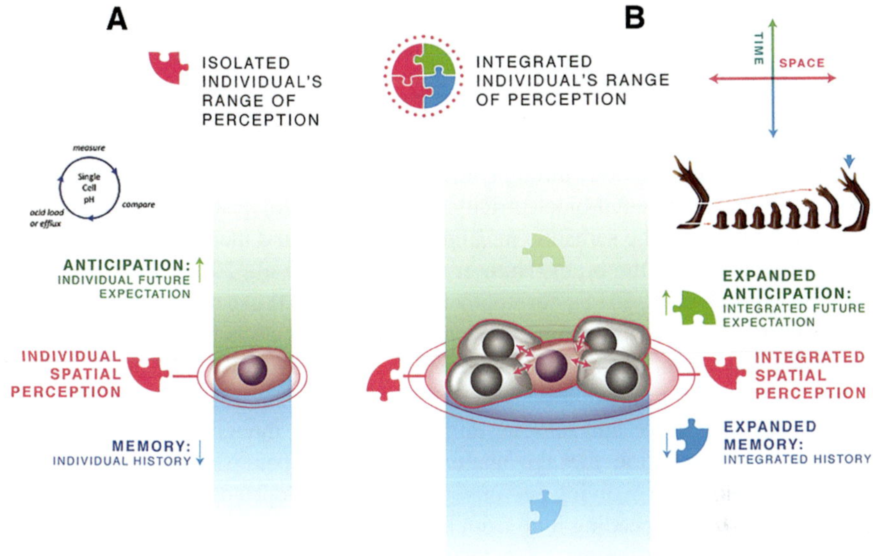

FIGURE 2.8 Scaling of the cognitive light cone by bioelectric networks. The cognitive light cone refers to the size, in space and time, of the largest goal state a system is capable of pursuing (i.e., the size of its setpoints). Single cells pursue small setpoints, such as scalar values of physiological parameters like pH, over short ranges of size and memory/anticipation horizon (A). Joined into large-scale tissues, groups of cells can hold much larger and more complex setpoints for homeostatic and allostatic processes, which enables them to reliably pursue enormous construction projects such as rebuilding whole limbs (B).

1983), and it is now known that even bacterial biofilms already use electrical networks to join into a kind of proto-multicellularity to do things no individual bacterium can do (Lee et al., 2017; Liu et al., 2017; Prindle et al., 2015). The ion channels (and pumps) set each cell's voltage potential, which can propagate to other cells via gap junctions. However, many of the channels and the gap junctions are themselves voltage-sensitive, which means there is potential for complex feedback loops that establish multicellular regions of different voltages that change over time.

Cells, in turn, react to the bioelectric state of themselves and of their neighbors. Because the same voltage can be reached by very different ion balances, resting potential itself achieves a kind of generalization of stimulus: different past states (whether induced by sodium, potassium, chloride, or other ions) are interpreted as the same coarse-grained parameter – "voltage potential", which is a precursor to the formation of generalized categories in brainy organisms during learning. Cell voltage regulates second-messenger systems (and, ultimately, controls gene expression and cell movement) via a set of known transduction mechanisms, which include serotonergic machinery, calcium signaling, and others (Levin et al., 2017). What cells appear to react to is not absolute voltage states but *differences* between cells and their neighbors – *patterns* of bioelectric state across cell sheets, which often set up compartment boundaries during embryogenesis or regeneration (Emmons-Bell and

Hariharan, 2021; Emmons-Bell et al., 2020; Petsakou and Perrimon, 2023; Schotthofer and Bohrmann, 2020; Vandenberg et al., 2011; Weiss and Bohrmann, 2019; Woodruff and Telfer, 1973).

When large-scale patterns of electrical activation move rapidly through the brain, we call it the province of neuroscience. When large-scale patterns of electrical activation move much more slowly through other somatic and embryonic tissues, this is the province of developmental bioelectricity. Just as neurobiologists seek to do "neural decoding" to extract the semantic meaning of thoughts and memories from electrophysiological data of the brain (Huth et al., 2016; Lu et al., 2021; Naselaris et al., 2009; Nishimoto et al., 2011), developmental physiologists can now extract information from the bioelectric dynamics of patterning tissues (Adams and Levin, 2012; McMillen and Levin, 2024). We know that bioelectric patterns in the brain evolved to think about how to move a body in 3D space. What did ancient bioelectric networks, predating nerve and muscle, think about? They thought about moving the organism's configuration through anatomical morphospace (Fields et al., 2020). What evolution did, in evolving nervous systems, was co-opt a navigational problem-solving mechanism, speeding it up from minutes to milliseconds and pivoting it into a new problem space (Figures 2.3 and 2.8).

2.4.2 Developmental Electrophysiology and the Scaling of Cognitive Light Cones

A field of cells expressing ion channels and gap junction proteins is, in effect, an excitable medium in which complex and beautiful patterns can form from spontaneous symmetry breaking and amplification. Developmental biologists are used to Turing patterns and other reaction-diffusion dynamics in chemical media (Karig et al., 2018; Marcon and Sharpe, 2012; Nakamasu et al., 2009; Strier and Ponce Dawson, 2007), but the same kind of phenomenon happens at the physiological level. Crucially, this happens because channels and gap junctions open and close post-translationally. Just as with the action potentials in the brain, which don't require rapid changes of transcription or translation to implement each signaling spike, cells can change electrical properties with no changes in mRNA or protein. Consequences of this include the facts that the same bioelectric state can be caused by different ion channels' functions, while cells with identical ion channel complements can be in different bioelectric states (because of past physiological stimuli leading the channels to be open or closed – an inherent historicity and memory in this system). Thus, bioelectric states are not derivable from transcriptomic or proteomic (and of course not genomic) data – physiomics are needed to understand the bioelectric state of cells, just as electrophysiological imaging is needed in neuroscience.

Individual cells can store tiny setpoints – metabolic, pH, and other ways to represent goals at the level of a single cell. However, joined into large-scale electrical networks, they are able to represent enormous setpoints or bioelectric prepatterns that can scaffold whole organs or bodyplan axes (Figure 2.9). As brains' bioelectric patterns store memories for actions in 3D space, the body's bioelectric network stores pattern memories that guide cell behavior toward morphogenesis of specific forms (Pai et al., 2012). Examples include the bioelectric prepattern (Figure 2.9C) that

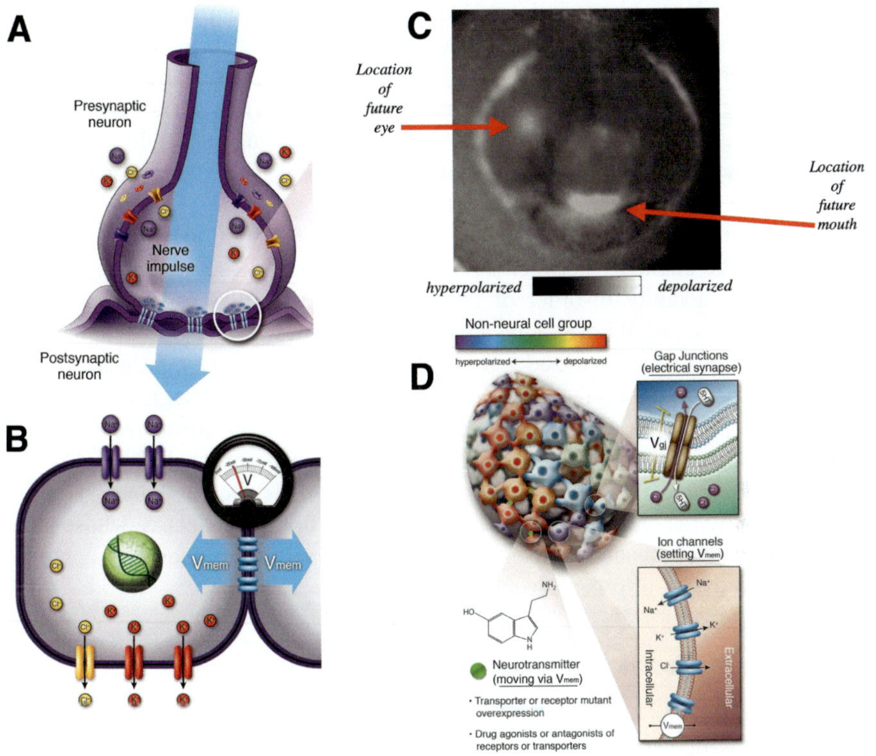

FIGURE 2.9 Developmental bioelectricity. In neurons, electrical signals are propagated across the network via ion channels that set resting potential and electrical synapses (gap junctions), which serve as context-sensitive conduits for that voltage to neighboring cells (A). The exact same system functions in the rest of the body, allowing most kinds of cells to couple into electrical networks (B). The result of voltage states propagating across a tissue of voltage-sensitive channels and gap junctions is complex self-organization of patterns, such as the voltage map of the early *X. laevis* embryo anterior neurectoderm (C, shown using voltage-sensitive fluorescent dye), which guides gene expression and cell behavior and often serves as a prepattern or scaffold for the subsequent anatomy, such as the face. (D) The information in these networks can be effectively modified (rewritten) by approaches taken from neuroscience, such as using drugs or optogenetics to target specific ion channels, pumps, and gap junctions.

Panels A, B, and D by Jeremy Guay of Peregrine Creative. Panel C taken with permission from Vandenberg et al. (2011).

precedes and functionally dictates the location of craniofacial structures (Vandenberg et al., 2011), and the planarian voltage gradient that directs the number of heads in flatworms (Durant et al., 2019). Three brief examples illustrate important features of this system, especially highlighting what information and tractable control points exist at the level of physiology as distinct from the genetics.

The first is the case of cancer. Transforming cells rapidly drop gap junctional communication with their neighbors (Krutovskikh and Yamasaki, 1997; Mesnil et al., 1995). This renders them incapable of remembering the grandiose morphogenetic setpoints they were working on, and they revert to their ancient unicellular goals of reproduction and survival – both behaviorally (metastasis) and transcriptionally (Bussey et al., 2017; Davies and Lineweaver, 2011). They are not more selfish than normal cells (as often modeled in game theory simulations); they *just have smaller selves*, in the sense that the size of an agent's goal states (the region of space-time it tries to manage toward specific setpoints – its cognitive light cone) defines the border between the Self and the outside world (Levin, 2019). Cancer results from a shrinkage of the boundary between a Self and its world (microenvironment), a kind of dissociative identity disorder of the morphogenetic collective intelligence. It reverses the processes of development (and evolution) in which cognitive light cones massively scaled up during the advent of multicellularity. It has now been shown that forcibly reinstating the correct bioelectric state (a kind of physiological integration therapy), despite the presence of strong oncogenic mutations, normalizes cells and prevents tumors in animal models (Chernet et al., 2016; Chernet et al., 2015; Chernet and Levin, 2013a, 2014). It doesn't kill cells or fix their genetic defect, but once joined back into the electrical network, they form part of the collective whose goals include morphogenesis and upkeep of whole organs, reinflating their tiny cognitive light cones to tissue- or organ-level.

2.4.3 The Bioelectric Interface for Large-Scale Pattern Control

Misexpression of ion channels (Chernet and Levin, 2013b) or their activation via optogenetics (Chernet et al., 2016) can thus regulate the size of the computational unit and its goal states. Other ways to exploit the natural bioelectric system that forms the decision-making medium of cellular collectives are to induce the formation of ectopic organs. The location of naive eyes, for example, is set by the presence of a specific pattern of resting potential within a set of patterns establishing the gene expression and morphogenesis of the vertebrate face (Figure 2.9C). A single ion channel misexpression can establish a similar pattern in aberrant locations in a frog embryo, resulting in the production of ectopic eyes (Figure 2.10A, Pai et al., 2012); other such voltage patterns can induce brain and other structures (Blackiston et al., 2009). Several key pieces of information are revealed in that experiment. 1) The bioelectric state is instructive – it can be used to create novel coherent structures, not just cause damage. 2) The bioelectric state can encode large-scale structures: in keeping with the notion of bioelectric networks operating with generalizations (primitive concepts) of single-cell states, this particular voltage pattern *means "eye"* – no individual cell knows what an eye is, but the collective interprets that pattern to stand for a specific organ and executes accordingly (other patterns have been found to indicate other outcomes). We didn't have to talk to stem cells or micromanage the structure of an eye; we specified a very high-level prompt to which the competent material reacted (as is the key ability of all embodied cognitive agents – high-level concepts make the chemistry work to implement abstract goals). 3) The material is self-scaling: when only a few cells are targeted with the intervention, they perform a secondary instruction and recruit un-manipulated cells to join in (Figure 2.10B). 4) The system actually contains a battle of goal states: while the injected cells try

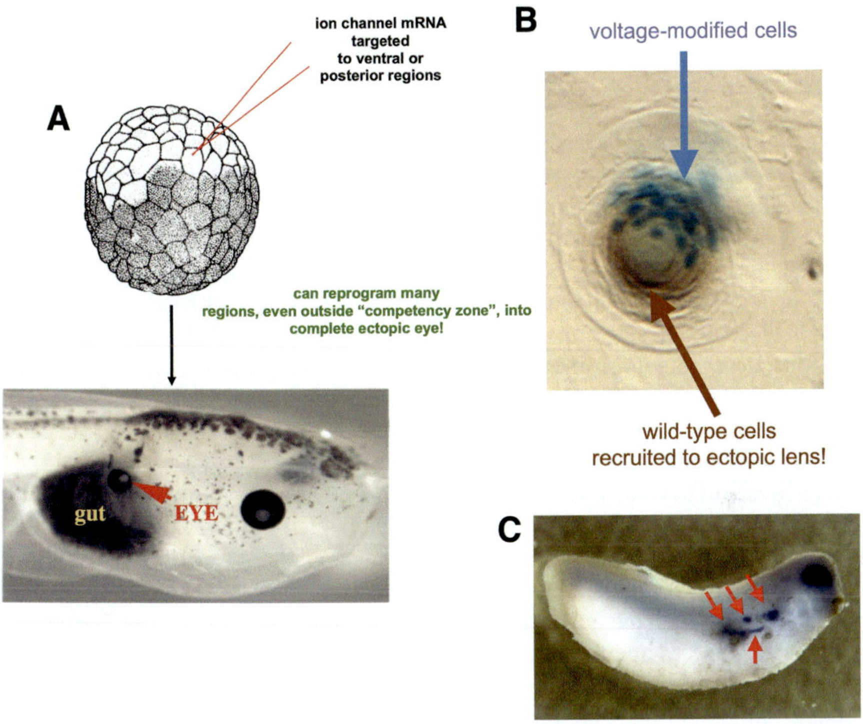

FIGURE 2.10 Example: organ-level communication with cellular collectives. When ion channel mRNA is injected into an ectopic region of an embryo, setting up an ectopic prepattern that matches the native eye prepattern from the face (Figure 2.9C), a new eye is induced (A). Sectioning reveals that only a few of their cells were actually injected: the rest of the organ is made from cells that joined the morphogenetic task by being secondarily recruited (B). This is a dynamic process, as revealed when early embryos are stained using in situ hybridization for an early eye marker – Rx1 (red arrowheads reveal ectopic eye spots, C). The majority of these ectopic eye spots will, in fact, not make eyes because surrounding cells normalize their voltage (a cancer suppression mechanism). This reveals the dynamic nature of the instructions mediated by these physiological signals, and the nature of the message – at the level of whole organs, not gene states or cell types – that is being communicated; the specific structure and the molecular events needed, including secondary interactions between cells, are all downstream of the primary voltage pattern trigger and don't need to be micromanaged by the bioengineer.

to make their neighbors join in the "eye" project, the surrounding cells, executing a cancer suppression mechanism (resisting rogue voltage states in the microenvironment), try to do the opposite – normalize the injected cells' voltage to that of the local average. It is an active area of investigation for biomedical applications of bioelectric signaling to understand how to make such bioelectric messages maximally convincing to cell groups, to efficiently rewrite setpoints.

A final example concerns the repair of birth defects. Notch is an important neurogenesis gene, and when a dominant form of the Notch-ICD mutant is injected into frog embryos, severe defects of the brain result, via disruption of the normal bioelectric prepattern that sets the regionalization of the neural tube. Recently, a computational model of this process was created, which could be asked: what ion channel could be opened or closed to get the pattern back to normal? The model suggested HCN2 – a very interesting channel with the property that it acts as a kind of "sharpen filter" on bioelectric prepatterns (Pai et al., 2020; Pai and Levin, 2022; Pai et al., 2018). Overexpressing HCN2, or opening existing HCN2 channels with drugs, restored the normal shape, gene expression, and even learning rates of Notch mutant-affected animals' brains, revealing that at least in some cases, hardware defects can be fixed "in software". And, consistent with the point above about bioelectric states being invisible to conventional omics, note that the number of heads in bioelectrically-induced two-head worms, the absence of tumors in KRAS (Kirsten rat sarcoma viral oncogene homolog)-injected animals, or the shape of brains in Notch-mutant animals could not be inferred from their genetics (which point to different and incorrect predictions).

The basic theme of this section can be summarized as follows. We now know quite a bit about the mechanisms responsible for cell groups' ability to reach specific anatomical outcomes – the mapping of genotype (which sets ion channel properties) to phenotype (anatomy) is mediated by the computational properties of bioelectric networks. Bioelectricity is not yet another piece of biophysics that operates in morphogenesis, nor are the regenerative, reparative, and organ-inducing abilities of molecular bioelectric perturbations some sort of external modulation applied to the system. They only work because they exploit the native interface by which cells become an organism. The bioelectric network implements the proto-cognitive medium in tissues that stores the homeodynamic setpoints: the goal states toward which cells will build. This is important in general because it fleshes out how collective intelligence of morphogenesis is implemented, and specifically because it offers an interface by which regenerative medicine workers and bioengineers can communicate with (not micromanage) the morphogenetic Self.

2.5 NOVEL BODIES, NOVEL MINDS, SAME HARDWARE

Framing morphogenesis as targeted navigation of anatomical space, with context-sensitive (but finite) capacity to forge new paths to the setpoint when circumstances demand, raises two related questions. What do problem-solving systems do when their standard goals cannot be achieved? And, where do the setpoints originate – what actually sets the target states for these homeodynamic processes? Bioengineering and synthetic morphology (Doursat and Sanchez, 2014; Doursat et al., 2013; Ebrahimkhani and Levin, 2021; Kamm et al., 2018; Sample et al., 2019) are natural extensions of developmental biology that extend our understanding of growth and form along those two lines of inquiry.

The standard answer to the origin of goal states is evolution: selection over eons in specific environments has ensured that the default hardware produced by embryogenesis has very specific anatomical setpoints. Indeed, we know that evolution is very good at managing both internal (morphogenetic) and external (shapes of bird nests, etc.) targets of dynamical processes (Gottlieb, 1999; Khuong et al., 2016; Turner, 2011). But, much as neuroscientists ask about the plasticity of nervous systems in handling novel circumstances, we can ask: what would be the morphogenetic, physiological, transcriptional, and behavioral properties of novel living beings that have never existed on Earth and have never faced selection at the level of the organism? When cell groups find themselves unable to reach the default target morphology, what will they do?

Unfortunately, we do not yet have any frameworks in developmental biology to predict this in advance. A more basic limitation (Figure 2.11A) is observed in the inability to predict outcomes in chimeras (Nanos and Levin, 2021). For example, axolotl embryo cells (which create larval legs) can be mixed with frog embryo cells (which do not have larval legs) to make a viable embryo. Despite the availability of both frog and axolotl genomes, we have no way of predicting whether such Frogolotls will have legs, and if they do, whether these legs will consist only of axolotl cells or will recruit frog cells as well. Like in the chimeric neoblast case in planaria (Figure 2.11B), what is missing despite the high-resolution molecular data is the knowledge of how collective intelligences (groups of cells) make decisions and how those decisions change when the different members of the collective are seeking

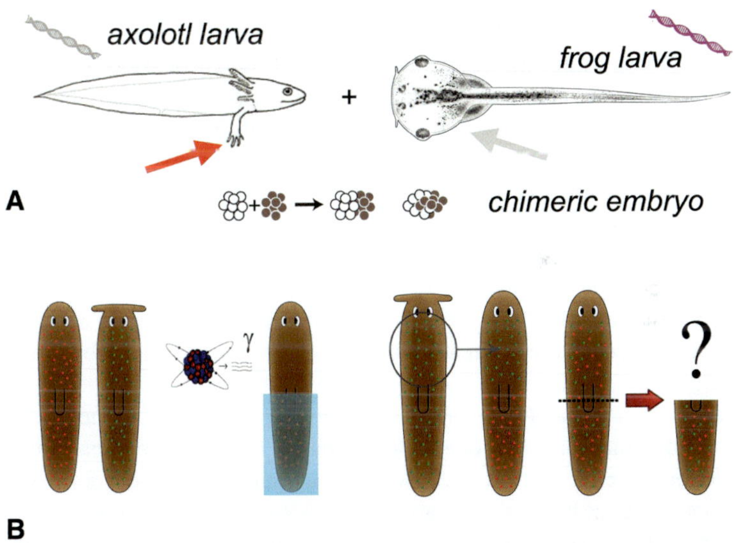

FIGURE 2.11 Limits of current predictive models. Despite having full access to the genomes of the axolotl and frog, no existing model allows us to predict whether chimeric "frogolotl" embryos will have legs (like axolotls) or not (like frog larvae): the molecular information does not directly reveal decisions made by groups of cells at a larger scale (A). Similarly, in planaria, despite decades of molecular analysis, no model makes a prediction on what kind of head will form after neoblast stem cells from species of worms with different head shapes are combined into a partially irradiated host: will one head shape dominate? Will the result be a composite? Or will the head never stop remodeling, as neither set of cells' complete target morphology is ever reached?

Images in A taken with permission from Zahn et al. (2017). Image in B taken with permission from Lobo et al. (2012).

different setpoints. Novel computational frameworks, using multiscale models from the collective intelligence field (Pezzulo and Levin, 2015), not CA, suitable for single-layer systems made of stupid components, are needed to better understand, predict, and control biological outcomes. And a deep question is, what does it mean to *understand* and predict morphogenetic outcomes as something different from simply running a fast simulation and observing what happens (Wolpert, 1994).

But, at least in chimeras, we have two known evolutionarily-established setpoints, even if we don't know whether the result will be one dominant one, an intermediate case, or an oscillation. What happens when cells find themselves in entirely novel circumstances? This question can likely be studied in many organoids, spheroids, assembloids, embryoids, etc. (Haase and Freedman, 2020; Panoutsopoulos, 2020; Vogt, 2021), if attention were paid to how they navigate their various problem spaces while remaining still in 3D space. However, two stark examples of this phenomenon are provided by model systems whose conventional motility makes it much more obvious that their properties as novel proto-organisms need to be studied.

The first are "Xenobots" (Figure 2.12), made from epithelial frog embryo cells. These self-assemble into spherical constructs that move, change direction, and even assemble copies of themselves from loose cells in their environment (kinematic self-replication) (Blackiston et al., 2023; Blackiston et al., 2021; Kriegman et al., 2020, 2021). They have a unique transcriptome with hundreds of genes differently expressed than the cells do in vivo, including a cluster of hearing-related genes, which led to the discovery of their ability to respond to acoustic stimuli in ways that embryos do not (Pai et al., 2025). The second are Anthrobots (Bhattaram et al., 2025; Gumuskaya et al., 2025; Gumuskaya et al., 2023), made from human adult tracheal epithelial cells. These have four different motility modes with known transition probabilities between behaviors, several discrete morphotypes, over 9000 differentially-expressed genes (about half the genome!), a partially reset epigenetic aging clock (they are younger than the cells they came from), and the ability to heal neural wounds in their environment. None of these capabilities were predicted in advance prior to their discovery.

Anthrobots reveal that the remarkable properties of Xenobots are not some unique feature of amphibian or embryonic cells: somatic, 100% *Homo sapiens* cells from elderly donors take on a novel lifestyle when liberated from their normal environment. In both types of biobots, no scaffolds, DNA editing, transgenes, or nanomaterials are added. Their novel properties derive from the *subtraction* of instructive cues provided in vivo that mask their intrinsic capabilities in forcing them to boring epithelial fates. Allowing them to reboot their multicellularity shows the plasticity and novel setpoints that the reliability of development obscures. Studies are underway to determine their proto-cognitive capacities (i.e., learning abilities and other problem-solving capacities).

Crucially, it is not enough, after such living constructs are made, to observe their surprising properties and note that they are compatible with the ciliated nature of their parent tissues, etc. Our goal, if we want to understand evolution, collective intelligence, and the roadmap to regenerative medicine, is not merely to record "emergent" outcomes after the fact, but to understand the option space from which

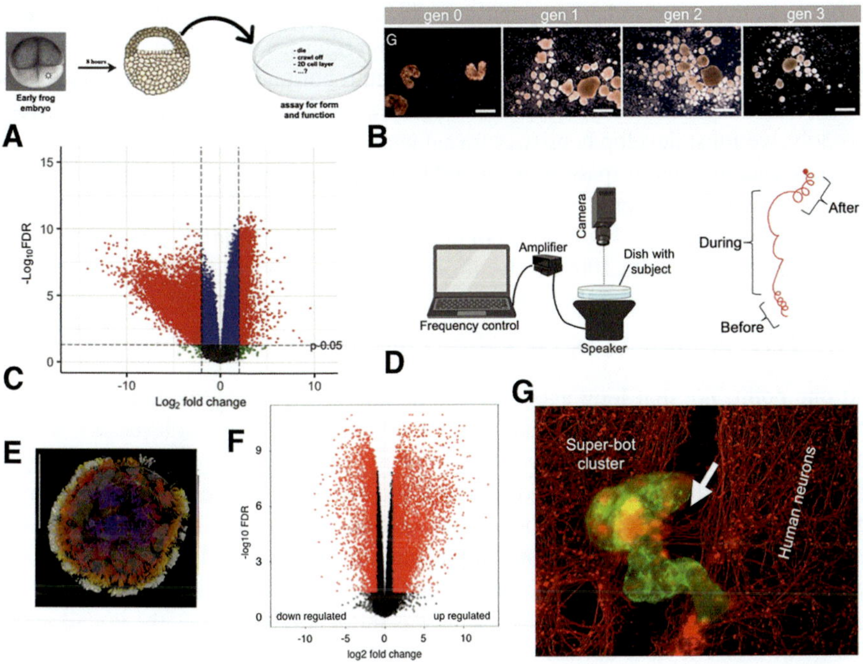

FIGURE 2.12 Same genetic hardware, different form and function. (A) Epithelial cells from a frog embryo combine to form "Xenobots", which move and can replicate by combining loose epithelial cells into the next generation of Xenobots (B). Xenobot cells express hundreds of genes differently than they do in vivo (C), including a cluster of genes related to sound perception, which enables them (unlike embryos) to respond to sounds (D). Likewise, epithelial cells taken from adult human donors form Anthrobots (E) – motile proto-organisms that express >9000 genes differently from the cells do in vivo, and have capabilities such as healing neural wounds (G; Anthrobots in green, forming a cluster to knit across a wound made in human iPS-derived neurons in red). Images in A and B taken with permission from Kriegman et al. (2021). In both cases, no genetic editing, scaffolds, synthetic circuits, or other means were used to induce these complex and novel behaviors in living systems that have never existed in this configuration before and were not selected for their numerous new capabilities, specific transcriptomes, etc.

Images in C and D taken with permission from Pai et al. (2025). Images in E–G taken with permission from (Gumuskaya et al., 2025; Gumuskaya et al., 2023).

such forms and behaviors are drawn, and to be able to predict and guide specific novel setpoints in advance. Moreover, there have never been any Anthrobots or Xenobots, and there's never been selection for any of their transcriptional, morphological, and behavioral properties. Whence their very specific traits? How much specificity between past environments and the resulting traits should we expect in an evolutionary origin story? Of course, there are many ways for traits to be established other than from direct selection for them, but it is not enough to say that these capacities were somehow acquired at the same time that human and frog environments drove selection for standard frog and human bodies, gene expression cascades, etc. The next opportunities in this field are to develop rigorous theories of how and when the computational cost for designing good Anthrobot and Xenobot capacities was paid (since they were not paid in the same way as human and frog traits were). More broadly, we must develop heuristics for understanding how the agential materials of life navigate the option space of form and behavior in truly novel circumstances that provided evolutionary guidance at the lowest levels (reuse of cells and molecules that do have a long history) but offer no tight historical information on how to be a multicellular organism with 100% conventional parts but a novel lifestyle.

2.6 FUTURE IMPACTS OF DEVELOPMENTAL BIOLOGY

At one point, not that long ago, the way to control a computer was by physically rewiring it. In other words, one had to interact with it at the lowest level. It is laughable nowadays to suggest that switching among applications on one's laptop should require a soldering iron. That is because if the hardware is good enough, it can be controlled with stimuli and reprogrammed, not micromanaged. This gave rise to the information technology revolution. I argue that the life sciences today, with all of the emphasis on genomic editing, single-molecule approaches, protein engineering, etc., are roughly where computer engineering was in the 1940s and 1950s: it's all about the hardware. The next decades will complement this essential body of knowledge with new tools focused on understanding not only the software of life but its essential agentic nature. Developmental biology plays a central role in a number of other fields (Figure 2.13).

Living matter is not just a computational material, offering reprogrammability and plasticity of outcome without changing the DNA. It is more than that – it is an agential material, composed of multilevel autonomous subsystems at many scales of organization that have agendas and the ability to behavior-shape other subsystems toward setpoints in diverse spaces. This means that as useful as the software–hardware paradigm is in rising beyond molecular determinism, it is limited; no current machine metaphor is sufficient to interact with life at its fullest. More appropriate formalisms, based on systems that dynamically improvise the meaning of their memories (Levin, 2024c) and erase the boundaries between active agents and passive information (Fields and Levin, 2025), are being developed.

2.6.1 Biomedicine

One of the key areas that will be impacted by these advances is regenerative medicine (Levin, 2024b). We currently have increasing control over molecular events, but definitive treatments for birth defects, missing or damaged organs, aging, and cancer elude us. A major limitation is the lack of *cures*. With the exception of surgery, antibiotics, and single-gene diseases amenable to gene therapy, there are very few therapies that solve a problem permanently: typically, it comes back with a vengeance when the drug is stopped (or earlier, when the tissue habituates to it). Massive differences across patients in efficacy and side effects arise, in part, because cells

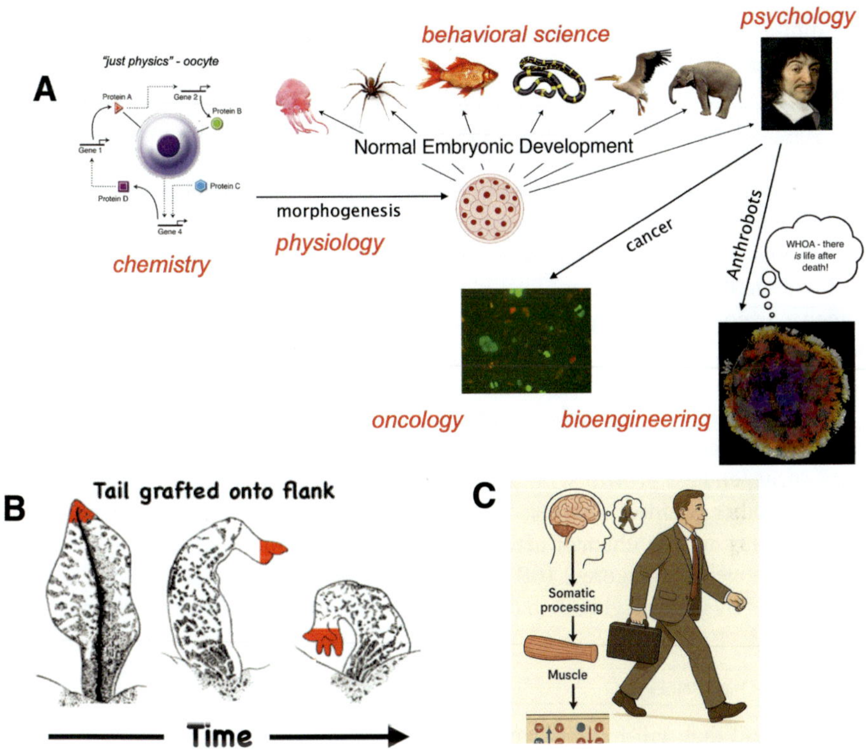

FIGURE 2.13 Morphogenesis and cognition both feature multiscale control. (A) One of the most remarkable aspects of developmental biology is that it illustrates how a system slowly progresses not just in complexity, but continuously traverses disciplines that are often taken to treat categorically distinct kinds of subjects. Starting as an unfertilized oocyte that is thought to be suitably handled by chemistry, a being self-assembles and journeys through developmental physiology, behavioral science, and perhaps psychology, with possible detours including oncology and bioengineering (which can allow fragments to persist long after their human donor has become deceased). (B) When a tail is grafted onto the flank of an amphibian, it slowly remodels into a limb. Consider the cells marked in red – tail tip cells, sitting at the end of a tail, which nevertheless become fingers. There is no local damage or injury, and yet their molecular signals and cellular architecture change in accordance with a high level, to them abstract and unknowable, body-wide target morphology, which forces modification toward a large-scale outcome more appropriate for the species. (C) This kind of top-down control of molecular events by higher-level goal states is not just for rare cases of mind-body medicine (such as changes in gene transcription and molecular signaling cascades by meditation or mental expectation of drug activity in placebo experiments; Black et al., 2019; Buric et al., 2017; Colloca and Benedetti, 2005, 2006; Dada et al., 2018; Lui et al., 2010). It is also precisely what happens every day, when the bioelectric system of the body enables the highly abstract goals of a human being to change the chemistry of ions crossing muscle cell membranes during voluntary motion. The top-down control of chemistry by cognitive constructs is a central feature of nervous systems, but they coopted this basic trick from far more ancient bioelectrical systems first used to control morphogenesis.

Panels in A by Jeremy Guay. Panel in B modified after Farinella-Ferruzza (1956). Panel C by GPT-5.

and multicellular collectives have not bought into the new regime and often resist (Lagasse and Levin, 2023). New computational models of multiscale autonomous cybernetic systems (Csermely et al., 2020; Keresztes et al., 2025; Veres et al., 2024) will enable strategies that do not attempt to force specific molecular states but rewrite setpoints, which the cells will pursue autonomously.

Part of the research program is to develop *convincing* strategies for effectively manipulating the encoded goal states of physiological, transcriptional, and many other subsystems. Cells, with a long evolutionary history of being subject to hacking and exploitation attempts by other biota, likely have many ways to resist signals that they perceive as coming from outside. Getting the cells' and tissues' buy-in, so that they don't try to contravene the treatment but rather continue to maintain health after the intervention is withdrawn, requires a better understanding of their setpoints, priors (physiological memories), and reprogrammability. One example of this is chemically-induced leg regeneration in amphibians, where a 24-hour intervention leads to 1.5 years of appendage growth with no further manipulations needed (Murugan et al., 2022). Another example is bioelectric eye induction (Pai et al., 2012), where cells misexpressing an ion channel actively recruit their neighbors to participate in the novel organogenesis (Figure 2.10B), or ignore the exogenous bioelectric state entirely if persuaded by their neighbors to do so (as part of a cancer suppression mechanism).

2.6.2 Bioengineering

Another field that will be massively impacted by developmental biology is bioengineering. As we move from a focus on life-as-it-is to life-as-it-can-be (Langton, 1995), an understanding of the plasticity and capability of living material at all scales is crucial. Synthetic biology – rewiring the phenotypes of single cells (Krzyszton et al., 2021; McNerney et al., 2021; Sole and Macia, 2014) – will expand to synthetic morphology (Aydin et al., 2019; Davies, 2017; Doursat and Sanchez, 2014; Doursat et al., 2013; Ebrahimkhani and Levin, 2021; Ho and Morsut, 2021; Kamm et al., 2018; Pagan-Diaz et al., 2019; Sample et al., 2019; Velazquez et al., 2018), as we learn to communicate larger-scale goal states. Ultimately, this will lead to a functional anatomical compiler (Davies and Levin, 2023) – a tool that will allow scientists and engineers to probe and access all possible regions of the latent space of form and function.

Bioengineering will benefit from a dissolution of artificial walls between the sciences of the body and those of the mind (Rouleau and Levin, 2023), not only because of useful new living machines but also because of the increasing development of practical tools (Abramson and Levin, 2021) for mutually enriching feedback. Indeed, the interoperability of life makes possible a very wide variety of instrumentized *ex vivo* constructs (Ando and Kanzaki, 2020; Bakkum et al., 2007; Clawson and Levin, 2023; Potter et al., 2004; Potter et al., 2003; Selberg et al., 2018; Tsuda et al., 2009; Warwick et al., 1998) – mergers of life with passive or AI-powered electronic components, which will need the insights of developmental biology and developmental psychology to guide both applications and ethics.

In many ways, communicating with cells and tissues is analogous to the search for extraterrestrial intelligence challenge (Grimaldi and Marcy, 2018; Maccone, 2011, 2017; Waldrop, 2011): these are somewhat alien intelligences who operate in a problem space difficult for us to visualize, and have capabilities and setpoints we do not yet understand. One of the most exciting tools to assist with this challenge is AI. The role of AI in this domain is not limited to data mining or analysis (Villoutreix, 2021). It holds the potential to be used as a *translator interface* to enable effective communication with these systems. Notice that this is similar to how language-using brain regions allow one human to, albeit indirectly, access and communicate with the private physiological information implemented by neuronal machinery in another human's brain. The full potential of this technology is when it is used as an "agential translation layer" – a sophisticated problem-solving system motivated to understand the beings on both sides of the equation and improve their interaction.

2.6.3 Cognitive Science and Philosophy of Mind

> Developmental biologists don't concern themselves with the mind-body problem.
>
> —anonymous reviewer

If developmental biologists don't concern themselves with the mind-body problem, who possibly could? Developmental biology, expanded by modern tools and approaches that transcend traditional disciplinary boundaries, is uniquely placed to provide scientific advances in an area previously left to philosophers: the origin and nature of embodied minds. Mechanisms of scaling of cognition from chemistry to psychology, the shape of the morphospace of possible embodied minds, the role of reductionist strategies in science (Rijssenbeek, 2025), machine and computational metaphors for mind and life (Boldt, 2018), the mechanisms of top-down causation (Figure 2.14), and the relationships between scientists/engineers and their increasingly autonomous materials are all crucial areas in which developmental biology can contribute essential ideas, model systems, and best of all, actual data that constrain and enable better philosophy.

It is especially timely now, with the arrival of AI in everyday life (Rouleau and Levin, 2024). The ubiquitous conversations about the difference between "machines" and "real beings" (Barwich and Rodriguez, 2024; Levin, 2024a; McShea, 2012; McShea, 2013; Nicholson, 2019) lead to deep questions about future relationships with not only software agents but with much more challenging mergers of human bodies and technology in cyborgs (Pio-Lopez, 2021) (Figure 2.15). The vast majority of these discussions are missing basic aspects of developmental biology and will be greatly enhanced by a wider appreciation of the unique science that addresses the origins of everything that makes us special (in the widest possible sense of "us", Levin, 2025).

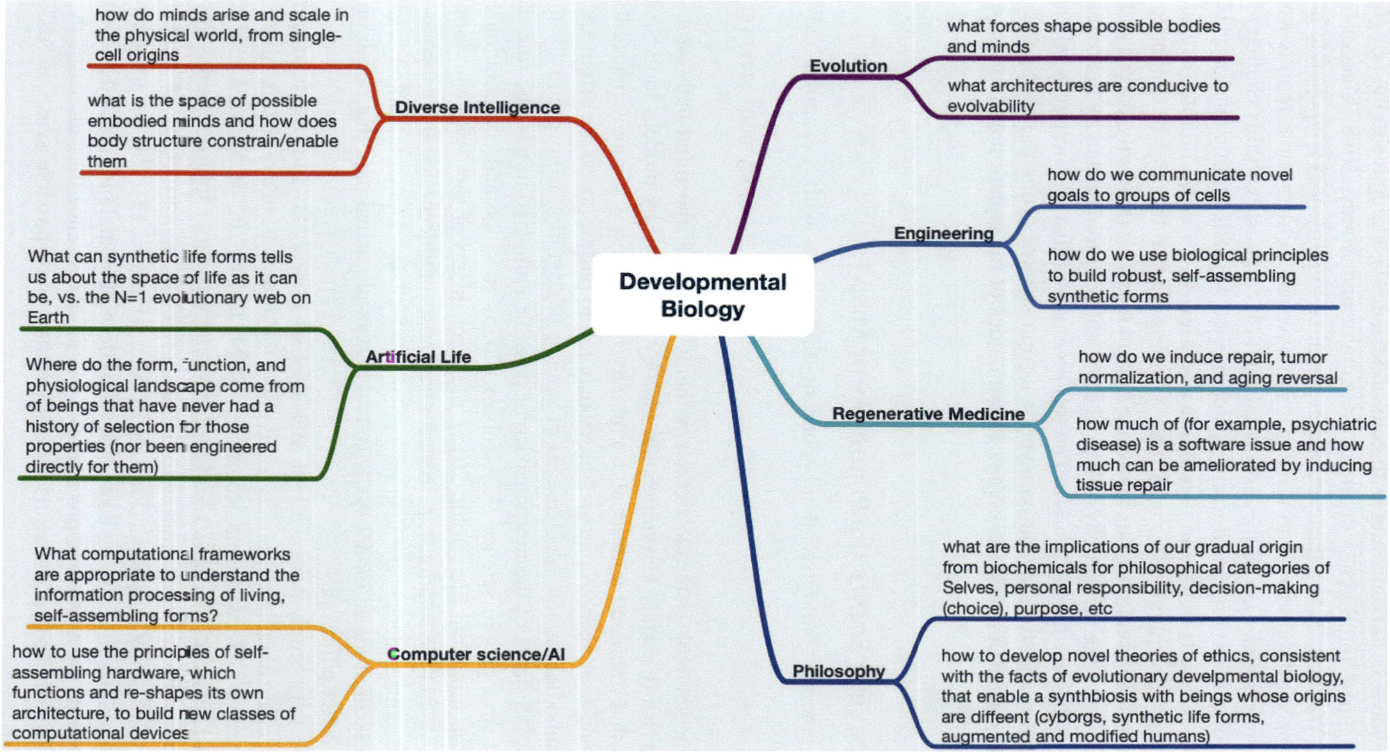

FIGURE 2.14 A mind map of developmental biology's essential contributions to other basic and applied fields. As shown in this mind map, the study of morphogenesis impacts efforts including diverse intelligence, artificial life, computer science/AI, evolution, engineering, regenerative medicine, and foundational issues in philosophy.

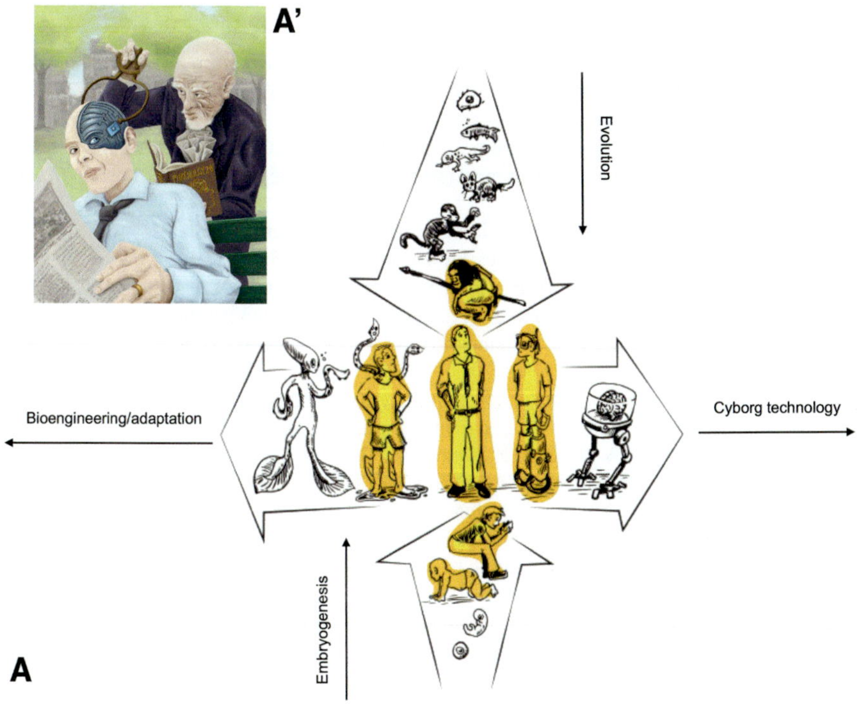

FIGURE 2.15 The future of developmental biology. (A) Understanding the multiscale, self-assembling properties of bodies and minds is not only about the continuum of increasing agency (symbolized by the yellow glow) of ontogenetic and phylogenetic change (vertical). It is also essential to understand the forthcoming changes, biological and technological, which will expand the natural living form in ways we can barely imagine. The early tests of social and ethical frameworks, which must be grounded in real biology, will not be in the relatively easy distinctions between software AI's and living forms, but by the inevitable appearance of composite beings – cyborgs that defy our prescientific, outdated categories of "life" vs. "machine" (panel A' schematizes futile attempts to figure out if a being is "man" or "machine", by measuring the precise amount of engineered vs. evolved tissue). Developmental biology, in its essential embrace of the development of mindful beings from a drop of chemicals in the oocyte (i.e., an implicit acceptance that we and the inanimate world are part of one continuum), is uniquely and ideally placed to help humanity through this transition.

Images by Jeremy Guay of Peregrine Creative.

ACKNOWLEDGMENTS

I thank Richard Watson, Scott Gilbert, Lev Beloussov, Katrina Schleisman, Mark Solms, Alexey Tolchinsky, Chris Fields, Tomas Pollak, Pamela Lyon, and many members of the Levin Lab, as well as numerous members of the developmental biology and diverse intelligence communities, for many useful discussions. I thank Igor Adameyko and Gerhard Schlosser for helpful comments on the manuscript. The chapter is dedicated to Alexander Gurwitsch, an early pioneer of key concepts in this area.

REFERENCES

Abramson, C.I., Levin, M., 2021. Behaviorist approaches to investigating memory and learning: A primer for synthetic biology and bioengineering. *Commun Integr Biol* 14, 230–247.

Adams, D.S., Levin, M., 2012. General principles for measuring resting membrane potential and ion concentration using fluorescent bioelectricity reporters. *Cold Spring Harbor protocols* 2012, 385–397.

Adar, O., Shakargy, J.D., Ilan, Y., 2025. The constrained disorder principle: Beyond biological allostasis. *Biology (Basel)* 14. https://www.mdpi.com/2079-7737/14/4/339

Akberdin, I.R., Ozonov, E.A., Mironova, V.V., Omelyanchuk, N.A., Likhoshvai, V.A., Gorpinchenko, D.N., Kolchanov, N.A., 2007. A cellular automaton to model the development of primary shoot meristems of Arabidopsis thaliana. *J Bioinform Comput Biol* 5, 641–650.

Albantakis, L., Marshall, W., Hoel, E., Tononi, G., 2019. What caused what? A quantitative account of actual causation using dynamical causal networks. *Entropy (Basel)* 21.

Albantakis, L., Marshall, W., Hoel, E.P., Tononi, G., 2017. What caused what? An irreducible account of actual causation. arXiv arXiv:1708.06716.

Ando, N., Kanzaki, R., 2020. Insect-machine hybrid robot. *Curr Opin Insect Sci* 42, 61–69.

Auletta, G., Ellis, G.F., Jaeger, L., 2008. Top-down causation by information control: From a philosophical problem to a scientific research programme. *J R Soc, Interface / the R Soc* 5, 1159–1172.

Aydin, O., Zhang, X., Nuethong, S., Pagan-Diaz, G.J., Bashir, R., Gazzola, M., Saif, M.T.A., 2019. Neuromuscular actuation of biohybrid motile bots. *Proc Natl Acad Sci U S A* 116, 19841–19847.

Bakkum, D.J., Gamblen, P.M., Ben-Ary, G., Chao, Z.C., Potter, S.M., 2007. MEART: The semi-living artist. *Front Neurorobot* 1, 5.

Baldwin, J.M., 1896. A new factor in evolution. *Am Nat* 30, 441–445.

Baluska, F., Miller, W.B., Jr., 2018. Senomic view of the cell: Senome versus Genome. *Commun Integr Biol* 11, 1–9.

Baluska, F., Miller, W.B., Reber, A.S., 2022a. Cellular and evolutionary perspectives on organismal cognition: From unicellular to multicellular organisms. *Biol J Linn Soc* 139, 503–513.

Baluska, F., Miller, W.B., Slijepcevic, P., Reber, A.S., 2025. Sensing, feeling and sentience in unicellular organisms and living cells. *Biosystems* 247, 105374.

Baluska, F., Reber, A.S., Miller, W.B., Jr., 2022b. Cellular sentience as the primary source of biological order and evolution. *Biosystems* 218, 104694.

Barwich, A.-S., Rodriguez, M.J., 2024. Rage against the what? The machine metaphor in biology. *Biol Philos* 39, 14.

Bassel, G.W., 2018. Information processing and distributed computation in plant organs. *Trends Plant Sci* 23, 994–1005.

Beane, W.S., Morokuma, J., Adams, D.S., Levin, M., 2011. A Chemical genetics approach reveals H,K-ATPase-mediated membrane voltage is required for planarian head regeneration. *Chem Biol* 18, 77–89.

Beisson, J., Sonneborn, T.M., 1965. Cytoplasmic inheritance of the organization of the cell cortex in Paramecium Aurelia. *Proc Natl Acad Sci U S A* 53, 275–282.

Berry, M.J., 2nd, Brivanlou, I.H., Jordan, T.A., Meister, M., 1999. Anticipation of moving stimuli by the retina. *Nature* 398, 334–338.

Bhattaram, D., Golestan, K., Zhang, X., Yang, S., Gong, Z., Brody, S.L., Horani, A., Webster-Wood, V.A., Farimani, A.B., Ren, X., 2025. AggreBots: Configuring CiliaBots through guided, modular tissue aggregation. *Sci Adv* 11, eadx4176.

Biswas, S., Clawson, W., Levin, M., 2023. Learning in transcriptional network models: Computational discovery of pathway-level memory and effective interventions. *Int J Mol Sci* 24, 285. https://doi.org/10.3390/ijms24010285

Biswas, S., Manicka, S., Hoel, E., Levin, M., 2021. Gene regulatory networks exhibit several kinds of memory: Quantification of memory in biological and random transcriptional networks. *iScience* 24, 102131.

Black, D.S., Christodoulou, G., Cole, S., 2019. Mindfulness meditation and gene expression: A hypothesis-generating framework. *Curr Opin Psychol* 28, 302–306.

Blackiston, D., Kriegman, S., Bongard, J., Levin, M., 2023. Biological robots: Perspectives on an emerging interdisciplinary field. *Soft Robot.* https://doi.org/10.1089/soro.2022.0142

Blackiston, D., Lederer, E.K., Kriegman, S., Garnier, S., Bongard, J., Levin, M., 2021. A cellular platform for the development of synthetic living machines. *Sci Robot* 6, eabf1571.

Blackiston, D., Shomrat, T., Levin, M., 2015. The stability of memories during brain remodeling: A perspective. *Commun Integrat Biol* 8, e1073424.

Blackiston, D., Shomrat, T., Nicolas, C.L., Granata, C., Levin, M., 2010. A second-generation device for automated training and quantitative behavior analyses of molecularly-tractable model organisms. *PLoS One* 5, e14370.

Blackiston, D.J., Levin, M., 2013. Ectopic eyes outside the head in Xenopus tadpoles provide sensory data for light-mediated learning. *J Exper Biol* 216, 1031–1040.

Blackiston, D.J., McLaughlin, K.A., Levin, M., 2009. Bioelectric controls of cell proliferation: Ion channels, membrane voltage and the cell cycle. *Cell Cycle* 8, 3519–3528.

Blackiston, D.J., Vien, K., Levin, M., 2017. Serotonergic stimulation induces nerve growth and promotes visual learning via posterior eye grafts in a vertebrate model of induced sensory plasticity. *npj Regener Med* 2, 8.

Boldt, J., 2018. Machine metaphors and ethics in synthetic biology. *Life Sci Soc Policy* 14, 12.

Boyle, E.A., Li, Y.I., Pritchard, J.K., 2017. An expanded view of complex traits: From polygenic to omnigenic. *Cell* 169, 1177–1186.

Bubenik, A., 1966. *Das Geweih.* Paul Parey Verlag, Hamburg.

Bubenik, A.B., Pavlansky, R., 1965. Trophic responses to trauma in growing antlers. *J Exp Zool* 159, 289–302.

Buckley, C.L., Lewens, T., Levin, M., Millidge, B., Tschantz, A., Watson, R.A., 2024. Natural induction: Spontaneous adaptive organisation without natural selection. *Entropy (Basel)* 26.

Buric, I., Farias, M., Jong, J., Mee, C., Brazil, I.A., 2017. What is the molecular signature of mind-body interventions? A systematic review of gene expression changes induced by meditation and related practices. *Front Immunol* 8, 670.

Bussey, K.J., Cisneros, L.H., Lineweaver, C.H., Davies, P.C.W., 2017. Ancestral gene regulatory networks drive cancer. *Proc Nat Acad Sci U S A* 114, 6160–6162.

Calvo, P., Baluska, F., Sims, A., 2016. "Feature Detection" vs. "Predictive Coding" models of plant behavior. *Front Psychol* 7, 1505.

Cannon, W.B., 1929. Organization for physiological homeostasis. *Physiol Rev* 9, 399–431.

Cannon, W.B., 1932. *The wisdom of the body*, Rev. and enl. ed. Norton, New York.

Cervera, J., Levin, M., Mafe, S., 2021. Morphology changes induced by intercellular gap junction blocking: A reaction-diffusion mechanism. *Biosystems* 209, 104511.

Chernet, B.T., Adams, D.S., Lobikin, M., Levin, M., 2016. Use of genetically encoded, light-gated ion translocators to control tumorigenesis. *Oncotarget* 7, 19575–19588.

Chernet, B.T., Fields, C., Levin, M., 2015. Long-range gap junctional signaling controls oncogene-mediated tumorigenesis in *Xenopus laevis* embryos. *Front Physiol* 5, 519.

Chernet, B.T., Levin, M., 2013a. Endogenous voltage potentials and the microenvironment: Bioelectric signals that reveal, induce and normalize cancer. *J Clin Exp Oncol Suppl* 1.

Chernet, B.T., Levin, M., 2013b. Transmembrane voltage potential is an essential cellular parameter for the detection and control of tumor development in a Xenopus model. *Dis Models Mechan* 6, 595–607.

Chernet, B.T., Levin, M., 2014. Transmembrane voltage potential of somatic cells controls oncogene-mediated tumorigenesis at long-range. *Oncotarget* 5, 3287–3306.

Chuang, J.S., Frentz, Z., Leibler, S., 2019. Homeorhesis and ecological succession quantified in synthetic microbial ecosystems. *Proc Natl Acad Sci U S A* 116, 14852–14861.

Ciszak, M., Masi, E., Baluska, F., Mancuso, S., 2016. Plant shoots exhibit synchronized oscillatory motions. *Commun Integr Biol* 9, e1238117.

Clawson, W.P., Levin, M., 2023. Endless forms most beautiful 2.0: Teleonomy and the bioengineering of chimaeric and synthetic organisms. *Biol J Linn Soc* 139, 457–486.

Cliff, O.M., Lizier, J.T., Wang, X.R., Wang, P., Obst, O., Prokopenko, M., 2017. Quantifying long-range interactions and coherent structure in multi-agent dynamics. *Artif Life* 23, 34–57.

Colditz, I.G., 2020. A consideration of physiological regulation from the perspective of Bayesian enactivism. *Physiol Behav* 214, 112758.

Colloca, L., Benedetti, F., 2005. Placebos and painkillers: Is mind as real as matter? *Nat Rev Neurosci* 6, 545–552.

Colloca, L., Benedetti, F., 2006. How prior experience shapes placebo analgesia. *Pain* 124, 126–133.

Cooke, J., 1979. Cell number in relation to primary pattern formation in the embryo of *Xenopus laevis*. I: The cell cycle during new pattern formation in response to implanted organisers. *J Embryol Exper Morphol* 51, 165–182.

Cooke, J., 1981. Scale of body pattern adjusts to available cell number in amphibian embryos. *Nature* 290, 775–778.

Craver, C.F., Bechtel, W., 2007. Top-down causation without top-down causes. *Biol Philos* 22, 547–563.

Csermely, P., Kunsic, N., Mendik, P., Kerestely, M., Farago, T., Veres, D.V., Tompa, P., 2020. Learning of signaling networks: Molecular mechanisms. *Trends Biochem Sci* 45, 284–294.

Csete, M., Doyle, J., 2004. Bow ties, metabolism and disease. *Trends Biotechnol* 22, 446–450.

Dada, T., Mittal, D., Mohanty, K., Faiq, M.A., Bhat, M.A., Yadav, R.K., Sihota, R., Sidhu, T., Velpandian, T., Kalaivani, M., Pandey, R.M., Gao, Y., Sabel, B.A., Dada, R., 2018. Mindfulness meditation reduces intraocular pressure, lowers stress biomarkers and modulates gene expression in glaucoma: A randomized controlled trial. *J. Glaucoma* 27, 1061–1067.

Davies, J., 2017. Using synthetic biology to explore principles of development. *Development* 144, 1146–1158.

Davies, J., Levin, M., 2023. Synthetic morphology with agential materials. *Nat Rev Bioengineer* 1, 46–59.

Davies, P.C.W., Lineweaver, C.H., 2011. Cancer tumors as Metazoa 1.0: Tapping genes of ancient ancestors. *Phys Biol* 8, 015001.

Davis, D., Cano, J., Miller, G., Loftus, E., 2024. The Multiple Roles of Emotion in Interpretation and Memory of Sexual Consent. *Top Cogn Sci* 16, 644–660.

Davis, G.V., de Souza Moraes, T., Khanapurkar, S., Dromiack, H., Ahmad, Z., Bayer, E.M., Bhalerao, R.P., Walker, S.I., Bassel, G.W., 2023. Toward uncovering an operating system in plant organs. *Trends Plant Sci.* 29, 753. https://doi.org/10.1016/j.tplants.2023.11.006

Dewan, E.M., 1976. *Consciousness as an emergent causal agent in the context of control system theory*, in: Globus, G., Maxwell, G., Savodnik, I. (Eds.). Plenum Press, New York, pp. 179–198.

Dopazo, H., Gordon, M.B., Perazzo, R., Risau-Gusman, S., 2001. A model for the interaction of learning and evolution. *Bulletin of Mathematical Biology* 63, 117–134.

Doursat, R., Sanchez, C., 2014. Growing fine-grained multicellular robots. *Soft Robotics* 1, 110–121.

Doursat, R., Sayama, H., Michel, O., 2013. A review of morphogenetic engineering. *Nat Comput* 12, 517–535.

Doyle, J.C., Csete, M., 2011. Architecture, constraints, and behavior. *Proc Natl Acad Sci U S A* 108 Suppl 3, 15624–15630.

Durant, F., Bischof, J., Fields, C., Morokuma, J., LaPalme, J., Hoi, A., Levin, M., 2019. The role of early bioelectric signals in the regeneration of planarian anterior/posterior polarity. *Biophys J* 116, 948–961.

Ebrahimkhani, M.R., Levin, M., 2021. Synthetic living machines: A new window on life. *iScience* 24, 102505.

Ellis, G.F.R., 2009. Top-Down Causation and the Human Brain. *Underst Complex Syst*, 63–81.

Ellis, G.F.R., Bloch, C.S., 2011. Top-down causation, adaptive selection, and their consequences. *Hum Dev* 54, 93–100.

Emmons-Bell, M., Durant, F., Tung, A., Pietak, A., Miller, K., Kane, A., Martyniuk, C.J., Davidian, D., Morokuma, J., Levin, M., 2019. Regenerative adaptation to electrochemical perturbation in planaria: A molecular analysis of physiological plasticity. *iScience* 22, 147–165.

Emmons-Bell, M., Hariharan, I.K., 2021. Membrane potential regulates Hedgehog signalling in the Drosophila wing imaginal disc. *EMBO Rep* 22, e51861.

Emmons-Bell, M., Yasutomi, R., Hariharan, I.K., 2020. Membrane potential regulates Hedgehog signaling and compartment boundary maintenance in the Drosophila wing disc. bioRxiv, 2020.2006.2002.128892.

Fankhauser, G., 1945a. The effects of changes in chromosome number on amphibian development. *Quart Rev Biol* 20, 20–78.

Fankhauser, G., 1945b. Maintenance of normal structure in heteroploid salamander larvae, through compensation of changes in cell size by adjustment of cell number and cell shape. *J Exp Zool* 100, 445–455.

Farinella-Ferruzza, N., 1953. Risultati di trapianti di bottone codale di urodeli su anuri e vice versa. *Riv. Biol.* 45, 523–527.

Farinella-Ferruzza, N., 1956. The transformation of a tail into a limb after xenoplastic transformation. *Experientia* 15, 304–305.

Fields, C., Bischof, J., Levin, M., 2020. Morphological coordination: A common ancestral function unifying neural and non-neural signaling. *Physiology* 35, 16–30.

Fields, C., Levin, M., 2017. Multiscale memory and bioelectric error correction in the cytoplasm–cytoskeleton-membrane system. *Wiley Interdisciplinary Rev: Syst Biol Med* 10, e1410-n/a.

Fields, C., Levin, M., 2020. Does Evolution Have a Target Morphology? Organisms. *Journal of Biological Sciences* 4, 57–76.

Fields, C., Levin, M., 2022. Competency in navigating arbitrary spaces as an invariant for analyzing cognition in diverse embodiments. *Entropy (Basel)* 24, 819. https://doi.org/10.3390/e24060819

Fields, C., Levin, M., 2025. Thoughts and thinkers: On the complementarity between objects and processes. *Phys Life Rev.* 52, 256–273.

Frank, S.A., 2007. Maladaptation and the paradox of robustness in evolution. *PLoS One* 2, e1021.

Frank, S.A., 2009. The common patterns of nature. *J Evol Biol* 22, 1563–1585.

Frank, S.A., 2019a. Evolutionary design of regulatory control. I. A robust control theory analysis of tradeoffs. *J Theor Biol* 463, 121–137.

Frank, S.A., 2019b. Evolutionary design of regulatory control. II. Robust error-correcting feedback increases genetic and phenotypic variability. *J Theor Biol* 468, 72–81.

Gallo, D.A., 2010. False memories and fantastic beliefs: 15 years of the DRM illusion. *Mem Cognit* 38, 833–848.
Gawne, R., McKenna, K.Z., Levin, M., 2020. Competitive and coordinative interactions between body parts produce adaptive developmental outcomes. *BioEssays* 42, e1900245.
Gottlieb, G., 1999. *Probabilistic epigenesis and evolution*. Clark University Press, Worcester, Mass.
Gray, R.S., Roszko, I., Solnica-Krezel, L., 2011. Planar cell polarity: Coordinating morphogenetic cell behaviors with embryonic polarity. *Dev Cell* 21, 120–133.
Grimaldi, C., Marcy, G.W., 2018. Bayesian approach to SETI. *Proc Natl Acad Sci U S A* 115, E9755–E9764.
Grossberg, S., 1978. Communication, memory, and development, in: Rosen, R., Snell, F. (Eds.), *Progress in Theoretical Biology*. Academic Press, New York.
Gumuskaya, G., Davey, N., Srivastava, P., Bender, A., Pio-Lopez, L., Hazel, D., Levin, M., 2025. The morphological, behavioral, and transcriptomic life cycle of anthrobots. *Adv Sci (Weinh)* 12, e2409330.
Gumuskaya, G., Srivastava, P., Cooper, B.G., Lesser, H., Semegran, B., Garnier, S., Levin, M., 2023. Motile living biobots self-construct from adult human somatic progenitor seed cells. *Adv Sci (Weinh)* 11, e2303575.
Guo, L., Accorsi, A., He, S., Guerrero-Hernández, C., Sivagnanam, S., McKinney, S., Gibson, M., Sánchez Alvarado, A., 2018. An adaptable chromosome preparation methodology for use in invertebrate research organisms. *BMC Biol* 16, 25.
Gyurkó, D.M., Veres, D.V., Módos, D., Lenti, K., Korcsmáros, T., Csermely, P., 2013. Adaptation and learning of molecular networks as a description of cancer development at the systems-level: Potential use in anti-cancer therapies. *Semin Cancer Biol* 23, 262–269.
Haase, K., Freedman, B.S., 2020. Once upon a dish: Engineering multicellular systems. *Development* 147.
Harrison, L.A., Gracias, A.J., Friston, K.J., Buckwalter, J.G., 2025. Resilience phenotypes derived from an active inference account of allostasis. *Front Behav Neurosci* 19, 1524722.
Hartl, B., Levin, M., 2025. What does evolution make? Learning in living lineages and machines. *Trends Genet* 41, 480–496.
Hartl, B., Risi, S., Levin, M., 2024. Evolutionary Implications of Self-Assembling Cybernetic Materials with Collective Problem-Solving Intelligence at Multiple Scales. *Entropy-Switz* 26, 532.
Hatzikirou, H., Deutsch, A., 2008. Cellular automata as microscopic models of cell migration in heterogeneous environments. *Curr Top Dev Biol* 81, 401–434.
Hinton, G.E., Nowlan, J., 1987. How learning can guide evolution. *Complex Sys* 1, 495–502.
Ho, C., Morsut, L., 2021. Novel synthetic biology approaches for developmental systems. *Stem Cell Rep* 16, 1051–1064.
Hoel, E., Levin, M., 2020. Emergence of informative higher scales in biological systems: A computational toolkit for optimal prediction and control. *Commun Integr Biol* 13, 108–118.
Hoel, E.P., 2017. When the map is better than the territory. *Entropy-Switz* 19.
Hoel, E.P., 2018. Agent above, atom below: How agents causally emerge from their underlying microphysics, in: Aguirre, A., Foster, B., Merali, Z. (Eds.), *Wandering Towards a Goal: How Can Mindless Mathematical Laws Give Rise to Aims and Intention?* Springer International Publishing, Cham, pp. 63–76.
Hoel, E.P., Albantakis, L., Marshall, W., Tononi, G., 2016a. Can the macro beat the micro? Integrated information across spatiotemporal scales. *Neurosci Conscious* 2016, niw012.
Hoel, E.P., Albantakis, L., Marshall, W., Tononi, G., 2016b. Can the macro beat the micro? Integrated information across spatiotemporal scales. *Neurosci Conscious* 2016, niw012.

Hoel, E.P., Albantakis, L., Tononi, G., 2013a. Quantifying causal emergence shows that macro can beat micro. *Proc Natl Acad Sci U S A* 110, 19790–19795.

Hoel, E.P., Albantakis, L., Tononi, G., 2013b. Quantifying causal emergence shows that macro can beat micro. *Proc Natl Acad Sci U S A* 110, 19790–19795.

Holtfreter, J., 1955. Transformation of a tail into a limb or gill-like structures. *Journal of Experimental Zoology* 129, 623–648.

Huang, S., Ernberg, I., Kauffman, S., 2009. Cancer attractors: A systems view of tumors from a gene network dynamics and developmental perspective. *Semin Cell Dev Biol* 20, 869–876.

Huth, A.G., Lee, T., Nishimoto, S., Bilenko, N.Y., Vu, A.T., Gallant, J.L., 2016. Decoding the semantic content of natural movies from human brain activity. *Front Syst Neurosci* 10, 81.

James, W., 1890. *The principles of psychology*. H. Holt and company, New York,.

Johnston, I.G., Bassel, G.W., 2018. Identification of a bet-hedging network motif generating noise in hormone concentrations and germination propensity in Arabidopsis. *J R Soc Interface* 15, 20180042.

Juarrero, A., 2009. Top-Down Causation and Autonomy in Complex Systems. *Underst Complex Syst*, 83–102.

Kamm, R.D., Bashir, R., Arora, N., Dar, R.D., Gillette, M.U., Griffith, L.G., Kemp, M.L., Kinlaw, K., Levin, M., Martin, A.C., McDevitt, T.C., Nerem, R.M., Powers, M.J., Saif, T.A., Sharpe, J., Takayama, S., Takeuchi, S., Weiss, R., Ye, K.M., Yevick, H.G., Zaman, M.H., 2018. Perspective: The promise of multi-cellular engineered living systems. *Apl Bioeng* 2, 040901.

Karig, D., Martini, K.M., Lu, T., DeLateur, N.A., Goldenfeld, N., Weiss, R., 2018. Stochastic Turing patterns in a synthetic bacterial population. *Proc Natl Acad Sci U S A* 115, 6572–6577.

Keresztes, D., Kerestely, M., Szarka, L., Kovacs, B.M., Schulc, K., Veres, D.V., Csermely, P., 2025. Cancer drug resistance as learning of signaling networks. *Biomed Pharmacother* 183, 117880.

Khuong, A., Gautrais, J., Perna, A., Sbai, C., Combe, M., Kuntz, P., Jost, C., Theraulaz, G., 2016. Stigmergic construction and topochemical information shape ant nest architecture. *Proc Natl Acad Sci U S A* 113, 1303–1308.

Klein, B., Hoel, E., 2019. Uncertainty and causal emergence in complex networks, arXiv e-prints.

Koshland, D.E., 1983. The bacterium as a model neuron. *Trends Neurosci* 6, 133–137.

Kouvaris, K., Clune, J., Kounios, L., Brede, M., Watson, R.A., 2017. How evolution learns to generalise: Using the principles of learning theory to understand the evolution of developmental organisation. *PLoS Comput Biol* 13, e1005358.

Kriegman, S., Blackiston, D., Levin, M., Bongard, J., 2020. A scalable pipeline for designing reconfigurable organisms. *Proc Natl Acad Sci U S A* 117, 1853–1859.

Kriegman, S., Blackiston, D., Levin, M., Bongard, J., 2021. Kinematic self-replication in reconfigurable organisms. *Proc Natl Acad Sci U S A* 118, e2112672118.

Krutovskikh, V., Yamasaki, H., 1997. The role of gap junctional intercellular communication (GJIC) disorders in experimental and human carcinogenesis. *Histol Histopathol.* 12, 761–768.

Krzyszton, R., Wan, Y., Petreczky, J., Balazsi, G., 2021. Gene-circuit therapy on the horizon: Synthetic biology tools for engineered therapeutics. *Acta Biochim Pol.* 68, 377–383.

Lagasse, E., Levin, M., 2023. Future medicine: From molecular pathways to the collective intelligence of the body. *Trends Mol Med* 29, 687–710.

Langton, C.G., 1995. *Artificial life: An overview*. MIT Press, Cambridge, MA.

Lee, D.D., Prindle, A., Liu, J., Suel, G.M., 2017. SnapShot: Electrochemical Communication in Biofilms. *Cell* 170, 214–214 e211.

Lehotzky, D., Zupanc, G.K.H., 2019. Cellular automata modeling of stem-cell-driven development of tissue in the nervous system. *Dev Neurobiol* 79, 497–517.

Levin, M., 2019. The computational boundary of a "Self": Developmental bioelectricity drives multicellularity and scale-free cognition. *Front Psychol* 10, 2688.

Levin, M., 2021. Bioelectric signaling: Reprogrammable circuits underlying embryogenesis, regeneration, and cancer. *Cell* 184, 1971–1989.

Levin, M., 2022a. Technological approach to mind everywhere: An experimentally-grounded framework for understanding diverse bodies and minds. *Front Syst Neurosci* 16, 768201.

Levin, M., 2022b. Technological approach to mind everywhere: An experimentally-grounded framework for understanding diverse bodies and minds. *Front Syst Neurosci* 16, 768201.

Levin, M., 2023. Bioelectric networks: The cognitive glue enabling evolutionary scaling from physiology to mind. *Anim Cogn* 26, 1865–1891.

Levin, M., 2024a. AI: A Bridge toward Diverse Intelligence and Humanity's Future. preprint.

Levin, M., 2024b. The Multiscale wisdom of the body: Collective intelligence as a tractable interface for next-generation biomedicine. *BioEssays* 47, e202400196.

Levin, M., 2024c. Self-improvising memory: A perspective on memories as agential, dynamically reinterpreting cognitive glue. *Entropy (Basel)* 26, 481. https://doi.org/10.3390/e26060481

Levin, M., 2025. Ingressing Minds: Causal Patterns Beyond Genetics and Environment in Natural, Synthetic, and Hybrid Embodiments. preprint.

Levin, M., Pezzulo, G., Finkelstein, J.M., 2017. Endogenous bioelectric signaling networks: Exploiting voltage gradients for control of growth and form. *Annu Rev Biomed Eng* 19, 353–387.

Liu, J., Martinez-Corral, R., Prindle, A., Lee, D.D., Larkin, J., Gabalda-Sagarra, M., Garcia-Ojalvo, J., Suel, G.M., 2017. Coupling between distant biofilms and emergence of nutrient time-sharing. *Science* 356, 638–642.

Lizier, J.T., 2014. JIDT: An Information-Theoretic Toolkit for Studying the Dynamics of Complex Systems. *Front Robot AI* 1, 11. https://doi.org/10.3389/frobt.2014.00011

Lizier, J.T., Heinzle, J., Horstmann, A., Haynes, J.D., Prokopenko, M., 2011. Multivariate information-theoretic measures reveal directed information structure and task relevant changes in fMRI connectivity. *J Comput Neurosci* 30, 85–107.

Lobo, D., Beane, W.S., Levin, M., 2012. Modeling planarian regeneration: A primer for reverse-engineering the worm. *PLoS Comput Biol* 8, e1002481.

Lobo, D., Solano, M., Bubenik, G.A., Levin, M., 2014. A linear-encoding model explains the variability of the target morphology in regeneration. *J R Soc, Interface / the R Soc* 11, 20130918.

Lu, H.Y., Lorenc, E.S., Zhu, H., Kilmarx, J., Sulzer, J., Xie, C., Tobler, P.N., Watrous, A.J., Orsborn, A.L., Lewis-Peacock, J., Santacruz, S.R., 2021. Multi-scale neural decoding and analysis. *J Neural Eng* 18, 045013.

Lui, F., Colloca, L., Duzzi, D., Anchisi, D., Benedetti, F., Porro, C.A., 2010. Neural bases of conditioned placebo analgesia. *Pain* 151, 816–824.

Lutz, H., 1949. Sur la production experimentale de la polyembryonie et de la monstruosite double chez les oiseaux. *Archs Anat Microsc Morph Exp* 38, 79–144.

Maccone, C., 2011. A mathematical model for evolution and SETI. *Orig. Life Evol. Biosph.* 41, 609–619.

Maccone, C., 2017. Evo-SETI: A Mathematical Tool for Cladistics, Evolution, and SETI. *Life (Basel)* 7.

Marcon, L., Sharpe, J., 2012. Turing patterns in development: What about the horse part? *Curr Opin Genet Dev* 22, 578–584.

Marder, E., Goaillard, J.M., 2006. Variability, compensation and homeostasis in neuron and network function. *Nature reviews. Neuroscience* 7, 563–574.

Markus, M., Bohm, D., Schmick, M., 1999. Simulation of vessel morphogenesis using cellular automata. *Math Biosci* 156, 191–206.

Marr, D., 1982. *Vision: A computational investigation into the human representation and processing of visual information*. W.H. Freeman, San Francisco.

Martinez-Corral, R., Liu, J., Prindle, A., Suel, G.M., Garcia-Ojalvo, J., 2019. Metabolic basis of brain-like electrical signalling in bacterial communities. *Philos Trans R Soc Lond B Biol Sci* 374, 20180382.

Mathews, J., Chang, A.J., Devlin, L., Levin, M., 2023. Cellular signaling pathways as plastic, proto-cognitive systems: Implications for biomedicine. *Patterns (N Y)* 4, 100737.

Matsushita, Y., Kaneko, K., 2020. Homeorhesis in Waddington's landscape by epigenetic feedback regulation. *Phys Rev Res* 2, 023083.

Maturana, H.R., Varela, F.J., 1980. *Autopoiesis and Cognition the Realization of the Living, Boston Studies in the Philosophy of Science*. Springer Netherlands,, Dordrecht, p. 1 online resource (180 pages).

McClelland, J.L., Botvinick, M.M., Noelle, D.C., Plaut, D.C., Rogers, T.T., Seidenberg, M.S., Smith, L.B., 2010. Letting structure emerge: Connectionist and dynamical systems approaches to cognition. *Trends Cogn Sci* 14, 348–356.

McClelland, J.L., Rogers, T.T., 2003. The parallel distributed processing approach to semantic cognition. *Nat Rev Neurosci* 4, 310–322.

McEwen, B.S., 1998. Stress, adaptation, and disease. Allostasis and allostatic load. *Ann N Y Acad Sci* 840, 33–44.

McMillen, P., Levin, M., 2024. Optical estimation of bioelectric patterns in living embryos. *Methods Mol Biol* 2745, 91–102.

McNerney, M.P., Doiron, K.E., Ng, T.L., Chang, T.Z., Silver, P.A., 2021. Theranostic cells: Emerging clinical applications of synthetic biology. *Nat Rev Genet* 22, 730–746.

McShea, D.W., 2012. Upper-directed systems: A new approach to teleology in biology. *Biol Philos* 27, 663–684.

McShea, D.W., 2013. Machine wanting. *Stud Hist Philos Biol Biomed Sci* 44, 679–687.

Melanitis, N., Nikita, K.S., 2019. Biologically-inspired image processing in computational retina models. *Comput Biol Med* 113, 103399.

Mesnil, M., Krutovskikh, V., Omori, Y., Yamasaki, H., 1995. Role of blocked gap junctional intercellular communication in non-genotoxic carcinogenesis. *Toxicol Lett* 82–83, 701–706.

Miller, W.B., Jr., Baluska, F., Reber, A.S., 2023. A revised central dogma for the 21st century: all biology is cognitive information processing. *Prog Biophys Mol Biol* 182, 34–48.

Miller, W.B., Jr., Baluska, F., Torday, J.S., 2020. Cellular senomic measurements in Cognition-Based Evolution. *Prog Biophys Mol Biol* 156, 20–33.

Miller, W.B., Jr., Torday, J.S., Baluska, F., 2018. Biological evolution as defense of 'self'. *Prog Biophys Mol Biol* 142, 54–74.

Mintz, B., 1971. Genetic mosaicism in vivo: Development and disease in allophenic mice. *Fed Proc* 30, 935–943.

Mintz, B., Illmensee, K., 1975. Normal genetically mosaic mice produced from malignant teratocarcinoma cells. *Proc Natl Acad Sci U S A* 72, 3585–3589.

Moore, D.G., Valentini, G., Walker, S.I., Levin, M., 2018. Inform: Efficient Information-Theoretic Analysis of Collective Behaviors. *Front Robot AI* 5, 60.

Morgan, C.L., 1903. Other minds than ours, in: Scott, W. (Ed.), *An Introduction to Comparative Psychology*, p. 59. The Walter Scott Publishing Co.

Murugan, N.J., Vigran, H.J., Miller, K.A., Golding, A., Pham, Q.L., Sperry, M.M., Rasmussen-Ivey, C., Kane, A.W., Kaplan, D.L., Levin, M., 2022. Acute multidrug delivery via a wearable bioreactor facilitates long-term limb regeneration and functional recovery in adult Xenopus laevis. *Sci Adv* 8, eabj2164.

Nakamasu, A., Takahashi, G., Kanbe, A., Kondo, S., 2009. Interactions between zebrafish pigment cells responsible for the generation of Turing patterns. *Proc Natl Acad Sci U S A* 106, 8429–8434.

Nanos, V., Levin, M., 2021. Multi-scale Chimerism: An experimental window on the algorithms of anatomical control. *Cells Dev* 169, 203764.

Naselaris, T., Prenger, R.J., Kay, K.N., Oliver, M., Gallant, J.L., 2009. Bayesian reconstruction of natural images from human brain activity. *Neuron* 63, 902–915.

Nicholson, D.J., 2019. Is the cell really a machine? *J Theor Biol* 477, 108–126.

Nishimoto, S., Vu, A.T., Naselaris, T., Benjamini, Y., Yu, B., Gallant, J.L., 2011. Reconstructing visual experiences from brain activity evoked by natural movies. *Curr Biol: CB* 21, 1641–1646.

Nishimura, O., Hosoda, K., Kawaguchi, E., Yazawa, S., Hayashi, T., Inoue, T., Umesono, Y., Agata, K., 2015. Unusually Large Number of Mutations in Asexually Reproducing Clonal Planarian Dugesia japonica. *PLoS One* 10, e0143525.

Noble, D., 2010. Biophysics and systems biology. *Philos Trans A Math Phys Eng Sci* 368, 1125–1139.

Noble, D., 2011. The aims of systems biology: Between molecules and organisms. *Pharmacopsychiatry* 44(Suppl 1), S9–S14.

Noble, D., 2012. A theory of biological relativity: No privileged level of causation. *Interface Focus* 2, 55–64.

Noble, D., 2022. Modern physiology vindicates Darwin's dream. *Exp Physiol* 107, 1015–1028.

Odling-Smee, F.J., 1995. Niche construction, genetic evolution and cultural change. *Behavioural processes* 35, 195–205.

Okasha, S., 2012. Emergence, hierarchy and top-down causation in evolutionary biology. *Interface Focus* 2, 49–54.

Otgaar, H., Dodier, O., Garry, M., Howe, M.L., Loftus, E.F., Lynn, S.J., Mangiulli, I., McNally, R.J., Patihis, L., 2023. Oversimplifications and Misrepresentations in the Repressed Memory Debate: A Reply to Ross. *J Child Sex Abus* 32, 116–126.

Oviedo, N.J., Morokuma, J., Walentek, P., Kema, I.P., Gu, M.B., Ahn, J.M., Hwang, J.S., Gojobori, T., Levin, M., 2010. Long-range neural and gap junction protein-mediated cues control polarity during planarian regeneration. *Dev Biol* 339, 188–199.

Pagan-Diaz, G.J., Ramos-Cruz, K.P., Sam, R., Kandel, M.E., Aydin, O., Saif, M.T.A., Popescu, G., Bashir, R., 2019. Engineering geometrical 3-dimensional untethered in vitro neural tissue mimic. *Proc Natl Acad Sci U S A* 116, 25932–25940.

Pai, V.P., Aw, S., Shomrat, T., Lemire, J.M., Levin, M., 2012. Transmembrane voltage potential controls embryonic eye patterning in *Xenopus laevis*. *Development* 139, 313–323.

Pai, V.P., Cervera, J., Mafe, S., Willocq, V., Lederer, E.K., Levin, M., 2020. HCN2 Channel-Induced Rescue of Brain Teratogenesis via Local and Long-Range Bioelectric Repair. *Front Cell Neurosci* 14, 136.

Pai, V.P., Levin, M., 2022. HCN2 channel-induced rescue of brain, eye, heart and gut teratogenesis caused by nicotine, ethanol and aberrant notch signalling. *Wound Repair Regen.* 30, 681.

Pai, V.P., Pietak, A., Willocq, V., Ye, B., Shi, N.Q., Levin, M., 2018. HCN2 Rescues brain defects by enforcing endogenous voltage pre-patterns. *Nat Commun* 9, 998.

Pai, V.P., Pio-Lopez, L., Sperry, M.M., Erickson, P., Tayyebi, P., Levin, M., 2025. Basal Xenobot transcriptomics reveals changes and novel control modality in cells freed from organismal influence. *Commun Biol* 8, 646.

Panoutsopoulos, A.A., 2020. Organoids, Assembloids, and Novel Biotechnology: Steps Forward in Developmental and Disease-Related Neuroscience. *Neuroscientist*, 1073858420960112.

Petsakou, A., Perrimon, N., 2023. Bioelectric regulation of intestinal stem cells. *Trends Cell Biol* 33, 555–567.

Pezzulo, G., Levin, M., 2015. Re-membering the body: Applications of computational neuroscience to the top-down control of regeneration of limbs and other complex organs. *Integr Biol (Camb)* 7, 1487–1517.
Pezzulo, G., Levin, M., 2016. Top-down models in biology: Explanation and control of complex living systems above the molecular level. *J R Soc Interface* 13.
Piaget, J., 1976. *Behaviour and evolution*. Pantheon Books, New York.
Pinet, K., Deolankar, M., Leung, B., McLaughlin, K.A., 2019. Adaptive correction of craniofacial defects in pre-metamorphic *Xenopus laevis* tadpoles involves thyroid hormone-independent tissue remodeling. *Development* 146, dev175893.
Pinet, K., McLaughlin, K.A., 2019. Mechanisms of physiological tissue remodeling in animals: Manipulating tissue, organ, and organism morphology. *Dev Biol* 451, 134–145.
Pio-Lopez, L., 2021. The rise of the biocyborg: Synthetic biology, artificial chimerism and human enhancement. *New Genetics and Society* 40, 599–619.
Potter, A.M., DeMarse, T.B., Bakkum, D.J., Booth, M.C., Brumfield, J.R., Chao, Z., Wagenaar, D.A., 2004. Hybrots: Hybrids of living neurons and robots for studying neural computation. In *Proceedings of Brain Inspired Cognitive Systems* (pp. 1–5). Stirling, UK.
Potter, S.M., Wagenaar, D.A., Madhavan, R., DeMarse, T.B., 2003. Long-term bidirectional neuron interfaces for robotic control, and in vitro learning studies. *P Ann Int Ieee Embs* 25, 3690–3693.
Power, D.A., Watson, R.A., Szathmary, E., Mills, R., Powers, S.T., Doncaster, C.P., Czapp, B., 2015. What can ecosystems learn? Expanding evolutionary ecology with learning theory. *Biol Direct* 10, 69.
Prindle, A., Liu, J., Asally, M., Ly, S., Garcia-Ojalvo, J., Suel, G.M., 2015. Ion channels enable electrical communication in bacterial communities. *Nature* 527, 59–63.
Rago, A., Kouvaris, K., Uller, T., Watson, R., 2019. How adaptive plasticity evolves when selected against. *PLoS Comput Biol* 15, e1006260.
Reber, A.S., Baluska, F., Miller, W.B., Slijepcevic, P., 2024a. The sensual cell: Feeling and affect in unicellular species. *Biosystems* 238, 105197.
Reber, A.S., Miller, W.B., Jr., Slijepcevic, P., Baluska, F., 2024b. The CBC theory and its entailments: why current models of the origin of consciousness fail. *EMBO Rep* 25, 8–12.
Rijssenbeek, J., 2025. Synthetic biology: Supporting an anti-reductionist view of life. *Synthese* 205, 52.
Roediger, H.L., McDermott, K.B., 1995. Creating false memories: remembering words not presented in lists. *J Exp Psychol* 21, 803.
Rosen, R., 1985. *Anticipatory systems: Philosophical, mathematical, and methodological foundations*, 1st ed. Pergamon Press, Oxford, England, New York.
Rosenblueth, A., Wiener, N., Bigelow, J., 1943. Behavior, purpose, and teleology. *Philos Sci* 10, 18–24.
Rouleau, N., Levin, M., 2023. The multiple realizability of sentience in living systems and beyond. *eNeuro* 10. https://doi.org/10.1523/ENEURO.0375-23.2023
Rouleau, N., Levin, M., 2024. Discussions of machine versus living intelligence need more clarity. *Nat Mach Intell* 6, 1424–1426.
Rubin, H., 1985. Cancer as a dynamic developmental disorder. *Cancer Res* 45, 2935–2942.
Rubin, H., 2006. What keeps cells in tissues behaving normally in the face of myriad mutations? *BioEssays* 28, 515–524.
Rubin, H., 2007. Ordered heterogeneity and its decline in cancer and aging. *Adv Cancer Res* 98, 117–147.
Rubin, H., Chow, M., Yao, A., 1996. Cellular aging, destabilization, and cancer. *Proc Natl Acad Sci U S A* 93, 1825–1830.

Saló, E., Abril, J.F., Adell, T., Cebrià, F., Eckelt, K., Fernandez-Taboada, E., Handberg-Thorsager, M., Iglesias, M., Molina, M.D., Rodriguez-Esteban, G., 2009. Planarian regeneration: Achievements and future directions after 20 years of research. *Int J Dev Biol* 53, 1317–1327.

Sample, M., Boulicault, M., Allen, C., Bashir, R., Hyun, I., Levis, M., Lowenthal, C., Mertz, D., Montserrat, N., Palmer, M.J., Saha, K., Zartman, J., 2019. Multi-cellular engineered living systems: Building a community around responsible research on emergence. *Biofabrication* 11, 043001.

Schotthofer, S.K., Bohrmann, J., 2020. Analysing bioelectrical phenomena in the Drosophila ovary with genetic tools: Tissue-specific expression of sensors for membrane potential and intracellular pH, and RNAi-knockdown of mechanisms involved in ion exchange. *BMC Dev Biol* 20, 15.

Schulkin, J., Sterling, P., 2019. Allostasis: A brain-centered, predictive mode of physiological regulation. *Trends Neurosci* 42, 740–752.

Selberg, J., Gomez, M., Rolandi, M., 2018. The potential for convergence between synthetic biology and bioelectronics. *Cell Syst* 7, 231–244.

Sennesh, E., Theriault, J., Brooks, D., van de Meent, J.W., Barrett, L.F., Quigley, K.S., 2022. Interoception as modeling, allostasis as control. *Biol Psychol* 167, 108242.

Shao, X., Chen, C., Loftus, E.F., Xue, G., Zhu, B., 2022. Dynamic changes in neural representations underlie the repetition effect on false memory. *Neuroimage* 259, 119442.

Shapiro, J.A., 2011. *Evolution: A view from the 21st century*. FT Press Science, Upper Saddle River, NJ.

Shapiro, J.A., 2022. Engines of innovation: Biological origins of genome evolution. *Biol J Linn Soc*.

Shreesha, L., Levin, M., 2023. Cellular competency during development alters evolutionary dynamics in an artificial embryogeny model. *Entropy-Switz* 25, 131.

Shreesha, L., Levin, M., 2024. Stress sharing as cognitive glue for collective intelligences: A computational model of stress as a coordinator for morphogenesis. *Biochem Biophys Res Commun* 731, 150396.

Sole, R., Macia, J., 2014. Synthetic biology: biocircuits in synchrony. *Nature* 508, 326–327.

Strier, D.E., Ponce Dawson, S., 2007. Turing patterns inside cells. *PloS one* 2, e1053.

Thorpe, W.H., 1945. Animal learning and evolution. *Nature* 156, 46–46.

Topham, A.T., Taylor, R.E., Yan, D., Nambara, E., Johnston, I.G., Bassel, G.W., 2017. Temperature variability is integrated by a spatially embedded decision-making center to break dormancy in Arabidopsis seeds. *Proc Natl Acad Sci U S A* 114, 6629–6634.

Tsuda, S., Artmann, S., Zauner, K.-P., 2009. The Phi-Bot: A Robot controlled by a slime mould, in: Adamatzky, A., Komosinski, M. (Eds.), *Artificial Life Models in Hardware*. Springer London, pp. 213–232.

Tung, A., Sperry, M.M., Clawson, W., Pavuluri, A., Bulatao, S., Yue, M., Flores, R.M., Pai, V.P., McMillen, P., Kuchling, F., Levin, M., 2024. Embryos assist morphogenesis of others through calcium and ATP signaling mechanisms in collective teratogen resistance. *Nat Commun* 15, 535.

Turing, A.M., 1950. Computing machinery and intelligence. *Mind* 59, 433–460.

Turing, A.M., 1952. The chemical basis of morphogenesis. *Philos T Roy Soc B* 237, 37–72.

Turner, J.S., 2011. Termites as models of swarm cognition. *Swarm Intell-Us* 5, 19–43.

Van Valen, L., 1973. A new evolutionary law. *Evol Theory* 1, 1–30.

Vandenberg, L.N., Adams, D.S., Levin, M., 2012. Normalized shape and location of perturbed craniofacial structures in the Xenopus tadpole reveal an innate ability to achieve correct morphology. *Dev Dyn* 241, 863–878.

Vandenberg, L.N., Morrie, R.D., Adams, D.S., 2011. V-ATPase-dependent ectodermal voltage and pH regionalization are required for craniofacial morphogenesis. *Dev Dyn* 240, 1889–1904.

Varley, T.F., Hoel, E., 2022. Emergence as the conversion of information: A unifying theory. *Philos Trans A Math Phys Eng Sci* 380, 20210150.

Velazquez, J.J., Su, E., Cahan, P., Ebrahimkhani, M.R., 2018. Programming morphogenesis through systems and synthetic biology. *Trends Biotechnol* 36, 415–429.

Veres, T., Kerestely, M., Kovacs, B.M., Keresztes, D., Schulc, K., Seitz, E., Vassy, Z., Veres, D.V., Csermely, P., 2024. Cellular forgetting, desensitisation, stress and ageing in signalling networks: When do cells refuse to learn more? *Cell Mol Life Sci* 81, 97.

Villoutreix, P., 2021. What machine learning can do for developmental biology. *Development* 148.

Vogt, N., 2021. Assembloids. *Nat Methods* 18, 27.

Waldrop, M.M., 2011. The search for alien intelligence: SETI is dead--long live SETI. *Nature* 475, 442–444.

Wang, X.R., Miller, J.M., Lizier, J.T., Prokopenko, M., Rossi, L.F., 2012. Quantifying and tracing information cascades in swarms. *PLoS One* 7, e40084.

Warwick, K., Nasuto, S. J., Becerra, V. M., Whalley, B. J., 1998. Experiments with an in-vitro Robot Brain, in: Cai, Y. (Ed.), *Computing with Instinct*. Springer.

Watson, R.A., Levin, M., Buckley, C.L., 2022. Design for an individual: connectionist approaches to the evolutionary transitions in individuality. *Frontiers in Ecology and Evolution* 10. https://doi.org/10.3389/fevo.2022.823588

Watson, R.A., Mills, R., Buckley, C.L., Kouvaris, K., Jackson, A., Powers, S.T., Cox, C., Tudge, S., Davies, A., Kounios, L., Power, D., 2016. Evolutionary connectionism: algorithmic principles underlying the evolution of biological organisation in Evo-Devo, Evo-Eco and evolutionary transitions. *Evol Biol* 43, 553–581.

Watson, R.A., Szathmary, E., 2016. How can evolution learn? *Trends Ecol Evol* 31, 147–157.

Webster, M., 2019. Morphological homeostasis in the fossil record. *Semin Cell Dev Biol* 88, 91–104.

Weiss, I., Bohrmann, J., 2019. Electrochemical gradients are involved in regulating cytoskeletal patterns during epithelial morphogenesis in the Drosophila ovary. *BMC Dev Biol* 19, 22.

Wibral, M., Lizier, J.T., Vogler, S., Priesemann, V., Galuske, R., 2014. Local active information storage as a tool to understand distributed neural information processing. *Front Neuroinform* 8, 1.

Wigner, E.P., 1960. The unreasonable effectiveness of Mathematics in the natural sciences. *Communications on Pure and Applied Mathematics* 13. https://doi.org/10.1142/1104

Wolpert, L., 1994. Do we understand development? *Science* 266, 571–572.

Woodruff, R.I., Telfer, W.H., 1973. Polarized intercellular bridges in ovarian follicles of the cecropia moth. *J Cell Biol* 58, 172–188.

Yang, C.Y., Bialecka-Fornal, M., Weatherwax, C., Larkin, J.W., Prindle, A., Liu, J., Garcia-Ojalvo, J., Suel, G.M., 2020. Encoding membrane-potential-based memory within a microbial community. *Cell Syst* 10, 417–423 e413.

Yokawa, K., Kagenishi, T., Pavlovič, A., Gall, S., Weiland, M., Mancuso, S., Baluška, F., 2018. Anaesthetics stop diverse plant organ movements, affect endocytic vesicle recycling and ROS homeostasis, and block action potentials in Venus flytraps. *Ann Bot* 122, 747–756.

Zahn, N., Levin, M., Adams, D.S., 2017. The Zahn drawings: New illustrations of Xenopus embryo and tadpole stages for studies of craniofacial development. *Development* 144, 2708–2713.

Section II

Computational Approaches to Development

3 New Approaches to an Old Problem

The Enigma of Cells Making Decisions

Igor Adameyko

3.1 ON HOW WE STUDY CELL LINEAGES AND DEPICT EMBRYONIC DEVELOPMENT

The question of how cells choose their fate during development has fascinated biologists for centuries. Understanding fate selection is central to deciphering how a single fertilized egg gives rise to the vast diversity of cell types in a multicellular organism. To tackle this question, the research community devised various techniques to track, manipulate, and analyze cell fate decisions in different organisms. These methodologies have evolved from simple observational techniques to sophisticated genetic tools, each providing new insights into the mechanisms of development.

In the late 19th and early 20th centuries, prominent researchers like Wilhelm Roux and Hans Driesch used basic embryology techniques to observe fundamentals of embryonic development (Sander, 1992). Roux's studies on frog embryos, in which he destroyed one cell of a two-cell stage embryo, suggested that early development followed a deterministic path (Sander, 1997). On the opposite, Driesch's sea urchin experiments surprisingly revealed that cells retained plasticity, compensating with robust responses for removed parts of the embryo (Huneman, 2013; Sander, 1992). These early experiments hinted at the complexity of fate selection and the balance between determinism and plasticity in development (Wellmann, 2022).

By the mid-20th century, the introduction of fluorescent dyes enabled more precise fate mapping. Our colleagues of that time injected or soaked cells in dyes that were retained for long enough to allow tracking over several cell divisions. This method provided crucial insights into lineage relationships and the spatial organization of tissues, particularly in amphibians and birds. In parallel, scientists began using cell transplantation to directly test how a cell's environment influences its fate (Liberali and Schier, 2024). A landmark discovery came from Spemann and Mangold's organizer experiment in 1924, in which they transplanted a small region of an amphibian embryo (the Spemann organizer) to another embryo (De Robertis, 2006). This graft triggered the formation of an entire secondary body axis, demonstrating that fate selection is not solely intrinsic but also influenced by external

DOI: 10.1201/9781003560784-5

signals. These studies laid the foundation for modern developmental biology by demonstrating the role of extrinsic cues, such as morphogens, in directing cell differentiation.

While transplantations and dye-based studies provided valuable insights, they were limited in their ability to track cells over extended developmental time frames and with high precision. The introduction of genetic lineage tracing revolutionized the field by allowing researchers to label and follow specific cells and their progeny through multiple generations (Liberali and Schier, 2024). One of the most powerful techniques emerged in the form of Cre-lox recombination, which was adapted from bacteriophage site-specific recombination systems (Joyner and Zervas, 2006). A variant of this method, inducible Cre systems (CreERT2), added another layer of control enabling researchers to activate genetic labeling at precise time points (Feil et al., 1997). This was particularly important for studying dynamic fate decisions and understanding how cells transition between different states.

Modern approaches now integrate molecular barcodes into cells, allowing massive parallel fate mapping at single-cell resolution (Erickson et al., 2024; McKenna et al., 2016; Raj et al., 2018). By tracking changes in gene expression alongside lineage tracing, these techniques provide insights into the molecular drivers of fate selection. In addition, CRISPR (clustered regularly interspaced short palindromic repeats)-based lineage tracing introduces CRISPR-induced mutations or edits into DNA, which act as heritable markers that can be used to reconstruct cell lineage trees computationally (Li et al., 2025). Traditional and old methods of fate mapping and lineage tracing, such as dye injection or genetic markers in small cell populations, have provided valuable insights but lacked the ability to capture the full complexity of cellular dynamics across an entire organism. Modern clonal tracing, particularly when performed in a massively parallel fashion, enables us to dissect developmental processes at an unprecedented scale. Such massive parallel clonal data are essential for understanding multicellular development because they provide a high-resolution, quantitative perspective on how cells choose fates, proliferate, differentiate, and contribute to tissue formation over time. One of the key contributions of such large-scale clonal analysis is its ability to reveal how multipotent progenitor cells resolve into distinct fates at the level of a cell collective within the context of patterned tissue (Erickson et al., 2024).

The concept of collective multipotency represents a rather recent shift in our understanding of how developmental potential is organized within cell populations (Erickson et al., 2023). Traditionally, multipotency has been described as a property of individual cells – a single progenitor possessing the ability to differentiate into multiple cell types. However, collective multipotency reframes this view, proposing that it is the ensemble of cells, rather than isolated individuals, that collectively retains the full spectrum of developmental options. In this framework, even if individual cells are already partially biased to particular fates, the group as a whole functions as a multipotent unit. This idea becomes especially apparent in embryonic systems where cell fate decisions are spatially and temporally distributed across dynamic populations. For instance, in the developing neural crest, a classic example of a multipotent population, it is increasingly evident that individual cells may already carry lineage biases (Erickson et al., 2023; Soldatov et al., 2019). Yet, the

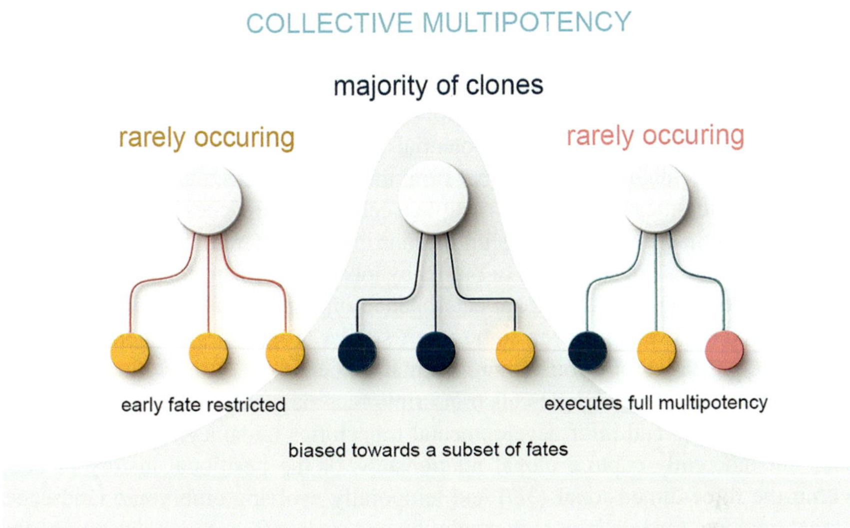

FIGURE 3.1 The concept of collective multipotency. The multipotent progenitors (or stem cells) are shown in white, whereas the differentiated progeny is shown in different colors depending on specific fate. Some progenitors are early fate-restricted and will give rise to only one fate, whereas a few others fully exercise their multipotency – they will actually give rise to all available fates. Finally, the majority of progenitors will be biased toward a specific subset of fates, depending on extrinsic signals, epigenetic priming, timing, and other factors. They will not give rise to all fates, or, even if they do, the proportions of resulting cell types may fluctuate.

collective migration and differentiation of the group leads to the generation of a wide array of derivatives, including neurons, glia, pigment cells, and facial cartilage. No single cell may generate all these fates alone, but the population collectively does (Figure 3.1). At the same time, every individual cell retains the full potential but with different propensities to give rise to its various potential fates (Baggiolini et al., 2015). What enables this collective form of multipotency is a combination of intercellular communication, functional redundancy, and environmental responsiveness. Cells in a collectively multipotent population are often in constant dialogue via chemical, mechanical, or electrical signals. These local interactions modulate fate decisions in real time, allowing the population to adapt and reallocate developmental roles in response to perturbations. If certain lineages are underrepresented – due to injury, mutation, or spatial constraints – other cells within the group can compensate, displaying remarkable plasticity. Moreover, this flexibility is not static but highly contextual. The spatiotemporal patterning of signals – such as gradients of morphogens or mechanical forces – contributes to the emergent behavior of the system (Kalcheim, 2000). In this sense, collective multipotency is not simply a sum of the fates of its constituent cells but a property that arises from the dynamic integration of signals and cellular states within a cohesive group.

This paradigm carries important implications. It cautions against overly reductionist interpretations of lineage potential based on single-cell assays, and instead emphasizes the importance of population-level dynamics in morphogenesis and regeneration. It also suggests that robust developmental outcomes may depend less on the rigid competence of individual cells and more on the adaptive, self-organizing properties of cell groups. Ultimately, collective multipotency illustrates a broader principle in biology: developmental potential, like many complex traits, emerges not solely from individual components but from their interactions, their history, and the context in which they operate. In principle, such new approaches allow us to determine how multiple individual progenitors contribute to different tissues and whether their fate decisions are stochastic or biased by local microenvironments.

Furthermore, figuring out clonal heterogeneity might help reveal how variability is introduced in development. In some cases, similar progenitors can generate divergent cell types based on subtle environmental influences or intrinsic fluctuations in gene expression. While single-cell transcriptomics has transformed our ability to classify cell types and infer developmental trajectories (Adameyko et al., 2024), it does not inherently capture clonal relationships or the positional history of cells within the three-dimensional (3D) and temporally evolving embryonic landscape. Clonal tracking, especially at high resolution and scale, offers a complementary and indispensable lens through which we can begin to understand how spatial and temporal patterning is orchestrated across the embryo. Indeed, morphogenesis is not merely about generating the right types of cells; it is equally about ensuring that correct proportions of these cells are allocated to specific regions, at the right time, and in coordination with tissue movements, shape changes, and environmental signals. Subtle but reproducible variations in clonal contributions can reveal hidden layers of positional information and regulatory logic. These patterns are not easily accessible by transcriptomic profiling alone, as they are encoded in lineage history and spatial distribution, rather than static molecular states (Erickson et al., 2024). By labeling thousands of clones and mapping their descendants, one can analyze how specific regions of an embryo are populated and whether developmental biases lead to uneven distributions of cell types or distorted proportions during body shaping.

3.2 INTRACELLULAR COMPUTATIONS IN DEVELOPING EMBRYOS

Although we witness great progress in methods that help us to describe the lineage transitions with quantitative precision (Rodriguez-Fraticelli et al., 2018; Weinreb and Klein, 2020), we largely lack the mechanistic understanding of internal computational processes that direct cells along their trajectories and perform decision-making, such as in fate selection. Overall, we know what cells do during development to a significant extent, and yet we do not know why and how they decide on the next action. To elucidate this internal clockwork and reveal the mechanisms of intracellular computations and decision-making, we must deal with the extraordinary complexity of integrating intracellular context (epigenetic and other) with a multitude of sensory inputs, ranging from mechanical to chemical and electrical signals.

Older concepts of intracellular decision-making in embryonic development harnessed the idea of master regulator genes, which were tied to the emergence of

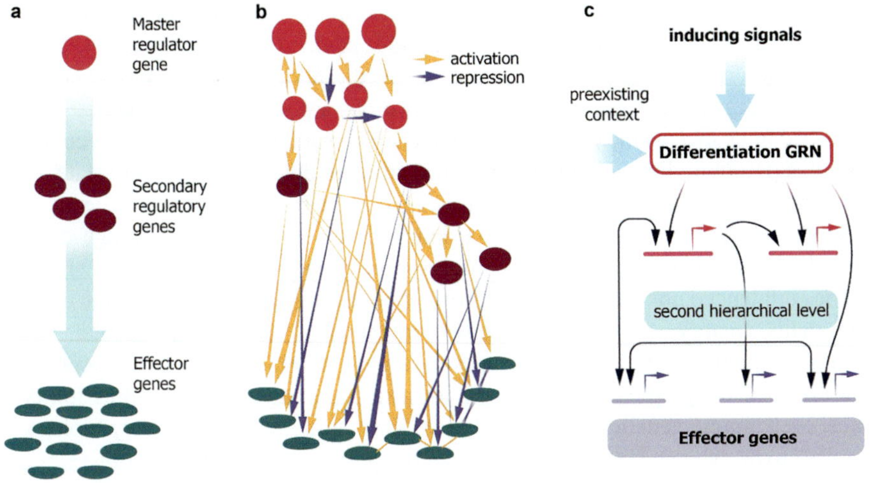

FIGURE 3.2 The historical shifts in our understanding of fate selection and cell identity control: from master regulators to GRNs. The concept of master regulators (left) evolved into the more context-aware concept of GRNs (right), which can be visualized in multiple different ways. a. Historical view of a cascade of transcription factors and effector genes regulated by a single all-powerful master regulator gene. It only includes unidirectional and positive activation of downstream genes. b. A more modern interpretation of the master regulator concept. It includes the option for multiple top-level regulatory genes, which can be influenced by each other as well as lower-level regulatory genes. This view accounts for both activation and repression. c. A contemporary view of a developmental GRN. The local top-level regulatory elements are themselves regulated by the pre-existing context-specifying GRNs as well as inductive signals that modulate signaling processes. The differentiation GRN initiates a self-reinforcing cascade of second-level transcription factors, which in turn activate a battery of effector genes, carrying out the morphological and structural processes of differentiation. GRN, gene regulatory network.

specific fates (Cai et al., 2020; Davis and Rebay, 2017) (see Figure 3.2). The support for the master regulator concept came from cell reprogramming experiments, particularly in vertebrates (Arumugam et al., 2020; Takahashi and Yamanaka, 2006; Wang et al., 2021). For example, the identification of MyoD (myoblast determination protein 1) as a transcription factor capable of converting fibroblasts into myoblasts provides striking evidence that single genes could override cellular identity (Rossi et al., 2023). However, that powerful concept soon started to melt and evolve into a concept of rather context-dependent gene regulatory networks (GRNs) (Badia-i-Mompel et al., 2023; Emmert-Streib et al., 2014). Thus, the once-dominant view of master regulator genes as solitary conductors of cellular fate has been refined into a more nuanced understanding of regulatory logic – where multiple players interact in a distributed fashion to produce robust yet flexible developmental outcomes (Karlebach and Shamir, 2008). As of today, the attempts to model such GRNs and understand how they mediate developmental transitions are progressing with a moderate degree of success, which is mostly attributed to the complexity and multitude of molecular players involved (Kamimoto et al., 2023). Also, the more stable states in development are often approximated as Lorenz attractors or valleys on a Waddington landscape (Mason et al., 2025; Sáez et al., 2022; Waddington, 1957 and 2014 (republished)). However, the mechanisms dissolving such attractors and driving cells from stable states into unstable transitions are yet to be discovered.

It becomes increasingly clear that the decisions made by a cell – such as committing to a particular lineage, initiating proliferation, or entering a quiescent state – are not simply outcomes of linear signaling cascades or binary switches. Instead, they emerge from dynamic, noisy molecular interactions, embedded within high-dimensional regulatory landscapes. These processes are deeply context-sensitive, with outcomes influenced not only by the identity and concentration of molecules but also by the cell's history, physical microenvironment, and fluctuating internal states (Kamenev et al., 2025). Recent advances in single-cell transcriptomics, epigenomics, and live-cell imaging have revealed the intrinsic variability in gene expression profiles, even among genetically identical cells. This variability is not merely biological "noise" to be averaged out, but rather a functional substrate for computation (Kramer et al., 2022). It provides the raw material for probabilistic decision-making and enables populations of cells to hedge their bets in unpredictable environments. In developmental systems, such "noise" may be buffered or amplified by local interactions or feedback from neighboring cells, further complicating the picture of how a cell interprets its surroundings and internal cues.

Indeed, a fundamental feature of both live neural circuits and gene regulation is the presence of slight variations among seemingly identical units. In neural circuits, even two neurons of the same type exhibit slight differences in synaptic connectivity, firing thresholds, and receptor expression. Similarly, in a population of cells, individual transcription factor concentrations, chromatin states, and signaling responses vary slightly (Erickson et al., 2023; Soldatov et al., 2019). While traditional models often depict cell differentiation as a linear and deterministic process, accumulating evidence suggests that subtle variations within seemingly homogeneous cell populations – referred to as microheterogeneity – can have profound consequences for fate determination, adaptability, and the emergence of complex multicellular structures (Carter and Zhao, 2021; Huang, 2009; Kamenev et al., 2025). Microheterogeneity refers to minor but possibly functionally significant variations in gene expression, protein levels, metabolic states, or signaling responsiveness among cells that are otherwise considered identical (Raj and Van Oudenaarden, 2008). These differences arise due to intrinsic stochastic fluctuations in gene expression, epigenetic modifications, and extrinsic influences such as microenvironmental signals and cell–cell interactions. Importantly, such heterogeneity does not represent mere noise but can provide an essential substrate for cellular decision-making, increasing the collective robustness and flexibility of developmental systems (Balázsi et al., 2011).

In many cases, cell fate determination relies on the presence of pre-existing heterogeneity, which biases cells toward specific differentiation pathways in response to external cues. For example, within progenitor pools, cells may exhibit distinct transcriptomic profiles that prime them for alternative fates before overt commitment (Soldatov et al., 2019). This has been demonstrated in hematopoiesis, neural development, and the formation of specialized epithelial structures, where subtle molecular differences predispose progenitors toward specific lineages (Rodriguez-Fraticelli et al., 2018; Weinreb and Klein, 2020).

The emergence of spatial patterns in developing tissues can, in some cases, originate from the inherent stochasticity in gene expression within individual progenitor cells (Losick and Desplan, 2008). Rather than being strictly guided by external morphogen gradients, certain progenitor populations exhibit fluctuating expression of gene modules that are permissive for multiple fates (Kelsh et al., 2021; Soldatov et al., 2019). These probabilistic differences can seed early asymmetries in the tissue (Symmons and Raj, 2016), initiating localized bias in cell identity that evolves into distinct domains (Thiery et al., 2023a). Such processes have been observed both *in vivo* and in controlled *in vitro* systems, particularly in studies of neural tube development, where regions like the floor plate and roof plate can begin to form even in the absence of canonical patterning signals (Lehr et al., 2023). For example, the neural tube can form a single, correctly positioned floor plate and roof plate domain even without the simultaneous presence of precise external gradients (Krammer et al., 2024). *In vitro* models have provided mechanistic insights into how such self-organization might occur. These include competitive interactions between emerging cell domains, spatially-dependent feedback based on cell positioning, and mutual inhibition between competing transcriptional programs. These dynamics promote the gradual segregation of cell populations and favor the expansion and stabilization of the largest emerging cluster. The result is a patterned structure arising not solely from external instructions but from interactions among cells exhibiting intrinsic variability and local feedback. This self-organizing behavior is particularly evident in regions of the embryo, such as its anterior portion, where the notochord – a source of SHH (Sonic Hedgehog signaling molecule) – is absent or inactive (Krammer et al., 2024). In such cases, floor plate-like domains can still arise, suggesting that SHH is not strictly necessary for the initial symmetry-breaking event. Instead, SHH, when present, appears to serve more as a refining signal, dampening the intrinsic variability among progenitor cells and enhancing the robustness and precision of the emerging pattern. By buffering noise, SHH helps ensure that a single, clearly defined domain emerges, preventing the coexistence of multiple, equally strong but improperly positioned cell groups.

These observations challenge the notion that pattern formation in embryonic tissues always and entirely depends on rigid morphogen hierarchies. Instead, they suggest that the capacity for symmetry breaking and biased domain formation may already exist in the form of latent, stochastic gene expression dynamics within progenitor cells. However, without extrinsic refinement, such as morphogen input or mechanical cues, the resulting patterns may be imprecise, prone to duplication, or misaligned (Abdel Fattah et al., 2023). Subtle differences in receptor levels, extracellular matrix properties, or sensitivity to diffusible signals can be amplified by these probabilistic processes, leading to divergent fate trajectories even in initially similar cells. In this context, stochastic variation contributes to – but does not override – structured fate selection processes. It increases the responsiveness of the system to contextual inputs, raising the probability of particular outcomes and supporting the robustness and adaptability of developmental patterning. This dynamic balance between noise and regulation allows biological systems to flexibly initiate structure while maintaining overall developmental fidelity.

One key mechanism contributing to microheterogeneity is stochastic gene expression, driven by the probabilistic nature of transcriptional activation and the asymmetric distribution of cellular components during division (Raj and Van Oudenaarden, 2008). These stochastic fluctuations can be amplified by feedback loops in signaling pathways, leading to bistable or multistable cell states that influence fate decisions. For instance, the balance between self-renewal and differentiation in stem cell populations often relies on the coexistence of distinct molecular states, with fluctuations enabling dynamic transitions between these states. A plethora of diverse epigenetic modifications (such as modified histones and DNA methylation) also contribute to meaningful microheterogeneity. Such differences can be heritable across cell divisions, contributing to the diversification of cell fates even in the absence of genetic changes (Carter and Zhao, 2021). The local microenvironment further reinforces heterogeneity through differential exposure to signaling molecules, mechanical forces, and metabolic gradients (Tassinari et al., 2023). Notably, developmental fields often contain morphogen gradients that induce graded responses in target cells, but within these fields, stochastic or microenvironment-driven variability can create distinct cellular niches with specialized functions (Rackauckas et al., 2018). In some cases, competitive chance-driven interactions between cells influence fate selection, as observed in lateral inhibition mechanisms mediated by Notch signaling (Galbraith et al., 2022).

The role of microheterogeneity extends beyond embryonic development into tissue homeostasis in adults. In the hematopoietic system, stem and progenitor cells exhibit varying degrees of lineage bias (Rodriguez-Fraticelli et al., 2018), which helps maintain a stable supply of differentiated cells under normal conditions while enabling rapid responses to stress or injury. From an evolutionary perspective, microheterogeneity provides a selective advantage by enabling cellular populations to explore a range of functional states. This bet-hedging strategy is particularly beneficial in fluctuating environments, where maintaining a pool of phenotypically diverse cells increases the likelihood of survival and adaptation. This connects to the role of microheterogeneity in disease contexts, particularly in cancer biology, where intratumoral heterogeneity of cancer cell phenotypes drives resistance to therapies. In line with this logic, the evolutionary conservation of microheterogeneity across diverse organisms, from simple metazoans to mammals, underscores its role in developmental plasticity. Furthermore, in artificial neural networks, controlled noise during training (such as dropout techniques) improves generalization and prevents overfitting (Li and Guo, 2018). Similarly, biological systems might utilize molecular noise as a functional tool (Balázsi et al., 2011; Huang, 2009; Rackauckas et al., 2018). Indeed, in early development, stochastic fluctuations in gene expression allow for alternative fate decisions, increasing the diversity of potential outcomes and enabling adaptability at the level of cell collectives.

Overall, rather than being a mere byproduct of biological noise, subtle variations contribute to a dynamic landscape that supports both robustness and flexibility in development. They shift our understanding from a model of strict molecular determinism toward one in which probabilistic processes guide developmental trajectories, with successive layers of refinement shaping these trajectories into consistent, genetically encoded outcomes (considered *de facto* deterministic as in the following example: He et al., 2012). Stochasticity in development generates essential heterogeneity in molecular responses and cell phenotypes, providing a rich and versatile starting material for the construction of the embryo (Figure 3.3). This initial diversity is not merely tolerated but actively harnessed through processes such as selection, feedback regulation, and local signaling interactions. These mechanisms progressively recruit and refine variable cell states, guiding them toward the formation of stable, functional structures that are both robust in their development and finely adapted to the specific demands of the organism's environment, just as learning algorithms reinforce optimal solutions in artificial intelligence (AI). Within this framework, master regulators and mechanisms with strong causative power act as key refining agents – imposing order on variability and integrating stochastic inputs into coherent developmental outcomes.

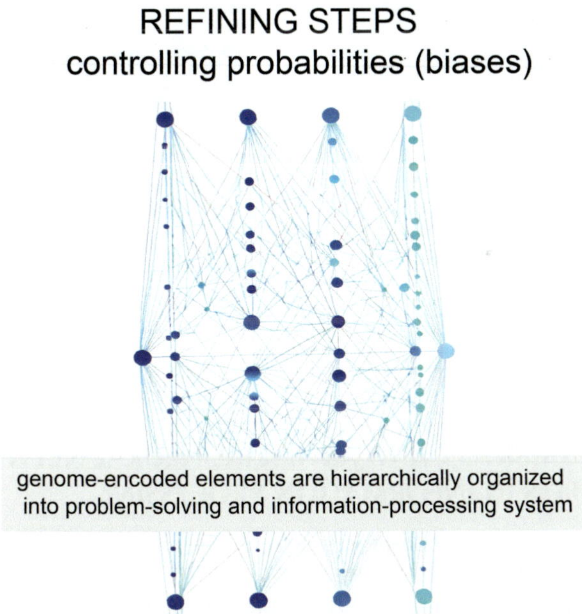

FIGURE 3.3 Taking advantage of noise and variation. The genome operates similar to a generative kernel, and, being a problem-solving system (Levin, 2023a; McMillen and Levin, 2024), it refines variation (introduced by noisy molecular and cellular processes) into developmental patterns resulting in near-deterministic outcomes. The upper part of the figure (blue graph with nodes) brings up a metaphor of a neural network operating based on analogous components in a cell (molecular machinery).

(Continued)

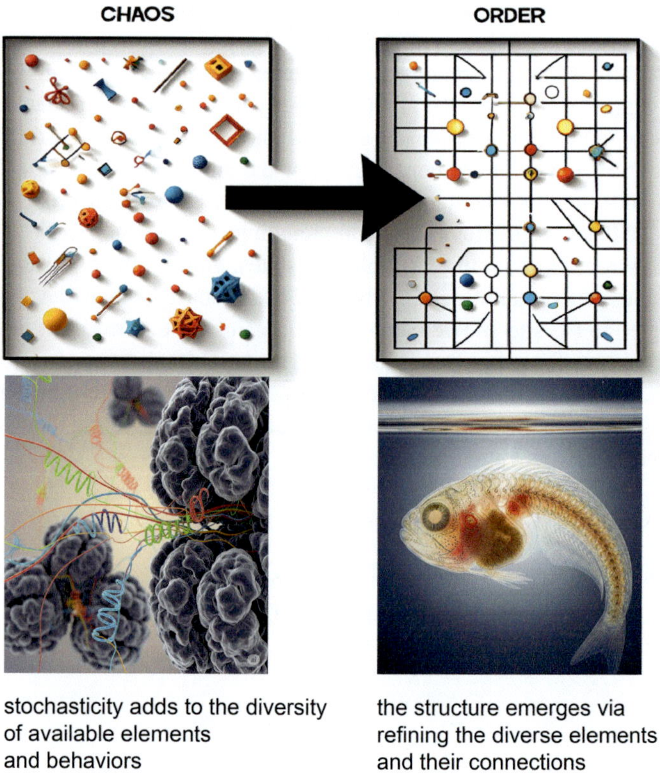

FIGURE 3.3 (Continued) The information about developmental processes might be (hypothetically) stored, compacted, and processed in such analogous networks during the construction of an organism.

3.3 CELLS, EMBRYOS, AND ARTIFICIAL NEURAL NETWORKS: PARALLELS IN INFORMATION PROCESSING

Individual cells are information-processing entities, integrating multiple inputs and producing coordinated outputs (Cardelli, 2005). While cell–cell interactions and live neural circuits are the most well-recognized biological information processors (Regot et al., 2011), the genome and its regulatory machinery also function as a highly complex computational system (Tkačik and Walczak, 2011). Cells interpret sensory inputs and adjust their behavior based on external signals and internal state. The regulatory mechanisms governing gene expression, chromatin accessibility, and signal transduction networks exhibit computational features akin to those found in artificial and biological neural networks. This parallel suggests that principles of information processing in neural network circuits could shed light on cellular decision-making (Marblestone et al., 2016), while concepts from genomic regulation might, in turn, inspire advances in AI and computational biology.

In computational terms, the genome can be considered a vast, dynamically regulated database, while transcription factors, chromatin modifications, and signaling molecules act as processing units that control gene expression and chromatin accessibility. The genome does not function as a static blueprint but as an adaptive and context-dependent system where different combinations of regulatory inputs determine specific transcriptional outputs (Grozinger et al., 2025). This mirrors how neural networks weigh multiple inputs to determine an appropriate action (Marblestone et al., 2016). More specifically, a cell likely operates through a probabilistic computing system, where biological information is processed in a way that embraces uncertainty, variability, and adaptability (Grozinger et al., 2025; Walczak et al., 2012). In this view, cells and genomes act not as deterministic machines but as networks of probabilistic decision-makers, where molecular interactions and cellular behaviors emerge from distributions of possible outcomes rather than fixed pathways. This perspective moves us beyond traditional genetic determinism, reshaping how we understand development, evolution, and cellular function.

In line with this logic, in a series of visionary works by Kevin J. Mitchell, Nick Cheney, Michael Levin, and their colleagues, a transformative rethinking of the genome is emerging – one that challenges the classical metaphor of the genome as a blueprint or deterministic program (Cardelli, 2005; Hartl and Levin, 2025; Levin, 2023a, 2023b; Mitchell and Cheney, 2025). Instead, these authors propose a model in which the genome is understood as a generative kernel: a compressed, abstract representation of organismal form and function that gives rise to complex biological structures through a contextual decoding process during development. This reimagining draws upon analogies with modern machine learning, particularly variational autoencoders (VAEs), neural cellular automata (NCAs), and diffusion models (DMs), to articulate a computational paradigm for understanding development, evolution, and morphogenesis.

In machine learning, a VAE encodes high-dimensional data into a low-dimensional latent space, preserving the essential features necessary to reconstruct similar outputs under various conditions (Kingma and Welling, 2019). Mitchell and Cheney argue that the genome operates in a similar fashion. Rather than encoding detailed anatomical instructions, the genome contains probabilistic constraints – tendencies and conditional rules that guide development through dynamic interactions with physics, epigenetic landscapes, and environmental inputs (Mitchell and Cheney, 2025). The genome, in this view, is not a linear script to be executed but a generative model, shaped by millions of years of evolutionary learning and fine-tuned to produce robust yet adaptable organisms across varying contexts. This generative model is implemented biologically through the architecture of GRNs, epigenetic regulators, non-coding RNAs, and chromatin organization. These components, instead of specifying outcomes explicitly, create attractor states in a high-dimensional gene expression landscape. Cells "choose" fates by traversing these landscapes under the influence of both intrinsic gene expression dynamics and extrinsic signals. The process is iterative, nonlinear, and sensitive to initial conditions – properties reminiscent of computation, rather than blueprint execution.

Mitchell and Cheney's metaphor of the genome as a generative kernel is expanded further by Hartl and Levin, who articulate the genome's role in a bowtie architecture.

Here, evolutionary history compresses information into a narrow latent space (the genome), from which development explosively re-expands this information into the rich variety of morphological and functional outcomes. The decoding process – development – is not fixed, but context-sensitive and competency-rich, shaped by physical, chemical, and bioelectric processes in the cellular environment (Hartl and Levin, 2025). Hartl and Levin presented the idea of multiscale competency architecture, suggesting that live organisms take advantage of problem-solving abilities. Each layer of organization (molecular, cellular, tissue, organ) contributes to computation in its own domain of state space. For example, cell collectives engage in emergent behaviors like morphogenesis, wound healing, and regeneration, all of which involve feedback, memory, and goal-directed action. Thus, morphogenesis itself is reframed not as the execution of a deterministic plan but as an emergent, computational process arising from distributed intelligence (Hartl and Levin, 2025).

The concept of collective intelligence, further elaborated in Levin and McMillen's work, ties this view together by showing that intelligence is not confined to brains or nervous systems (McMillen and Levin, 2024). Even non-neural tissues and single-cell systems like bacterial biofilms exhibit coordinated, intelligent responses to perturbation. For instance, planarian regeneration relies on bioelectric signals to restore correct anatomy; when bioelectrical cues are disrupted, abnormal yet coherent forms emerge, highlighting the computational nature of morphogenetic fields (Kingma and Welling, 2019).

Crucially, these generative models are not brittle. They are robust to noise, adaptive to perturbations, and flexible in function – key properties of evolvable systems. Because traits are encoded abstractly through latent representations and interpreted via dynamic developmental processes, small genetic changes can produce large or negligible effects depending on where in the system they occur. This allows evolution to shape specific traits without destabilizing the overall system, preserving function while experimenting with form. Moreover, the generative kernel view bridges genotype and phenotype by emphasizing the interpretive machinery of the cell. DNA sequences do not operate in isolation; they rely on cellular contexts, 3D genome organization, chromatin accessibility, and prior cellular states to express their meaning. Thus, gene expression is contextual computation, not simply command execution.

Ultimately, these works converge on a powerful new metaphor: the genome is not a script but a generative engine, encoding abstract potentialities that unfold into concrete realities through the dynamics of development (Figure 3.3). Morphogenesis is computation. Cells are decision-making agents. Evolution is learning. And biology, far from being a deterministic cascade of molecular events, is a layered symphony of interpretive, context-sensitive processes capable of producing the astonishing diversity and resilience of life.

3.3.1 Stochasticity, Probabilistic Biases, and *De Facto* Determinism in Developing Biological Systems

At this point, I would like to introduce a few remarks about the difference between random (purely stochastic), deterministic, and probabilistic processes in developmental biology, and also in evolution (which is largely related to fundamentals of development of form and function). In biological systems, many key processes – ranging from gene expression to cell fate decisions – exhibit some degree of randomness. However, not all randomness is equal. It is essential to distinguish between stochastic and probabilistic decision-making in the context of cell biology and developmental biology. Although both involve uncertainty, they differ fundamentally in the underlying mechanisms, the role of noise, and how outcomes are determined. Understanding this distinction requires both biological intuition and some mathematical grounding in the theory of stochastic processes.

In the strict mathematical sense, a stochastic process is one whose behavior is determined by random variables that evolve over time. In cellular systems, truly stochastic decision-making refers to fate choices that arise from intrinsic molecular noise – random fluctuations in molecular components, such as mRNA or protein concentrations (Elowitz et al., 2002; Galbraith et al., 2022; Losick and Desplan, 2008), arising from the inherent discreteness and low copy number of molecules in a cell. For example, consider a GRN where two transcription factors mutually inhibit each other and promote their own expression. If the initial levels of both are equal, and no external signals break the symmetry, the decision of which factor dominates may be made purely by stochastic fluctuations – minor differences in transcriptional bursting, random promoter activation, or asymmetric segregation during cell division. Mathematically, this can be modeled using Markov processes or stochastic differential equations, in which the future state of the system depends on a probability distribution that incorporates random noise. The cell's decision is unpredictable and can be visualized as a particle diffusing in a potential landscape with multiple valleys (attractors), with noise pushing it randomly into one valley or another.

This type of decision-making is memoryless and unbiased in the absence of external asymmetries. Each possible fate has a well-defined, fixed probability, and individual outcomes are governed by chance. Importantly, such processes are often non-deterministic at the single-cell level, although they can yield reproducible proportions at the population level purely as a result of averaging many random events (He et al., 2012).

A canonical biological example is the formation of persisting cells in bacteria such as *Escherichia coli* (Balaban et al., 2004). Here, a small subpopulation enters a dormant, antibiotic-tolerant state not because of an environmental cue but due to stochastic fluctuations in the levels of toxin–antitoxin systems. These fluctuations tip the balance randomly, allowing some cells to survive antibiotics despite being genetically identical to their neighbors. In evolutionary terms, this is a bet-hedging strategy that relies on noise to generate diversity in the absence of specific environmental guidance.

In contrast, probabilistic decision-making in cells refers to processes that still involve randomness but are biased by internal or external regulatory signals. Here, the system is not choosing blindly among outcomes; rather, it integrates molecular signals, regulatory networks, and environmental cues to alter the probability distribution over possible fates. The decisions are still not deterministic at the single-cell level, but the outcome probabilities are modulated by known factors. Mathematically, such systems can still be described by stochastic models, but with transition probabilities that depend on additional variables. For instance, in a birth–death process with state-dependent rates, the probability of transitioning from one state to another is influenced by factors such as signal gradients, receptor activation, or epigenetic modifications. This leads to biased random walks or non-homogeneous Markov processes, where the noise remains but the system has preferences – encoded in the probability weights – based on regulatory inputs.

A well-studied example comes from hematopoiesis, the process by which multipotent hematopoietic stem cells (HSCs) differentiate into various blood cell lineages. Although each HSC can theoretically give rise to all blood cell types, the probability of committing to a particular lineage is strongly influenced by cytokines, transcription factors, and the local bone marrow microenvironment. If PU.1 is upregulated, the cell becomes more likely to follow the myeloid lineage, but this does not guarantee that outcome. Rather, it skews the probabilities, often in a dose-dependent manner (Roeder and Glauche, 2006). This reflects a biased stochastic process: the cell's decision is still probabilistic, but the odds are shaped by a complex web of intrinsic and extrinsic factors.

In developmental biology, a similar phenomenon occurs in the vertebrate retina. Progenitor cells differentiate into photoreceptors, bipolar cells, or ganglion cells over time, with the distribution of cell types depending on both lineage history and environmental signaling. While the ultimate fate of any given progenitor is uncertain (Boije et al., 2015), statistical analyses show that the probabilities of different outcomes change as development proceeds and as feedback from neighboring cells accumulates. These are not random outcomes driven by noise alone – they are probabilistic outcomes guided by an evolving landscape of regulatory inputs.

This distinction between stochastic and probabilistically guided processes is more than semantic. It has deep implications for how we interpret experimental data, particularly in single-cell transcriptomics and lineage tracing. Apparent cell-to-cell variability may reflect stochastic noise or may indicate finely tuned regulatory programs that encode probabilistic rules. Experimental perturbations – such as knockouts or signal manipulations – can shift outcome probabilities (Erickson et al., 2024), revealing that many biological systems operate not through hard-wired, binary switches, but through soft probabilistic biases (Figure 3.3).

From a mathematical standpoint, recognizing whether a process is governed by homogeneous stochastic noise or by biased, signal-dependent randomness determines the modeling approach. Purely stochastic models rely on fixed transition matrices or constant diffusion coefficients, while probabilistically guided systems require dynamic, input-dependent parameters, often modeled using Bayesian inference, dynamic Bayesian networks, or nonlinear stochastic differential equations with external forcing terms.

Thus, stochastic decision-making in biology refers to random fate choices arising from intrinsic molecular noise, with little or no directionality. In contrast, probabilistic decision-making retains an element of randomness but is structured and biased by regulatory factors – both intrinsic (transcription factor levels) and extrinsic (morphogens or cell–cell signals). Both play critical roles in development, but they reflect fundamentally different architectures of decision-making: one dominated by chance and the other by guided uncertainty. Recognizing this distinction is essential for decoding the logic of biological systems and for constructing accurate models of cellular behavior.

The distinction between stochastic and probabilistic processes also plays a fundamental role in evolutionary biology, where randomness and determinism coexist in shaping the fate of populations, lineages, and phenotypes over time. Mutations, gene duplications, recombination events, and chromosomal rearrangements arise randomly with some biases, with examples including "mutagenesis hot spots in a genome" and mutation signatures introduced by biochemical processes and specific enzymes. That is, despite some biases, the generation of genetic variation is a largely stochastic process: mutations do not "know" what the environment will be, and their occurrence is governed by intrinsic molecular mechanisms such as polymerase errors or DNA damage – essentially random events at the biochemical level. Furthermore, once variation exists, its fate in a population can be heavily influenced by genetic drift, especially in small populations. Drift is a classic example of stochastic evolution: alleles can become fixed or lost entirely by chance, even if they confer no selective advantage. In such cases, the probability of fixation of a neutral allele is simply $1/2N$, where N is the population size – derived from a purely stochastic model of allele frequency changes over generations (the Wright–Fisher model) (Crow and Kimura, 1970). These stochastic processes can lead to divergent evolutionary paths in identical populations exposed to the same conditions, highlighting the role of historical contingency. For example, in Lenski's long-term evolution experiment with *E. coli*, some traits (like citrate metabolism) emerged only in certain lineages, not because of deterministic selection alone, but because a rare, stochastic mutational event opened a path for subsequent adaptation (Blount et al., 2008; Quandt et al., 2015).

At the same time, natural selection introduces a probabilistic bias: it increases the chances that beneficial variants will spread in the population. This is not deterministic – selection does not guarantee that a beneficial mutation will fix – but it increases the probability of fixation. Mathematically, in a population of effective size N, the fixation probability of a beneficial allele with selection coefficient s is approximately $2s$ (for small s), reflecting a probabilistically guided process where the odds are tilted in favor of the more fit variant (Crow and Kimura, 1970; Kimura, 1962).

Developmental biology adds another layer: developmental constraints and canalization can make some phenotypes more likely to arise than others, even if they are not the optimal solution. These biases in the genotype–phenotype map – shaped by GRNs, morphogenetic processes, and pleiotropy – introduce probabilistic structure to the space of phenotypic variation. Some variants are simply more accessible, due to how development funnels random genetic changes into certain morphologies or functions. This idea is central to the theory of evolvability: the capacity of a system to generate heritable phenotypic variation in a way that is non-uniform and structured (Le Rouzic et al., 2013; Pigliucci, 2008). Evolution is not a random walk in phenotype space; it is a biased exploration, where the biases are imposed by both selection and the architecture of developmental systems. The probabilistic aspect comes from how these constraints shape the distribution of possible outcomes.

Some organisms explicitly leverage stochasticity for evolutionary advantage, a strategy known as bet-hedging (Levy et al., 2012). In unpredictable environments, producing a diverse range of phenotypes – even if most are suboptimal – can ensure that at least some individuals survive. This is common in microbial systems, where genetically identical cells can stochastically adopt alternative phenotypes (spore formation, dormancy) to hedge against environmental collapse (Veening et al., 2008). Over evolutionary time, such strategies are shaped by selection acting on the probability distributions of phenotypes rather than deterministic outcomes. Understanding this duality is crucial for interpreting evolutionary trajectories, especially in the genomic era, where high-throughput data reveal vast amounts of variation. Distinguishing what is random noise from what is shaped by probabilistic forces – be they fitness-related, developmental, or ecological – allows us to disentangle the deep logic of evolutionary change.

Now let's return to our discussions of embryology, especially to how probabilistic or stochastic processes result in deterministic developmental outcomes. In developmental biology, determinism *de facto* refers to the empirically observed consistency and predictability of developmental outcomes, even though many of the underlying biological processes involve intrinsic noise and probabilistic decision-making (He et al., 2012). This concept stands in contrast to strict, theoretical determinism, in which every outcome is fully dictated by initial conditions and laws without any element of randomness. In biological systems, determinism *de facto* emerges as a systems-level property: while stochastic and probabilistic mechanisms operate at the molecular and cellular scales, the overall developmental program produces highly reproducible and robust macroscopic structures such as tissues, organs, and body plans.

Development is inherently hierarchical and involves numerous layers of regulatory control. At the lower levels – gene transcription, signaling cascades, and cell division – events are often subject to stochastic fluctuations. These fluctuations can stem from low molecular copy numbers, transcriptional bursting, variability in protein degradation, or noise in receptor–ligand interactions. However, developmental systems have evolved mechanisms to buffer, integrate, and canalize these variations. The result is that, despite noisy micro-level dynamics, the organism exhibits reliable macro-level outcomes. This emergent reproducibility is what we call determinism *de facto*.

One striking example of this phenomenon is vertebrate somitogenesis. The periodic formation of somites along the embryonic axis is governed by a segmentation clock, an oscillatory gene network that is inherently noisy at the level of individual cells. Yet, due to synchronization among cells and regulatory feedback mechanisms, somite boundaries form with remarkable precision and timing (Riedel-Kruse et al., 2007). Similarly, in early *Drosophila* embryogenesis, spatial patterning is controlled by morphogen gradients such as Bicoid. These gradients are subject to significant variability, yet they generate sharp, consistent expression domains across embryos due to robust thresholding and cross-regulation among downstream genes (Gregor et al., 2007).

Even in organisms often regarded as paradigms of deterministic development, such as *Caenorhabditis elegans*, the story is more nuanced. *C. elegans* has a nearly invariant cell lineage, with each embryonic division giving rise to predictable daughter cells. However, studies have shown that some variability – particularly in gene expression timing and levels – still occurs at the molecular level (Raj et al., 2010). What gives rise to the apparent determinism is not the absence of stochasticity, but the presence of mechanisms that buffer its effects. Regulatory networks, mechanical constraints, and feedback loops ensure that developmental trajectories converge on the same endpoint, even if the paths are subtly different. This empirical determinism is closely related to the concepts of canalization and robustness. Waddington's idea of canalization described how developmental pathways can buffer genetic and environmental variation, leading to a stable phenotype (Waddington, 1942). Modern systems biology has expanded this view, identifying architectural features such as redundancy, modularity, and negative feedback as sources of robustness. These features do not eliminate stochasticity; rather, they structure it, constrain it, and guide it toward consistent outcomes. Thus, determinism *de facto* is not a property of individual molecules or reactions, but a property of the developmental system as a whole.

Overall, determinism *de facto* in developmental biology is best understood as a pragmatic, emergent property. Developmental systems, through their organization and regulatory logic, produce consistent and functionally coherent outcomes despite being built on probabilistically operating components. This understanding reconciles the molecular randomness and probabilistic biasing in single-cell measurements with the striking regularity of organismal form. Determinism, in this view, does not arise from eliminating chance, but from mastering it.

3.3.2 Cell Is a Probabilistic Computer

At the core of this probabilistic operation lies the regulation of the genome itself. Genes are not simple on/off switches but dynamic elements that exist in states of fluctuating activity. Transcription factors, for instance, bind to DNA based on affinity, yet their actual binding events are influenced by a complex interplay of chromatin structure, molecular competition, and epigenetic marks (Crispatzu et al., 2021; Roeder and Glauche, 2006; Veening et al., 2008). Gene expression, therefore, is not a straightforward execution of instructions but a stochastic process where transcription occurs in bursts (Leyes Porello et al., 2023), with probabilities determining the frequency and intensity of expression rather than a binary system of activation or repression.

In artificial neural networks, each node receives numerous weighted inputs and generates a result based on an activation function. Transcription factors operate similarly: they integrate multiple signals from the environment (hormones, morphogens, mechanical forces) and combine these influences in a nonlinear manner to modulate gene expression, also accounting for the noise and multiple probabilistic biases. Some TFs act as activators, others as repressors, and their combinatorial effects determine the probability of whether a gene will be transcribed or silenced. This resembles the way neurons in a network sum excitatory and inhibitory inputs before deciding to fire.

Furthermore, in biological neural networks, plasticity – defined by synaptic modifications – allows circuits to store information, adapt, and refine responses over time. In genome regulation, chromatin accessibility plays an analogous role. Epigenetic modifications such as histone acetylation, DNA methylation, and nucleosome repositioning influence the accessibility of transcriptional machinery to DNA, effectively "storing" regulatory states and shifting probabilities of processes in the future (Carter and Zhao, 2021; Moris et al., 2016). Just as neurons retain memory traces through altered synaptic weights, cells maintain transcriptional states through epigenetic memory (Thiagalingam, 2020), ensuring that once a gene expression program is established, it remains stable until an external or internal cue modifies it.

Neural networks utilize feedback loops to reinforce specific activation patterns (Shao et al., 2025). Similarly, GRNs rely on feedback mechanisms – both positive and negative – to fine-tune responses. Positive feedback loops reinforce a particular cell fate or differentiation trajectory, ensuring commitment to a developmental path. Negative feedback introduces a regulatory balance, preventing excessive activation and maintaining homeostasis. This feedback architecture allows both neural and genomic networks to function with remarkable precision and adaptability, resulting in patterns that appear deterministic despite the probabilistic nature of many processes.

The concept of probabilistically-influenced computation becomes even more relevant when considering how cells make decisions (Soldatov et al., 2019). Whether a stem cell differentiates into one lineage or another is rarely dictated by a deterministic set of rules but is instead governed by the interplay of signaling gradients, transcriptional noise, and intracellular feedback loops (Kelsh et al., 2021; Rodriguez-Fraticelli et al., 2018; Soldatov et al., 2019). Each cell evaluates its internal state and external signals, assigning different weights to each input and reaching a decision probabilistically. This process mirrors the way artificial probabilistic computers function, where calculations do not follow a single predefined path but explore multiple possible solutions simultaneously, adjusting outcomes based on dynamic inputs.

A probabilistic computer, unlike classical deterministic systems, does not rely on fixed logical steps to arrive at an answer. Instead, it integrates uncertain data and computes likelihoods rather than absolute certainties (Manhaeve et al., 2021; Orhan and Ma, 2017). The logic of such a system involves multiple feedback mechanisms where different inputs can shift the probabilities of specific outcomes. In the case of biological systems, cells use transcriptional networks, protein interactions, and epigenetic modifications to weigh competing influences, allowing them to compute responses to environmental stimuli with built-in flexibility. This form of computation is particularly advantageous in noisy environments where rigid, deterministic decision-making would fail to adapt to fluctuations and uncertainties.

At the molecular level, chromatin accessibility might play a key role in probabilistic genome computation. Rather than genes being either permanently open or closed, they exist in states of varying probability of accessibility during periods of time, influenced by histone modifications, DNA methylation, and non-coding RNA interactions (Mansisidor and Risca, 2022). This means that the likelihood of a gene being transcribed during a certain time period is not absolute but varies within a defined range, allowing the system to respond dynamically to new conditions. Such a mechanism ensures that cells can balance stability with adaptability, maintaining core functions while also being capable of rapid response when necessary. Beyond gene regulation, the principles of probabilistic computation extend to cellular signaling pathways, where molecular signals do not elicit deterministic responses but instead modify the likelihood of particular cellular outcomes. For instance, in developmental biology, morphogen gradients influence cell fate not by providing strict instructions but by biasing the probabilities of different differentiation paths. This system allows populations of cells to collectively shape tissues and organs without requiring explicit programming at the individual cell level. Generally, the ability to operate within a framework of probability allows biological systems to ensure robustness in response to critical threats.

One of the most intriguing implications of this perspective is how probabilistic decision-making at the cellular level parallels neural network architectures in some computational systems. In neural networks, individual nodes integrate multiple weighted inputs before producing an output (Li and Guo, 2018), a process that resembles how cells integrate molecular signals before making transcriptional or behavioral decisions. Both systems function through a hierarchical interplay of probability-weighted interactions, allowing for learning, adaptation, and emergent complexity. Understanding biological computation as a probabilistic process challenges traditional views of genetic determinism and suggests that genomes do not act as rigid blueprints but as dynamic processors of information, constantly adjusting their outputs based on shifting probabilities. This perspective also aligns with evolutionary theory, emphasizing that variation and adaptability emerge from probabilistic mechanisms rather than preordained genetic instructions. Evolution itself, in this view, is not simply a matter of fixed genetic programs being acted upon by selection but a probabilistic exploration of phenotype space, where molecular and cellular randomness fuels the emergence of novel forms and functions, which I discuss below using the concept of evolutionary capacitance (Masel, 2013).

This concept allows organisms to accumulate hidden variation resulting from noise, which can later be tested and refined by natural selection – enhancing evolvability without compromising short-term fitness. The heat shock protein Hsp90 exemplifies this principle. As a molecular chaperone, Hsp90 stabilizes a wide array of client proteins, buffering the effects of destabilizing mutations. When Hsp90 is inhibited – either genetically or by stress – this buffering fails, and cryptic genetic variation is unmasked. In *Drosophila melanogaster*, for instance, reduction of Hsp90 function reveals a spectrum of phenotypes that are specific to the genetic background (Gangaraju et al., 2011; Rutherford and Lindquist, 1998). Many of these traits are heritable even after Hsp90 activity is restored, implying that selection can act on this revealed variation, a process called genetic assimilation. Such evolutionary capacitance finds its most direct and impactful expression during embryonic development, where the genome is translated into the phenotype through multilayered processes of cell differentiation, pattern formation, and morphogenesis. This is precisely where cryptic variation has the greatest chance to be revealed, tested, and potentially fixed by selection.

This phenomenon is not limited to *Drosophila*. In the beetle *Tribolium castaneum*, Hsp90 impairment produced a reduced-eye phenotype that became stably inherited, further supporting the idea that Hsp90 can toggle hidden genetic variation into the open (Aboelsoud and Kurtz, 2019). In a seminal work, Queitsch and coauthors demonstrated that HSP90 inhibition in various *Arabidopsis* plants led to a broad spectrum of morphological phenotypes that depended on underlying – but previously hidden – genetic differences (Queitsch et al., 2002). This finding mirrored earlier work in *Drosophila* and highlighted HSP90's conserved role in phenotypic buffering across kingdoms. Subsequently, Sangster and colleagues used recombinant inbred lines to show that HSP90 influences quantitative traits such as hypocotyl elongation in darkness: many HSP90-dependent alleles were frequent in wild populations and could be mapped by QTL (quantitative trait locus) methods (Sangster et al., 2008). These results confirm that HSP90 masks natural variation in environmentally responsive traits, maintaining developmental stability and canalization. Not surprisingly, in *E. coli*, overexpression of another chaperone, GroEL, also increases tolerance to mutations, effectively enlarging the space of viable genotypes (Fares et al., 2002). Even prion-like elements such as PSI+ (Sup35p – the *Saccharomyces cerevisiae* eukaryotic translation release factor, in a prion state) in yeast act as evolutionary capacitors by transiently disrupting translational fidelity, revealing alternative phenotypes under stress (Griswold and Masel, 2009).

Such capacitance functions play out especially when it matters most – under environmental stress, when adaptation is urgently needed. Stress-induced breakdown of robustness allows organisms to "test" previously hidden, noisy, and normally neutral variants that might offer an advantage. Viewed from this angle, the genome's role is not to dictate every detail of phenotype in a deterministic fashion. Instead, it establishes rules and constraints within which noisy elements – fluctuating molecular interactions, redundant genetic circuits, or variable developmental cues – are sculpted into coherent biological patterns. The genome filters noise, but doesn't necessarily erase it. Instead, it may act more like a lens that focuses variation toward useful outcomes. Through networks of feedback loops, epigenetic modifiers, and regulatory hierarchies, the genome converts molecular entropy into organismal order. Importantly, these refining mechanisms do not necessarily require precision at the individual level. Evolution acts on populations. Some individuals might fail due to noisy developmental outputs, but others may gain beneficial combinations of traits. Over time, natural selection filters the stochastic and probabilistically biased outputs of biological systems, locking in those rare but favorable outcomes. In this view, variation is not the antithesis of determinism – it is the engine of it, operating on evolutionary timescales. This leads us to a profound concept: determinism is emergent from noise. Systems that appear highly ordered – like live neural circuits or electronic neural networks, immune systems, or even embryonic development – do not function by eliminating variability. Instead, they depend on it in their origin and function. Random V(D)J recombination in the immune system, for instance, generates a repertoire of antibodies that is then refined through somatic selection. In a different situation, noisy gene expression in progenitors or stem cells can bias cell fate decisions before final commitment. This principle is also echoed in morphogen gradients, which pattern tissues not by prescribing exact positions but by offering probabilistic cues that cells interpret in context. In evolution, transcriptional noise allows regulatory sequences to test new configurations without fatal consequences. Once a configuration proves adaptive, it can be refined, stabilized, and integrated. Thus, noise precedes innovation; refinement follows success.

3.4 ON THE NATURE OF INFORMATION IN BIOLOGICAL SYSTEMS

The question of what constitutes "information" in biological systems – particularly in developmental biology – and how it is stored and interpreted has long provoked debate, blending rigorous formalism with philosophical controversy. Some thinkers might even like to link the "information" term to "the phlogiston of the 20th century," criticizing its vague, overused presence in biological discourse. And yet, when carefully defined and applied, information theory offers a powerful lens through which to interpret both natural and artificial systems, such as neural networks. Exploring how information is represented in artificial neural networks can help illuminate the way biological systems, especially during development, store and process information – though the analogy has both strengths and limitations.

In artificial neural networks, information is not stored as discrete symbols or localized instructions (Shreyas et al., 2023). Instead, it is embedded implicitly in the architecture of the network (Achille et al., 2019) – the pattern of connections between layers – and in the distribution of weights that determine the strength of influence one unit exerts on another (Montesinos López et al., 2022). These weights are adjusted through optimization processes, such as stochastic gradient descent, during training. After learning, the network has developed a representation of the data that allows it to perform a task – such as classification or prediction – with accuracy. But this representation is high-dimensional, distributed, and opaque; no single unit "contains" a specific concept. Rather, meaning emerges from patterns of activity spread across the network. The structure of the network constrains its representational capacity, and its final configuration reflects a compressed abstraction of the training data that is sufficient to accomplish the desired task (Achille et al., 2019).

This idea of distributed, implicit information parallels how biological systems operate, particularly during development. In developmental biology, we speak of genetic information in DNA, epigenetic information in chromatin states, morphogenetic information in gradients and tissue geometry, and dynamic information in signaling cascades and feedback loops. Much like a neural network, the developing embryo is not governed by a centralized instruction manual; instead, developmental outcomes emerge from the interaction of multiple distributed signals, often interpreted contextually and nonlinearly. DNA in this analogy is akin to the network's architecture: it encodes the potential for various outcomes, but does not by itself determine the exact course of development. The "weights" in the biological network might be likened to transcription factor concentrations, signaling molecule dynamics, and chromatin modifications, all of which modulate how the genomic architecture is interpreted. Cell fates and tissue structures do not unfold from a static code but result from dynamic processes in which local conditions, temporal sequences, and spatial constraints interact. Information, in this context, is not a message passed from gene to phenotype, but a set of evolving constraints that guide trajectories of differentiation and morphogenesis.

Indeed, all of that resembles how analog computers deal with information. Analog computers, including ancient devices (Arenas, 2025), operate on a fundamentally different principle from modern digital systems. Instead of encoding information in discrete binary units, analog computation stores and processes data through the continuous variation of physical quantities such as position, voltage, rotation, or pressure. This form of computation is deeply embedded in the structure and dynamics of the system itself – information is not represented symbolically, but physically.

The Antikythera mechanism, a Greek device constructed over two millennia ago, is one of the earliest known examples of an analog computer. It was designed to track astronomical cycles, such as the phases of the moon, eclipses, and the positions of planets (Arenas, 2025). These complex calculations were achieved not by following symbolic instructions but through a system of finely interlocked gears, each encoding specific celestial cycles through its size and placement. When the main crank was turned, the rotations of these gears translated the input into precise mechanical outputs, displaying future astronomical events. The mechanism did not store numbers or logic gates; rather, the relationships between the gears were themselves the computational substrate. Later, mechanical analog computers followed similar principles (Efstathiou et al., 2021). In the 19th and early 20th centuries, devices such as the differential analyzer used rotating shafts and wheels to solve differential equations (Hartree, 1940; Thomson III, 1876) – "On an integrating machine having a new kinematic principle," 1876). These machines did not perform symbolic manipulation of equations – instead, they physically modeled them. Rotational angles, resistances, or distances on sliding parts represented variables, and the mechanical interplay between components performed the computation. The output, often a drawn curve or the position of a dial, emerged naturally from the interactions of the system. Analog electrical computers used varying voltages to model continuous systems, providing real-time solutions to problems such as ballistics trajectories or aircraft control.

This physical and emergent nature of analog computation bears striking similarities to biological systems. Cells and organisms do not process information through explicit symbolic rules, and there is no central processor or symbolic script dictating outcomes. Instead, information is stored in patterns of molecular concentration, spatial gradients, and interaction networks (Figure 3.4).

In contrast, digital computers store information as sequences of discrete, atomic bits. They manipulate this information through well-defined logic gates and execute instructions sequentially, governed by rigid syntax and deterministic rules. Every operation in a digital system is symbolic and localized; meaning is constructed externally, by interpreting bit patterns through predefined coding schemes. Digital and symbolic systems are built on abstract, formal logic. Analog and biological systems compute through embodied interaction – their form and operations are inseparable (Hornischer, 2021). The Antikythera mechanism did not simulate astronomy symbolically; it was a mechanical instantiation of celestial motion. Similarly, a cell does not "run a program" in the digital sense – it is the program, dynamically enacting it through material processes. Understanding this distinction is crucial as we attempt to bridge biological and artificial systems. The insight that biology functions more like an analog computer than a digital machine shifts our focus from decoding symbolic instructions to interpreting emergent properties of material computation and information-processing.

Yet the use of the term "information" in biology is not unproblematic. In physics and in Shannon's formal information theory, information is defined as a reduction in

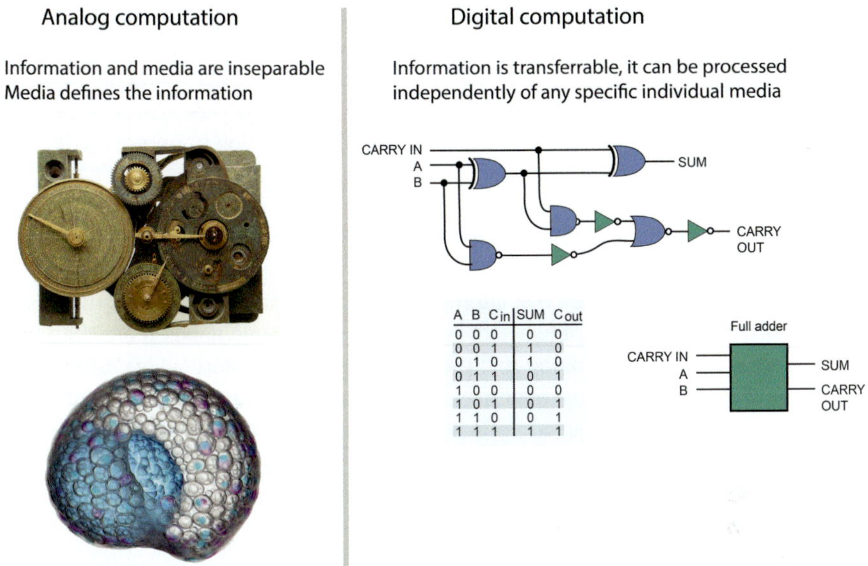

A	B	C_{in}	SUM	C_{out}
0	0	0	0	0
0	0	1	1	0
0	1	0	1	0
0	1	1	0	1
1	0	0	0	0
1	0	1	0	1
1	1	0	0	1
1	1	1	1	1

FIGURE 3.4 The examples of analog and digital computers. The live systems, as well as many engineered technical solutions, embed the information into their structure and operational dynamics, similar to how the diameter of the mechanical gears encodes simple mathematical rules. Such analog computation fully depends on the specific embodiment and media, which physically incorporate the information-processing principles.

uncertainty and is always relative to a known or assumed probability distribution over possible states (Chen, 2016). In developmental biology, however, the term often slips into functional or teleological language: for instance, when we say that "the genome contains the information to build an organism," we imply a semantic richness that classical information theory is explicitly agnostic to. Moreover, biological information is not "for" an external observer – it is used and interpreted internally, in the service of survival and reproduction, often shaped over evolutionary time. This leads to an important distinction: Shannon information measures the amount of uncertainty reduced by a signal (for instance, how much a morphogen gradient tells a cell about its position in a tissue), but says nothing about the meaning or utility of that information. Biological systems, by contrast, are fundamentally interested in meaning: a concentration of a signaling molecule has consequences only because the cell has evolved machinery to interpret it and respond appropriately (to guarantee survival and reproduction). Thus, biological information is inseparable from its interpretation and functional consequences.

Nonetheless, tools from information theory have been productively applied to biology. For example, mutual information has been used to measure how much positional information a morphogen gradient conveys to a developing cell, often in bits. In such cases, researchers have found that gradients like Bicoid in *Drosophila* carry about 1.5–2 bits of information, meaning they can resolve three to four distinct cell fates (Tkačik and Gregor, 2021). Signal transduction pathways, too, have been found to transmit surprisingly low amounts of information – typically fewer than 2 bits – prompting questions about how cells make reliable decisions in noisy environments (Grabowski et al., 2019; Topolewski and Komorowski, 2021). These findings suggest that biological systems often operate with minimal information, using strategies such as temporal integration, feedback, and cell–cell communication to improve effective resolution.

The unit of information in both artificial and biological contexts remains the bit – the amount of uncertainty reduced when distinguishing between two equally likely alternatives. In biology, a bit may represent a binary gene expression decision (on or off), a cell fate choice between two alternatives, or the presence or absence of a regulatory motif. However, the richness of biological systems demands more than bits alone can offer. Because meaning in biology is contextual, historical, and often nonlinear, the challenge is not just quantifying information, but understanding how it is integrated, interpreted, and acted upon across multiple scales and modalities. Thus, the analogy between neural networks and developmental systems highlights a shared principle: information can be distributed, implicit, and dynamically interpreted to generate complex outputs from simple inputs. Yet biology also requires a more expansive notion of information – one that incorporates semantics, dynamics, and the evolutionary embedding of interpretive machinery. Rather than viewing information as a substance or a static message, it may be more fruitful to treat it as a constraint on possible futures, a guide to unfolding structure in a system constantly in motion.

However, as biological systems and artificial neural networks increasingly appear to share principles of distributed, high-dimensional, and context-sensitive information processing, we face a key epistemological challenge: can we ever understand such systems in human-intuitive terms, using the traditional tools of logic, mathematics, and symbolic reasoning? Or are we confronting the limits of human cognitive models – models that evolved to deal with low-dimensional, rule-based environments? Indeed, the question of whether we can understand the distributed, high-dimensional nature of information in systems like artificial neural networks in human terms – using mathematics, logical gates, and Boolean logic – goes to the heart of a growing tension in science. Biological systems and artificial neural networks increasingly appear to operate not through localized, symbolic instructions, but via complex, implicit, and context-sensitive processes (Velik and Bruckner, 2008; Zhang and Sheng, 2024). This challenges the traditional human intuition that evolved to reason in low-dimensional, rule-based environments.

A useful extension of this view is to recognize that genomic information is often profoundly delocalized across the genome through the complex regulation landscape, rather than neatly contained in isolated, gene-like units. In this respect, the genome may be compared to a hologram: in a holographic diffraction film, information specifying a particular 3D point is not stored at one local site, but is distributed across many regions of the recording medium, such that numerous parts of the film contribute to the reconstruction of a given *xyz* coordinate in the final image. Likewise, biological form and function often do not arise from single, spatially localized informational elements, but from distributed interactions among regulatory sequences, chromatin states, epigenetic marks, gene networks, and cellular interpretive machinery acting across multiple scales. If this analogy is taken seriously, then understanding information in the genome requires moving beyond a purely local or atomistic concept of informational units. Instead, we may need to think of genomic information as encoded in relational, distributed, and context-dependent patterns, where the effective "unit" of information is not always a discrete sequence element, but sometimes a higher-order configuration whose meaning emerges only through its participation in a broader informational architecture.

Human understanding is deeply rooted in sequential logic, causal chains, and symbolic abstraction. These principles underlie classical computing and Boolean logic, where well-defined inputs pass through explicit gates like AND, OR, and NOT to produce deterministic outputs. While this framework is powerful in many domains, it becomes inadequate when applied to systems where information is not localized, not symbolically encoded, and not explicitly rule-based. Artificial neural networks, like biological regulatory systems, encode meaning across distributed patterns. A concept such as "cat" is not stored in a single unit or location, but emerges from activity patterns spread across the network's many layers and neurons. These patterns reside in high-dimensional vector spaces, shaped by nonlinear transformations and trained through stochastic processes like gradient descent (Welper, 2024). Unlike logical gates, which are explicitly interpretable, these systems are functionally opaque: we know what they do, but often cannot easily say how or why.

Nonetheless, efforts have been made to interpret or approximate such systems in more human-intuitive ways (Zhang and Sheng, 2024). The challenge, then, is not simply to use neural networks, but to reverse engineer them – to extract from them symbolic representations or compact mathematical descriptions that are comprehensible to the human mind. Understanding the logic and mathematics inside an artificial neural network involves identifying what kind of functions the network is computing, how those functions are implemented internally, and whether these functions can be expressed in a human-interpretable form (Arani et al., 2019). In some narrow settings – such as simple feedforward networks trained to perform arithmetic or logical operations – it is possible to map internal layers onto known mathematical operations. But for most modern networks, especially deep and highly overparameterized ones, the learned representations are not modular in the way traditional logic circuits are. Instead, the network approximates a complex function by layering many nonlinear transformations, each involving distributed interactions among thousands of parameters. The resulting computation is both efficient and flexible, but highly non-symbolic.

To address this, several approaches have been developed to extract symbolic structure or mathematical logic from trained networks (Arani et al., 2019). One of the most promising strategies is model distillation, where a complex network is approximated by a simpler, interpretable model, such as a decision tree, a set of logical rules, or a linear classifier (Baek and Tegmark, 2025). The idea is to treat the original network as an oracle and to generate a large number of inputs and outputs, from which a symbolic model can be learned. This distilled model may not capture every nuance of the original network, but it can offer insight into the dominant patterns or decision rules the network has internalized. In effect, it reconstructs a symbolic approximation of the network's function (Baek and Tegmark, 2025). Another approach is to probe the network with carefully designed inputs – a form of experimental mathematics. By analyzing how a network responds to systematically varied stimuli, one can infer which variables it is sensitive to, whether it obeys algebraic identities, and how it generalizes mathematical structure (Davis, 2021). For instance, by training a network on functional equations or algebraic systems and analyzing its learned representations, one can begin to tease out whether it has internalized symbolic structure like associativity, distributivity, or symmetry.

Advances in concept activation vectors and regions, together with representation analysis, allow us to trace how abstract concepts are encoded within the geometry of the network's latent space (Crabbé and van der Schaar, 2022; Grossmann et al., 2022). If a certain direction in the high-dimensional space corresponds to a concept such as "number of objects" or "presence of symmetry," we can begin to form a semantic map of the network's internal logic. Likewise, network dissection techniques aim to assign human-interpretable meanings to specific units or subnetworks, identifying neurons that respond to edge detection, curves, quantities, or other mathematical primitives. Yet even these tools fall short of true symbolic extraction. Neural networks do not compute via algebraic rules or formal logic; they approximate functions by statistical interpolation over vast parameter spaces. To reverse engineer them fully, we may need to develop a new branch of mathematics – a kind of algebra of distributed representations – that captures how high-dimensional transformations can implement logical or symbolic functions. In such a framework, we could begin to characterize the geometry of computation: how a specific function is "folded" into a layered architecture, how modularity arises from nonlinearity, and how symbolic reasoning can emerge from pattern recognition.

This becomes specifically important when we consider that nature may, in fact, compute in ways similar to artificial neural networks (even though this analogy has many limitations – artificial neural networks are not self-maintaining and don't have to survive and reproduce). Biological systems do not appear to rely on symbolic logic or centralized control. Evolution has not designed cells to follow Boolean circuits. Rather, biological computation unfolds through dynamic, adaptive interactions among genes, proteins, spatial gradients, and mechanical forces. Meaning in such systems is not a fixed instruction but an emergent property of process. The genome is not a codebook in the classical sense, but a framework within which interpretation, context, and history shape outcomes. This distributed and emergent style of computation mirrors the way artificial neural networks operate and raises the possibility that our standard models – grounded in discrete logic, linear causality, and symbolic abstraction – are mismatched to the systems we are trying to explain. Indeed, if biological systems, particularly neural and developmental systems, process information in a similar fashion – using distributed, context-sensitive representations rather than explicit symbolic rules – then the task of understanding biology becomes very complex. Traditional approaches to biological explanation rely on identifiable pathways, discrete gene functions, and linear cause–effect chains. But if biological systems compute in the same way as artificial neural networks, then many of these assumptions may be insufficient. For example, cell fate decisions might not be triggered by single genes but by complex threshold crossings in a high-dimensional regulatory space. GRNs might not contain symbolic logic gates but instead perform analog computation over diffuse transcriptional landscapes.

To interpret such systems in a symbolic way, we would need to combine high-throughput data acquisition (such as single-cell transcriptomics, epigenomics, and proteomics) with representation learning tools that identify the underlying structure of information flow in biological tissues. Borrowing from AI, we could use autoencoders or variational models to uncover the low-dimensional manifolds on which biological decision-making takes place. We would then use symbolic regression or program synthesis tools to approximate the learned functions with compact, interpretable expressions. In this way, we could translate distributed biological logic into symbolic hypotheses – rules that describe how inputs are transformed into outputs and how gene expression programs encode developmental strategies. Ultimately, solving this problem requires more than technical innovation – it demands a conceptual shift. We must recognize that nature, like artificial neural networks, may compute not with explicit symbols but with geometry, dynamics, and context. If we want to reverse engineer these systems, we must embrace a framework that allows for symbolic approximations of non-symbolic processes – a way of translating the implicit into the explicit. This may not yield perfect equivalence, but it may give us the tools to build symbolic models that are faithful to biological complexity while remaining comprehensible to human reasoning. In this vision, biology becomes not just a science of parts and interactions, but a science of information geometry, where symbolic understanding emerges from the deep structure of high-dimensional biological computation. The fusion of machine learning interpretability and systems biology could thus provide a path forward – not only for understanding artificial networks but for decoding the logic of life itself. Also, to grapple with this, we may need to go beyond Boolean logic and develop new conceptual and mathematical languages. Information theory, control theory, dynamical systems theory, and even abstract fields like category theory offer promising avenues for capturing systems where information is dynamic, modular, and context-sensitive. However, it is entirely possible that no fully intuitive, human-readable model will ever suffice. Our cognitive systems evolved for clarity and reductionism, not for navigating the probabilistic, high-dimensional landscape of biological or artificial information processing. Even so, we can still build bridges between this complexity and our understanding – through visualization, metaphor, and mathematical abstraction. While interpretative tools can offer approximations, we may need to cultivate a new form of intuition, one that embraces complexity, probability, and emergence not as obstacles to comprehension but as the very basis of how meaning is encoded in living systems.

However, all this reasoning does not mean that the problem of comprehending distributed neural network-like computations in biological systems (including developing embryos) is solely responsible for precluding us from making better progress in understanding the complexity of life. Counterintuitively, at the current level of our science, the problematic milestone might lie in an orthogonal realm perfectly presented by the recent semi-theoretical work by Jonas and Kording (Jonas and Kording, 2017). This study, "Could a Neuroscientist Understand a Microprocessor?" by Eric Jonas and Konrad Kording, explored a provocative question: if neuroscientists applied their standard analytical tools to a complex yet fully understood artificial information processing system – a microprocessor – would those tools yield real understanding? The motivation stems from the broader assumption in neuroscience that collecting larger, more detailed datasets, combined with increasingly sophisticated analysis methods, will eventually lead to a mechanistic understanding of the brain. However, neuroscience lacks a ground truth system where we can objectively verify whether such methods truly uncover meaningful explanations. The authors therefore chose a fully simulated and functionally understood system – a MOS 6502 microprocessor – as a testbed to evaluate whether neuroscientific methods are sufficient to explain how it works (Jonas and Kording, 2017).

The study approached this by treating the processor like a mini-nervous system. The authors applied a wide array of standard neuroscientific tools to it: connectomics to map the circuit architecture, lesion studies to assess the effect of removing individual transistors, tuning curve analysis to relate activity to behavior (screen brightness, computer game behavior), pairwise correlation and population synchrony analysis, local field potential analogs, Granger causality to infer functional connectivity, and dimensionality reduction methods like non-negative matrix factorization to identify low-dimensional structure in the global dynamics. Each of these techniques is commonly used in systems neuroscience to make inferences about neural circuits, especially in the context of complex behavior.

Despite the diversity and sophistication of the techniques, the authors found that none led to anything approximating a functional understanding of how the processor actually works. While some results mimicked real neuroscience data – such as discovering "tuning curves," "functional connectivity," or "oscillatory dynamics" – the apparent insights were largely superficial or misleading. For instance, some transistors appeared crucial to one game but not others, leading to spurious interpretations analogous to saying a neuron is specialized for one behavior, when in fact its function is much more general. Likewise, frequency analysis revealed power-law spectral content suggestive of self-organized criticality – a common claim in neuroscience – but which, in the case of the chip, was simply an artifact of its digital clock cycles.

The authors conclude that current methods in neuroscience may be fundamentally inadequate for reverse-engineering complex systems – even in the idealized case where all data are accessible and the ground truth is known. This casts doubt on the prevailing optimism that "more data plus better algorithms" will automatically yield deep insights into the brain or developing body. The inability to derive a functional, hierarchical, or mechanistic understanding from these methods – even in a system orders of magnitude simpler than the brain – suggests that neuroscience is limited not by data volume but by its conceptual and methodological frameworks.

What's missing, they argue, is a robust methodology for validating analytical tools against systems where we already do understand the ground truth. Without such validation, we cannot know whether the insights we generate about the brain are meaningful or illusory (Jonas and Kording, 2017). The microprocessor thus serves as a powerful control experiment – a known dynamical system with definable behaviors and internal architecture, allowing us to test whether our interpretive techniques are genuinely explanatory or merely descriptive.

To improve our chances of successfully reverse-engineering complex biological or artificial systems, several areas need development. First, biology requires tools that can explicitly detect and represent hierarchical structure, since both artificial processors and brains operate across many scales of organization. Second, there is a need for task-controlled experimental designs that isolate computational functions, akin to executing specific instructions on a chip rather than letting it run arbitrary code. Third, causal modeling frameworks must move beyond correlational proxies like Granger causality and better reflect the mechanisms underlying observed dynamics. Finally, the field must adopt a culture of cross-validation against synthetic systems where hypotheses can be objectively tested and refined. In other words, reverse engineering requires not just more data, but a principled way of navigating the enormous hypothesis space and confirming functional understanding. The central irony underscored by the study (Jonas and Kording, 2017) is that, given current neuroscience methods, we would fail to understand a reasonably simple system that we already know perfectly well. This implies that without substantial changes in strategy, the dream of reverse-engineering the living things might remain forever out of reach – not because it is inherently unknowable, but because our tools are not yet fit for the task.

3.4.1 The Philosophical Insights Into the Future of Understanding Intracellular Computations

The current philosophical perspective invites a deeper reflection on the co-evolution of human understanding and the natural phenomena it seeks to describe. Our scientific worldview does not arise in a vacuum, nor is it a neutral window onto reality. It is, rather, a tapestry woven from conceptual tools, analogies, and interpretive frameworks – many of which are contingent on the intellectual and technological resources available at a given historical moment. The idea that "we see only what we are prepared to see" becomes strikingly relevant: phenomena that have always been present in nature may remain invisible to us until we possess the right metaphors and methods to conceptualize them.

The example of electricity is particularly apt, not just for its historical significance, but for what it reveals about the interdependence between empirical discovery and theoretical innovation. The physiology of nerve conduction was utterly opaque before the development of concepts like voltage, current, and resistance. The advent of tools like the galvanometer and the oscilloscope didn't just reveal new phenomena – they transformed our cognitive architecture, providing new ways of parsing biological behavior. In this sense, theory is a prosthesis for perception: without the scaffold of electromagnetism, the action potential would remain a mystery.

As for today, we enter an interesting new time when a fruitful combination of information theory, probabilistic modeling, and machine learning is applied to biology. These frameworks allow us to reimagine biological systems not as static blueprints, but as dynamic, adaptive, noise-aware, and computation-like processes. This isn't to say that biology is literally a computer in the classical sense, but that biological systems share deep formal similarities with computational architectures. Cells integrate inputs, apply weighted responses, update internal states, and even perform decision-making in uncertain environments – activities that are strikingly familiar to those who study algorithms, control theory, or neural networks.

Crucially, this convergence does not imply reductionism. To view cells as information processors is not to deny their physicality or biochemical complexity. Rather, it is to gain a higher-order perspective – a systems view – in which structure, function, and behavior are interlinked by abstract rules of signal propagation, feedback control, and optimization. These principles are not imported arbitrarily from engineering; they emerge organically from the constraints of physics and evolution. A signaling pathway must manage noise. A regulatory network must stabilize developmental trajectories. Evolution, operating through selection, rewards systems that can process information efficiently under fluctuating conditions. In this way, biology and computation are not merely analogous – they are shaped by shared pressures in the landscape of possibility.

We can also consider this from an epistemological angle: each generation's scientific imagination is co-produced by its technologies. Just as the telescope expanded our understanding of the cosmos and the microscope opened a new world of the cell, today's deep learning systems are giving us new lenses to model complexity, nonlinearity, and emergence. For example, concepts like latent space, distributed representation, and gradient descent are not just engineering tricks – they may offer profound insights into how GRNs stabilize phenotypes or how neural circuits generalize experience. In turn, biology may teach AI about robustness, adaptability, and learning in decentralized, noisy, and energy-constrained systems – areas where engineered models still fall short.

This mutual inspiration between artificial and biological systems challenges older divisions between the natural and the artificial. It suggests a broader philosophical claim: computation is not a metaphor imposed on life but a real feature of dynamic systems operating under uncertainty. Biological systems compute – not in silicon and code, but in chemistry and time. Recognizing this doesn't reduce life to code; it elevates our models to better match the richness of living processes. In doing so, we open new paths not only for scientific discovery but for understanding the logic of life itself.

In this light, the history of science can be seen as a sequence of conceptual revolutions, each sparked by a new mode of thinking – mechanical, electromagnetic, evolutionary, and informational. The task before us now is to refine our computational literacy in biological contexts, to develop not just metaphors but precise, testable frameworks that capture the probabilistic, adaptive, and hierarchical nature of biological decision-making. And as we do, we may not only deepen our understanding of life but also evolve the very structures of thought through which we engage with the natural world.

3.5 BIASING AND PROBABILISTIC FEATURES EMBEDDED IN MECHANISMS OF FATE SELECTION

To fully grasp how dispersed information is integrated into a coherent developmental outcome, we must investigate the underlying mechanisms by which cells interpret, bias, and ultimately commit to specific fates. Such fate-related biases arise from a combination of inherited factors, such as epigenetic modifications, asymmetric distribution of molecules during cell division, and environmental signals that cells experience within their niche (Moris et al., 2016; Swift et al., 2023; Wang et al., 2018, 2024). These biases do not dictate a specific fate but rather constrain the decision-making landscape, making certain outcomes more likely than others (Erickson et al., 2024, 2023; Kastriti et al., 2022; Soldatov et al., 2019). They provide a framework within which subsequent signals and intracellular computations operate, ensuring that cell fate transitions remain within biologically meaningful boundaries. Different modes of this mechanism can be envisioned, ranging from systems that rely heavily on strong initial biases to those that depend more on dynamic, context-dependent decision-making. In some cases, early fate biases might be reinforced through positive feedback loops, locking cells into specific trajectories. In other scenarios, cells might retain multipotency for an extended period, with fate commitment occurring as a gradual, emergent process. The balance between these modes varies across species, tissues, and developmental stages, providing evolutionary flexibility in how organisms build their bodies.

Recently, research from the Adameyko lab has provided a more formal mechanistic framework for understanding the phenomenon of fate selection via competition of opposing genetic programs (embodying biases), revealing a three-step mechanism of cellular decision-making that unfolds through co-activation of competing programs (biasing), their repulsion (or competition), and commitment (Soldatov et al., 2019). This sequence of steps ensures that cells maintain flexibility before ultimately locking into a specific fate, balancing adaptability with developmental precision (Figure 3.5).

The first step in fate selection involves the co-activation of competing transcriptional programs, creating a state of multipotency where several differentiation trajectories remain viable. This phenomenon is observed in neural crest cells, mesenchymal progenitors, and stem cell populations where multiple lineage-specific gene networks are transiently expressed in parallel (Figures 3.6 and 3.7). The co-activation of

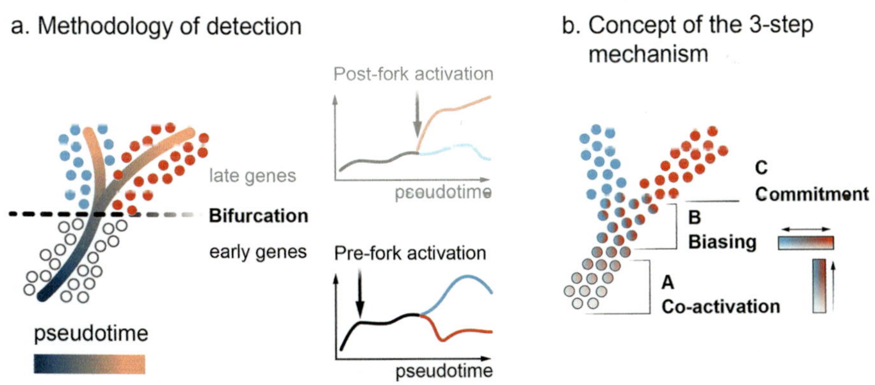

FIGURE 3.5 The three-step mechanism of fate selection. The three-step mechanism of fate selection includes the following stages: A. Co-activation of opposing gene expression modules – a cell becomes capable of responding to external or internal cues that influence its developmental fate. B. Biasing – certain extrinsic signals and chromatin states start to favor one lineage over others, making the cell more likely to adopt a specific fate. At this step, we observe the repulsion of opposing gene expression programs within individual cells and the formation of heterogeneity, which reflects differentially-biased cells within the entire population.

(Continued)

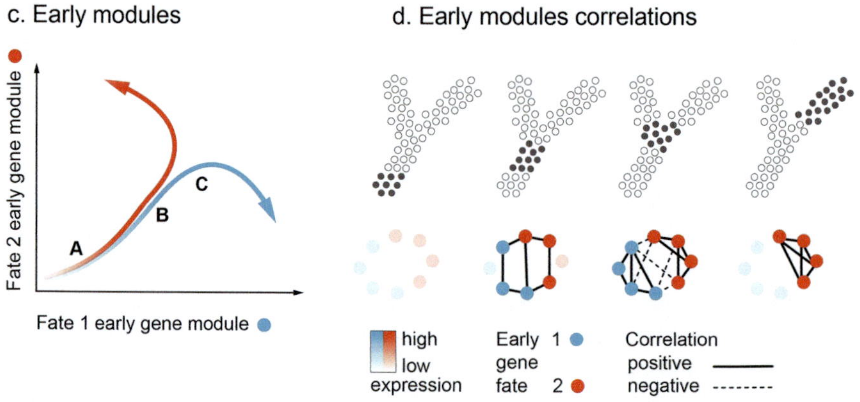

FIGURE 3.5 (Continued) C. Commitment – the cell irreversibly leans toward a specific future identity, turning on fate-specific genes and shutting down alternative programs.

The scheme is modified after Faure et al. (2023) and Soldatov et al. (2019).

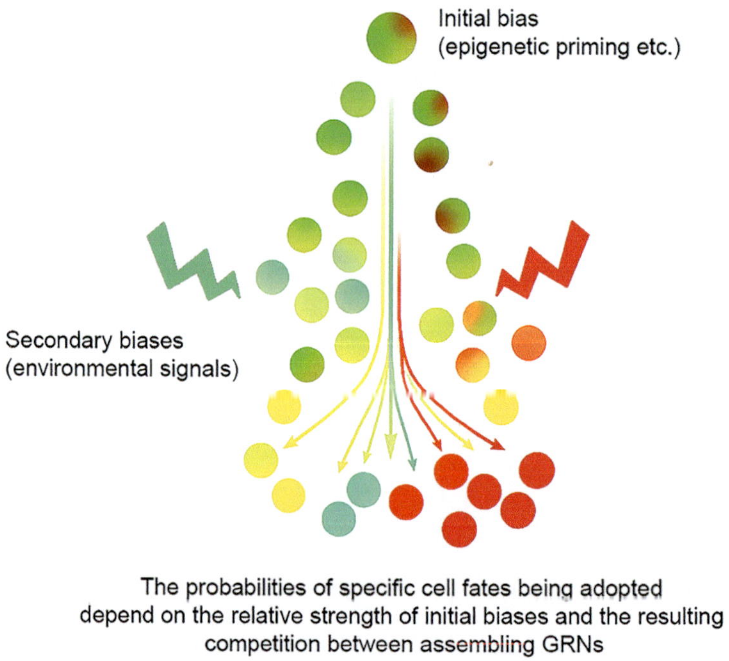

FIGURE 3.6 Biasing of multipotent progenitors. The scheme focuses on the progressive development of meaningful and future fate-related microheterogeneity due to incorporation of biases of different nature and strength.

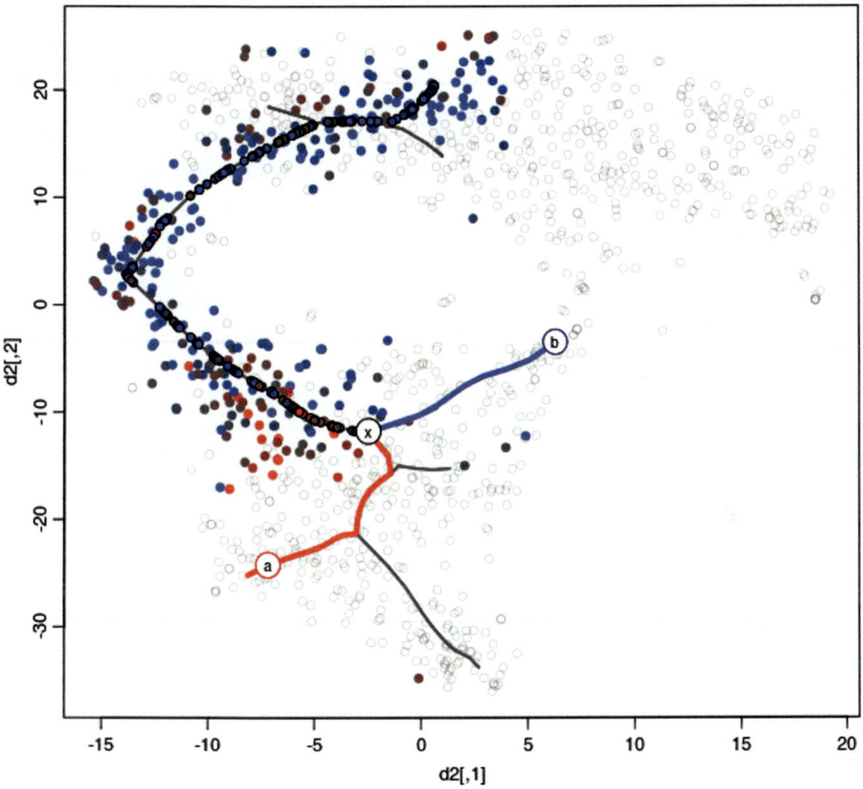

FIGURE 3.7 The actual picture shows biasing of migratory neural crest progenitors. This is the transcriptional tSNE (t-distributed stochastic neighbor embedding) from Soldatov et al. (2019), showing individual cell transcriptomes of migrating and differentiating murine neural crest cells. The red trajectory goes toward autonomic neuronal fates, whereas the blue trajectory ends with somatosensory neurons. Notice the mixing of "red" and "blue" gene expression modules prior to the first bifurcation and corresponding heterogeneity increase toward the fate split point.

competing transcriptional circuits ensures that a cell does not prematurely commit to a single fate, allowing it to "sample" potential identities before a final choice is made. Such dynamic biasing does not dictate an absolute fate but instead shifts the probability distribution of available choices. Some fates become more energetically favorable or more actively reinforced by external signals, making them more likely to be selected. This is akin to a probabilistic computational system where multiple inputs modify the weighting of different outputs before a final selection is made. For instance, neural crest-derived progenitors can receive subtle biasing cues that favor an autonomic neuronal cell fate over a somatosensory neuronal fate depending on local signaling environments (Figure 3.7). At this stage, heterogeneity among cells becomes a crucial factor.

At a molecular level, this phase is characterized by the simultaneous expression of markers from multiple lineages, often involving mutually inhibitory transcription factors that establish a dynamic equilibrium. For example, neural crest cells can transiently express melanocytic and neuronal markers before differentiating into either lineage. Similarly, mesenchymal progenitors can co-express osteogenic and adipogenic programs before selecting a final path. These early-stage gene expression conflicts create a metastable state where the cell remains poised for multiple fates. Importantly, this competition between fates is not random. The environment, intracellular signaling pathways, and metabolic states influence the stability of competing programs. For instance, oscillatory gene expression patterns – reminiscent of neural networks evaluating multiple inputs – allow cells to integrate signals over time before gradually tipping the balance toward a specific outcome. As the window of multipotency progresses, cells begin to experience competition and repulsion of opposing biasing transcriptional programs that gradually tilt the balance toward one fate over another. The microheterogeneity of such progenitors increases (Figures 3.5b and 3.6), as different cells enhance the mosaicism of specific fate biases within the population.

The final step in the process is commitment, where a cell decisively locks into a single fate and silences alternative transcriptional programs. This commitment occurs once a dominant transcriptional network stabilizes and reinforces itself through positive feedback mechanisms (Soldatov et al., 2019). The loss of competing fate determinants, either through active repression or chromatin remodeling, ensures that the decision becomes irreversible. Importantly, even at the commitment stage, some residual plasticity can remain under certain conditions. In regenerative processes, some cells can "unlock" previously silenced fates under extreme stress or injury. Understanding how cells transition from biasing to commitment can therefore provide insights into regenerative medicine and stem cell reprogramming strategies.

The process of cell fate determination can be metaphorically likened to the developmental trajectory of high school students exploring their future careers. In the early stages, much like pluripotent cells co-activating multiple transcriptional programs, students often cultivate a wide array of interests – dabbling in science, arts, sports, or literature – without immediate pressure to commit. This is a state of broad potential, where multiple directions are entertained simultaneously.

As time progresses, subtle biases begin to emerge. Some interests grow stronger, while others fade, influenced by personal experiences, peer dynamics, mentorship, and evolving self-identity. This mirrors the second step in cellular decision-making: a biasing or repulsion phase, where initially co-activated transcriptional programs begin to antagonize one another, nudging the cell toward one direction over others. Finally, just as students near graduation and commit to a college major or career path – often solidifying what had only been a latent preference – a cell reaches its final step: commitment to a specific fate, locking in a program while repressing alternative options. In both cases, what appears as a binary outcome (scientist vs. artist, neuron vs. muscle cell) is in fact the result of a dynamic, gradual process of weighing possibilities, shaped by both internal inclinations and external contexts.

A growing body of evidence suggests that poised bistable enhancers may serve as crucial regulatory elements in fate selection, particularly at points preceding bifurcations in differentiation trajectories (Crispatzu et al., 2021; Minoux et al., 2017). These enhancers exist in a metastable state, marked by the concurrent presence of activating (H3K4me1) and repressive (H3K27me3) histone modifications, positioning them as molecular "switchboards" capable of rapidly tipping toward activation or silencing in response to environmental cues or intracellular signals. Prior to cell fate bifurcations, such enhancers may act as sensitive integrators of upstream regulatory input, maintaining accessibility and transcription factor occupancy for competing lineage programs without committing to either. This poised state reflects the underlying plasticity of the progenitor cell, enabling it to remain responsive to divergent developmental cues. As differentiation proceeds and biasing signals accumulate, these enhancers may resolve into a monostable active or repressed state, thereby locking in one transcriptional program while excluding the alternative. In this way, ready-mode bistable enhancers could help to encode the cell's probabilistic landscape by representing the tipping points of the system, facilitating sharp transitions in gene expression following relatively subtle shifts in signaling context – thus functioning as molecular inflection points that mediate the transition from multipotency to commitment. Despite such sensitivity to signals and fluctuating states producing diverse outcomes and collective behaviors in multicellular systems, the robustness of development and higher levels of control will ensure the more or less genetically determined output at the level of high-level structure and function. Also, some factors might possess heavier weights in decision-making, thus creating situations that fall very close to a deterministic scenario or become *de facto* deterministic. In this case, everything depends on the details of the molecular mechanism involved.

Another emerging paradigm in developmental biology is the idea that cell fate decisions are governed by oscillatory dynamics, rather than being dictated by a linear, deterministic program (Meijer and Sonnen, 2024). This view is inspired by research on GRNs, signaling feedback loops, and dynamic cellular behaviors, which suggest that cell decisions often oscillate between competing fate options before committing to a particular outcome. Such oscillations introduce a temporal dimension to cell fate selection, allowing cells to integrate multiple signals over time before making an irreversible choice. At the heart of the oscillatory model is the idea that cells do not instantly commit to a single fate upon receiving differentiation cues. Instead, they go through cycles of fluctuating gene expression, switching between competing programs before stabilizing on a final choice (Kelsh et al., 2021). These fluctuations can be thought of as a form of "hesitation" in decision-making, allowing the cell to probe multiple potential fates before committing and increasing robustness by avoiding premature or erroneous decisions.

This transition from oscillatory multipotency to commitment resembles a bifurcation in dynamical systems theory, where a system with multiple stable states is pushed into one final attractor (Huang et al., 2007; Mojtahedi et al., 2016; Zhou et al., 2012). In practical terms, this means that cell fate is not simply a response to a singular signal but the result of accumulated oscillatory fluctuations resolving into a stable choice. The oscillatory model also has implications for understanding cell fate as a computational process. By acknowledging that cell fate decisions are dynamic, iterative, and influenced by fluctuating signals, the oscillatory model moves away from simplistic gene-centric views of differentiation. Instead, it supports a perspective where phenotypic outcomes arise from continuous feedback loops, increasing both the flexibility and robustness of development.

As a more precise example, Bayesian decision theory offers a framework (Knill and Pouget, 2004; Parmigiani, 2001) for understanding how biological systems, particularly cells during embryonic development, make decisions under uncertainty. Bayesian decision theory is a mathematical framework used to make optimal decisions under conditions of uncertainty. It combines Bayesian probability theory (Linden et al., 2014), which models how we update beliefs in light of new evidence, with decision theory (Cao, 2023), which provides principles for choosing actions that maximize expected utility or minimize expected loss. At its core, this theory integrates prior knowledge with current observations to evaluate the most probable outcome, a principle strikingly analogous to how cells assess and commit to specific fates. Originally developed within the realms of statistics and machine learning, Bayesian reasoning (Steib et al., 2025) has begun to inspire biological models by providing a structured way to represent how decisions emerge from dynamic, context-sensitive inputs. In the biological context, the concept of prior probability can be understood as the pre-existing internal biases within a cell – such as lineage history, chromatin accessibility, or epigenetic modifications – that make it more likely to adopt certain fates (Barua and Hatzikirou, 2023). These biases reflect the "memory" of the cell and shape its response landscape even before any new signal is received. In contrast, the likelihood corresponds to the impact of present cues: morphogens, local signaling, mechanical forces, and metabolic states. These environmental inputs update the cell's internal predictions, adjusting the probabilities of different fate outcomes in real time. When combined, these factors produce the posterior probability – the most refined prediction of which fate the cell is likely to adopt given both its history and its current state. This probabilistic integration has profound relevance for the study of fate decisions. Rather than following a rigid, deterministic path, many progenitor cells exist in a metastable state, co-expressing transcriptional programs for multiple potential fates. This co-activation phase, observable in neural crest cells, mesenchymal progenitors, and various stem cell systems, is reminiscent of how Bayesian models initially keep all hypotheses "open," awaiting further evidence to tilt the decision toward a specific outcome (Parmigiani, 2001). As new environmental inputs accumulate, the relative stability of competing GRNs shifts. Some programs are reinforced, while others are repressed, in a phase that can be likened to the recalibration of probabilities in Bayesian terms. Eventually, commitment occurs: a dominant program wins out, alternatives are silenced, and the cell enters a new, stable state. This final outcome is not an abrupt switch but the resolution of many probabilistic evaluations over time.

Applying Bayesian reasoning to developmental biology also reframes our understanding of heterogeneity, which enables populations of progenitor cells to explore a broader decision space, ensuring robust tissue patterning even under variable conditions. Just as dropout regularization improves generalization in neural networks by embracing noise, developmental systems may leverage biological noise to maintain plasticity and prepare for rapid adaptation during cell fate selection. This conceptual alignment between Bayesian decision theory (Parmigiani, 2001) and cell fate selection is not merely metaphorical – it suggests practical strategies for modeling and experimentation (Wilkinson, 2006). For instance, integrating prior lineage data with spatial transcriptomics could help assign probabilistic fate potentials to progenitor cells. Live-cell imaging of gene expression dynamics might capture temporal changes in "posterior" fate probabilities, especially in systems characterized by oscillatory expression or feedback-regulated commitment. Bayesian models could also assist in the design of reprogramming protocols, helping to predict which combinations of cues most efficiently convert a given cell into a desired fate by maximizing the posterior probability of that outcome. Importantly, this perspective does not imply that cells conduct formal Bayesian calculations, but rather that evolution has shaped cellular systems to approximate these logic structures. Feedback loops, bi-stable enhancers, and chromatin remodeling complexes may serve as biological analogs of Bayesian inference components, updating gene expression in response to internal states and external inputs. Even more intriguingly, probabilistic frameworks may help us model collective decisions in cell populations, such as the emergence of spatially patterned fates in developing tissues, as outcomes of distributed inference.

Overall, as research advances into the mechanisms underlying cell fate selection, numerous scientific and technical challenges remain. For instance, the interplay between stochastic fluctuations, their refinement, and biasing regulatory programs remains poorly understood, which is the core of the problem in modern developmental biology. While stochastic gene expression contributes to cell fate plasticity, pro-deterministic rules eventually push the system toward commitment. Finally, developmental biology has largely focused on cell fate selection during embryogenesis, but similar principles likely govern adult stem cell renewal, regeneration, and disease progression (cancer cell plasticity). Do all these systems operate under the same three-step mechanism (co-activation, biasing, commitment), or do alternative modes of decision-making exist?

3.6 HOW DO WE DESCRIBE AND CLASSIFY CELL STATES, CELL TYPES, AND TRANSITIONS BETWEEN THEM?

Understanding the dynamic mechanisms by which cells select their fates naturally leads to a critical question: once a fate is selected, how do we define and classify it? The field must grapple not only with the logic of fate decisions but also with the criteria we use to recognize and name distinct cellular states. This challenge lies at the heart of modern cell type taxonomy. Given the plasticity and modularity of gene expression programs (Melo et al., 2024; Verd et al., 2019; Wagner et al., 2007), traditional definitions of cell types might be revisited and refined. Also, as we trace the complex and often gradual transitions cells undergo from multipotency to commitment, a deeper challenge becomes apparent: how do we define and classify these intermediate, branching progenitor states within a coherent framework? Unlike terminally differentiated cells, whose identities often align with clear functional roles and stable gene expression programs, progenitor states exist in a continuum (Kastriti et al., 2022). They shift progressively, with overlapping and dynamically regulated transcriptional modules that resist sharp boundaries. This fluidity is particularly evident at branching points in differentiation trajectories, where competing fate programs are partially active, and gene expression changes incrementally rather than abruptly. Traditional classification systems – reliant on fixed markers or discrete cell identities – struggle to capture this complexity (Figure 3.8a). Instead, a new conceptual and technical framework is needed to describe cell states that are transitional, probabilistic, and embedded in a branching developmental landscape.

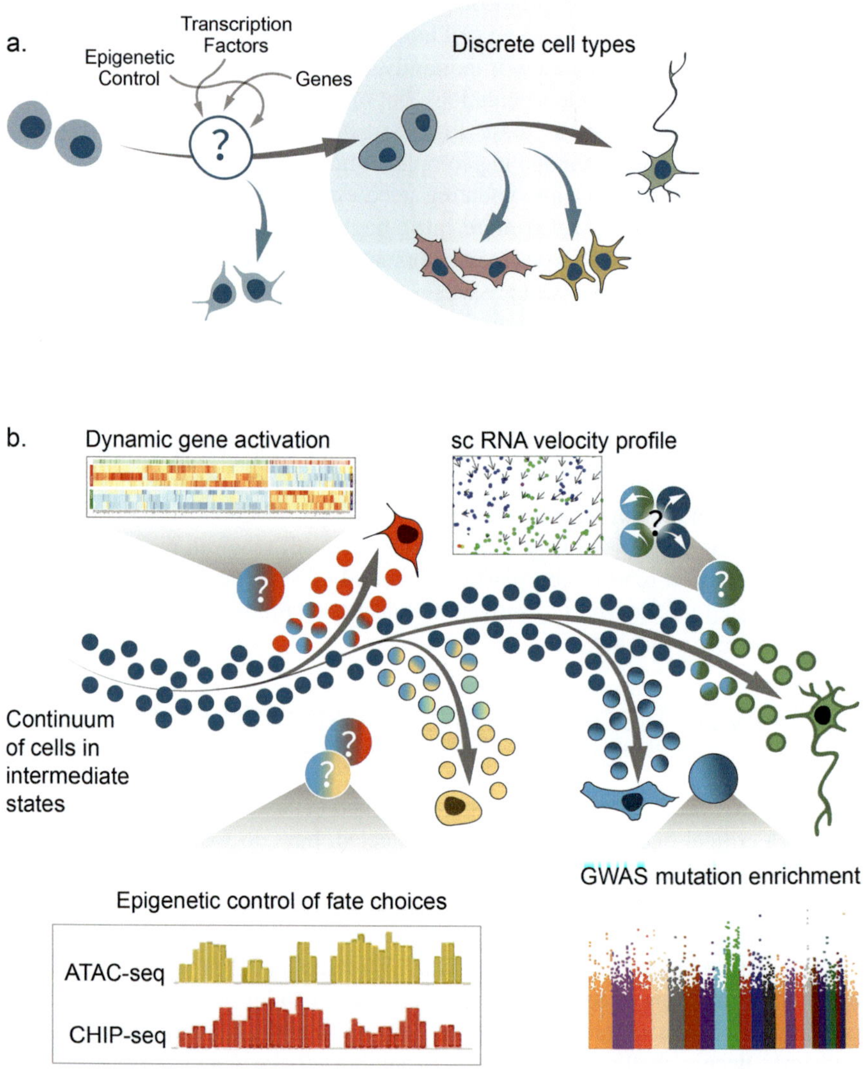

FIGURE 3.8 Modern approaches helping to understand the diversity of cell types and cell states in developmental and adult contexts. The older concepts about development had to rely on discrete pictures and snapshots of events (a) because of the lack of resolution in the experimental data. Now, this problem is largely solved by single-cell omics approaches (b) – single-cell transcriptomics, single-cell ATAC-seq, and spatial transcriptomics, all providing sufficient temporal and spatial resolution. The analysis toolkit also helps to make sense out of such data to align the individual transcriptomes into biologically meaningful sequences – developmental trajectories (Faure et al., 2023; La Manno et al., 2018).

3.6.1 Modular Structure of Gene Expression Helps to Improve Cell Type Taxonomy

In the realm of cell type classification (Clevers, 2017), consensus largely centers around one key aspect: the presence of a unique cellular function. This characteristic is perceived as pivotal, given that such distinct functions have been observed to undergo evolutionary changes, diversify, and become distributed among various cells within multicellular organisms. Consequently, they play a defining role in the evolutionary success of multicellular life forms. Thus, the existence of a unique and specific function constitutes a historically acknowledged and biologically significant attribute that forms an integral part of the traditional cell type definition.

It is important to note that a specific known function, while emblematic, often represents only a fraction of the myriad functions and processes transpiring within a cell. It is conceivable, therefore, to regard each cell as a repository of both unique and non-unique functions, with these functions being underpinned by transcriptional programs – essentially, the discrete modules organizing gene expression. Once evolved, these programs have the capacity to be transposed or "coopted" into other cell types sharing access to the same genetic makeup within a multicellular organism (McQueen and Rebeiz, 2020). This phenomenon challenges the classical concept of homology by introducing scenarios wherein the same function or cell type may emerge from unrelated developmental origins or distinct groups of progenitor cells through the co-option of a genetic program initially developed for cell types of different origin and locations. Such instances have led to a reevaluation and diversification of the concepts of "molecular" and "structural" homology, a matter that remains a subject of considerable debate in contemporary literature (Faunes et al., 2015; Rusin, 2023; Wagner, 2021).

However, it is essential to acknowledge that this function-centric definition of cell types poses several challenges. In the present scientific landscape, a substantial portion of cellular functions and compartmentalized processes remain inadequately characterized, and many of them may operate at various hierarchical levels or exhibit intricate interconnections. As a result, there is an imperative need for better methodologies to identify and delineate discrete cellular functions, which can be achieved by scrutinizing the molecular content of cells, including their transcriptional pro grams and epigenetic states. This is precisely where single-cell omics technologies have the potential to yield considerable insights (Zeng, 2022), a subject that will be elaborated upon in the subsequent discussion.

In fact, the functional attributes of an individual cell are multifaceted and are evident in various aspects such as cell morphology, proliferative or migratory capabilities, and the presence or absence of specific molecular components (Schlosser, 2021). Consequently, the categorization of cell types has evolved into a complex and sometimes disorderly process during earlier phases of exploring the cellular landscape within multicellular organisms. In these earlier periods, cell types were often identified based on criteria such as cell morphology, spatial positioning, function, or the capacity to selectively bind a limited set of chromogenic dyes.

The advent of molecular biology, which gave rise to techniques like immunohistochemistry and RNA hybridization, brought about sporadic and unstructured discoveries of marker genes that contributed to the ongoing refinement of cell type descriptions. Much like naturalists in previous centuries, biologists embarked on expeditions to identify new cell types armed with antibodies and *in situ* hybridization probes. During this phase, characterizing newly uncovered cell types was a relatively straightforward endeavor, involving the assignment of specific combinations of molecular markers to distinct cell populations. The number of such identified "cell species" hinged on the quantity of unique marker combinations discovered (Figure 3.8a). The apparent simplicity and unambiguous nature of this system, however, were rooted in the limited state of knowledge at the time.

Lately, with the use of unbiased and systematic methods, such as single-cell transcriptomics and ATAC-seq (assay for transposase-accessible chromatin with sequencing), the situation changed dramatically, as the number of descriptors ("markers") grew enormously, leading to a "near-complete" picture, which appeared rich and informative, and yet complex, vague, and uneven (Rood et al., 2025; The Tabula Muris Consortium et al., 2018). The single-cell multiomics approach provides a wealth of biologically relevant insights into cell typology and transitions between cell types (Figure 3.8b) (Adameyko et al., 2024). However, although it is extraordinarily informative and helpful for detecting transcriptional states with high precision, it comes with issues that are not so easy to fix. For instance, when cell typology is approached with combinations of thousands of markers, the resulting picture suggests the existence of cell types, which are not fully separable based on universal criteria.

Clustering of individual cell transcriptomes depends on many parameters and cannot be applied with the same settings to different datasets, producing similarly good results. Hierarchical clustering attributes individual cell transcriptomes to clusters, and although the clustering principles might be universal, the exact parameters and the number of "desired" clusters will not be universal or uniform across different data. Although this approach yields well-resolved groups of cells in an unbiased way, it is extraordinarily unsystematic in terms of applied parameters, which often creates debatable situations, such as whether individual transcriptomics belong to one cell type or should be split into two or three or more. The criteria of the sufficient difference are decidedly arbitrary. However, some are reasonable and grounded in biology. For example, one approach proposed – emphasized by Gerhard Schlosser in his chapter – is to use evolutionary comparisons, focusing on the identification of equivalent cell types across different species as a basis for defining cell identity.

Still, it might be possible to agree upon a set of criteria for more or less universal separation of cell types from single-species, single-cell multiomics datasets. However, in this case, the community must agree about placing weights on some specific molecular differences over others. A similar situation was encountered in the field of animal systematics and cladistics, when more important anatomical and embryological criteria had to be preferred over the others (Cuvier et al., 1985; Doyle, 2022; Richardson et al., 1997). An advantage of single-cell transcriptomics datasets is that the agreeable criteria will inevitably include large gene expression programs, including hundreds or thousands of genes, which will ensure a better biological grounding of the new systematics of cell types as compared to previous limited approaches. Similar to the recent emergence of at least 24–26 modern definitions of organismal species (Wilkins, 2011), the new depth of molecular knowledge due to the application of single-cell technologies generates a diversity of definitions of a cell type. There might be specific subsets of genes or arbitrary metrics in transcriptional space to define what a cell type is, and what is not.

All these approaches can be divided into two categories: those suggesting to use all genes and the entire spectrum of information unbiasedly, and those proposing placing specific weights on some groups of genes, for instance, DNA-binding proteins, including transcription factors and chromatin remodelers.

When pondering the advantages and disadvantages of these approaches, the following remark is necessary at the start of this discussion: there might be an impression in the community that a part of "the cell type resolution problem" is simply related to technical imperfections. However, the current technical proficiency has already created enough datasets with satisfactory depth to the extent that we can say that although "the cell type resolution problem" might be amplified by poor technology, the current state allows pondering a cell type problem assuming an ideal situation and implying "omni-knowledge" about transcriptional and epigenetic states. In the following section, I will largely theorize about such perfect situations.

The current, most popular approach to separating cell populations and grouping them into cell types based on individual cell transcriptomics relies on unbiased analysis of all expressed genes. This wealth of information and unbiased criteria cause a situation with higher but also lower resolution in some critical areas, as "the less important genes," for instance, of a stress response, cell cycle or circadian variation, or transcriptional noise, have the same weight as fate-controlling transcriptional factors. This creates some degree of fuzziness and loss of ultimate resolution of cell types, and provokes a comparison with zoological cladistics, where evolutionarily conserved and important anatomical features are weighted equally heavily as any other random organismal features.

Using such a completely unbiased approach, situational physiological processes or noise at the level of transcriptome might overshadow the importance of fate- and cell identity-related transcripts. At the same time, using the expression data for all genes in an unbiased way must be the most informative approach, as the entire available information about the system is taken advantage of. One of the possible solutions to this awkward situation is to switch from the unbiased analysis to a biased approach, where some additional weights are attributed to genes that are more informative about cell fates and types. This, however, requires a clear understanding of what is more relevant to a cell identity, and what is not. As for today, single-cell studies are still in the reasonably early explorative phase, and this knowledge is not yet robust, especially when it comes to non-canonical model organisms. Thus, the unbiased approach, although it comes with some problems, appears to be the best.

Is there any way for further improvement toward clearly separating and defining cell types in an unbiased way? In fact, there is a hope that, instead of sub-selecting only fate-related and other "meaningful" genes for further analysis in a biased way, adding additional layers of knowledge about gene expression regulation, hierarchy, and causality to an unbiased gene selection might improve the resolution of clusters and cell taxonomy.

One such informative layer is represented by coherent (or evolutionary stable) transcriptional programs, whereas another complementary approach might take advantage of gene expression modules, which are dedicated to some specific roles in cells. Different cell types have different structures of gene expression in terms of the combination of upregulated and downregulated modules (Melo et al., 2024; Verd et al., 2019; Wagner et al., 2007). Therefore, elucidation of such modules and their co-application to cell–cell comparisons will enable smarter analysis of cell populations and attribution of cells to cell types. Thus, a cell type can be defined as a unique combination of gene expression modules dedicated to specific roles (see Figure 3.9). The connection between gene expression and cell function, or morphology and position, can be better perceived at the level of transcriptional modules, which provide for implementation of any cellular function. Bearing the idea of modules corresponding to functions and other characteristics in mind, we can accept that such multiomics-based definition of a cell type converges and reconciles with more classical function- and feature-based definitions.

Nevertheless, do we really need such a reconciliation? The bioinformatics-centered view is largely self-sufficient and can bridge cell populations via label-transfer procedures or co-clustering of different datasets (Gonzalez-Ferrer et al., 2024). These approaches align heterogeneous cell populations without any need for a formal cell

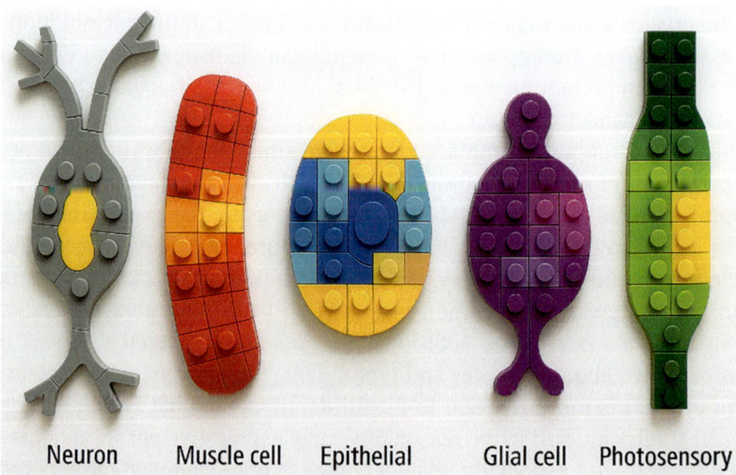

FIGURE 3.9 Cell types are made of functional modules (colorful blocks), which are organized into overlapping and unique sets. The detection of such compact, coherently evolved, and function-related modules is possible via comparative single-cell multiomics approaches (Fatieieva et al., 2025). A cell type can be defined as a container or collection of such modules, which might further undergo individual or co-dependent evolutionary changes.

type definition. The success of such a "computer language" raises a doubt regarding whether we really need to reinvent a cell type concept at all.

To advocate a need for a cell type concept as something biologically-grounded and not only as an idealistic framework to discretize nature, it is good to mention that transcriptional and epigenetic states of a cell represent only a part of its phenotype, as numerous processes proceed at a protein level, and are driven by cell morphology, cell–cell communication, and electrophysiology. For instance, the neuronal wiring profile is a key feature of a neuronal cell type.

Indeed, there is an unmet need to integrate multiple criteria, such as connectivity, position, physiology, cell shape, cell functions, gene expression, epigenetic profile, and migratory behavior, which altogether might discretize cells much better than clustering in transcriptional space. Thus, terminal cell types might emerge as discrete based on unique sets of functions and related phenotypical features, which will emerge as distinct and separable with more criteria included. At the same time, we need to get the best proxy to recognizing such integrated cell types. To bridge and integrate all these features with identified transcriptional states, we must develop a multilayered and coherent concept of a cell type, preferably the one that still allows recognizing cell types from transcriptional data only. Already today, a cell type might be defined as a unique combination of stable gene expression modules related to cell identity and implementation of cellular functions, being combined with specifics of cell morphology, contacts, and physiological outputs.

The challenge to this approach is rooted in a necessity to have excellent single-cell multiomics datasets covering a widest possible spectrum of cell populations of the body, as the identification of modules requires robust comparisons of transcriptional programs between all existing cell populations. A proxy to this approach is represented by the concept of a regulon, which includes a transcriptional activator with the downstream expression of a set of target genes in a specific cell (Bravo González-Blas et al., 2023). The elucidation of regulons is possible with SCENIC (Single-Cell rEgulatory Network Inference and Clustering) and DoRothEA (Discriminant Regulon Expression Analysis) tools and is based on the knowledge of DNA-binding sites of a wide array of transcription factors in combination with available chip-seq data (Bravo González-Blas et al., 2023; Garcia-Alonso et al., 2019; Müller-Dott et al., 2023). However, although a regulon represents a unit of structured gene expression, it is not a direct equivalent of a gene expression module dedicated to some function, as the module might be driven by several transcription factors in combination and, thus, might be composed of several coherent regulons.

Furthermore, the detection of gene expression modules is preferable over the detection of transcriptional factors driving them because of technical and transcriptional noise. Indeed, the detection of modules must be more robust, as those often include dozens or hundreds of genes instead of a small set of transcriptional controllers governing them. Any insights into a structured regulation of gene expression will inevitably assist better definitions of a cell type in the future, as it provides a more discrete picture because expression of modules is expected to be more crisp and discrete (being linked to cell functions, including unique ones), as compared to noisy expression of individual genes and their pleiotropic roles. Thus, the addition of the dimension, including the hierarchy of gene expression represented by nested transcriptional modules and regulons, might additionally improve the developing cell taxonomy. The use of single-cell ATAC-seq and other approaches addressing epigenetic states (Moris et al., 2016) will improve the predictions of such regulation and hierarchy.

Although this multilevel unbiased approach has good prospects for defining cell types from multiomics data (Bakken et al., 2017), there is another dilemma rooted in the necessity to separate dynamically changing cell states and more stable cell types. Indeed, how can we separate definitions of “a cell state” and “a cell type” (Morris, 2019; Zeng, 2022)? As discussed above, the cell type notion is grounded in a concept of stable cell identity and is often seen as a set of temporally persistent expressed genes and cellular functions specific to a concrete group of cells. However, within a cell type, a diverse spectrum of temporal fluctuations is possible, many of which are attributed to variations in cell function performance, phases of a cell cycle, stress response, circadian control, cell–cell interaction, neural and endocrine control, and many more. Splitting the modules related to cell identity and capacity to perform unique functions long term from the modules responsible for temporarily controlled physiological responses is important for formal separation of cell types and cell states. Thus, contrary to the fluctuating physiological state-related gene expression modules, the modules related to a cell type stay stable independently of the physiological status, and must be encoded in a persistent combination of transcription factors and epigenetic configurations. The identification of modules related to cell states requires a temporal analysis over different physiological contexts. Needless to say, numerous cell states with corresponding gene expression modules might be nested within a single cell type, reflecting the tunability of any specific function and housekeeping program.

Similar to how the cell state-related modules can be detected and attributed to temporal fluctuations in cell functions and housekeeping responses, a spatial variation in cell types can provide insight into the modules modifying cellular transcriptional programs depending on the position in an organism. As we know from different studies, the transcriptional status of a cell can be affected long term by location-defined cell–cell communication, developmental positional signals, and formation of a specific microenvironment niche. In that case, although the functions of a cell type may vary in their implementation across body territories, this variation is temporally stable and can account for a definition of a cell subtype (Erickson et al., 2024). The examples of cell subtypes related to spatial properties include oligodendrocytes and other positional adaptations in the brain, endothelial, and perivascular cells (Hilscher et al., 2022).

Therefore, the integration of spatial single-cell transcriptomics data becomes a key for inferring subtypes of specific cell types depending on the position together with correlated gene expression modules. Biased and unbiased single-cell spatial transcriptomics keep transforming the understanding of variation and microheterogeneity of cells with similar or same functions across the body territories (Haviv et al., 2025).

At last, it would be interesting to explore how cell type definitions, especially stemming from a single-cell multiomics approach, fit the concept of Waddingtonian landscape (Waddington, 1957, 2014 (republished), 1942). The landscape notion suggests placing cell states and types into some multidimensional space, similarly to how single-cell transcriptomes cluster in a space defined by differentially expressed genes and principal components. This implies that cells of a body occupy only specific locations in this space, being developmentally connected by transitions between these rather stable cell type occupancy areas. Temporally fluctuating cell states, as defined in the paragraph above, might represent an unstable cloud surrounding more stable cell identity centers. In this sense, a concept of cell types and subtypes resembles stable attractors in a Waddingtonian space, whereas the fluctuating cell states and transitions represent unstable areas and channels connecting the attractors (Mason et al., 2025). Stabilization of attractors and locking of cell type phenotypes might occur due to different molecular mechanisms, starting from stable GRN presence and epigenetic restrictions to genomic editing in cell type-specific way (Smith et al., 2009).

Altogether, this picture implies that a vast space of cell types and states stays unoccupied, while the occupied locations correlate with specific cell performances/phenotypes and are carved by evolution. The shift of cell phenotypes toward non-occupied locations in this space might correlate with the development of a pathology, such as malignant transformation or metabolic disease. Understanding the gene expression modules characterizing the new stable pathological attractors, as well as the trajectories toward the new attractor spaces, might yield better models of diseases at the level of cell types. Also, this approach might help to classify pathological cell types in the framework as normal healthy cell types, thus keeping the new cell taxonomy uniform and universal.

Finally, cell types might be primarily defined via the evolutionary framework, and not according to any minimal significant difference in gene expression with other closest clusters in a transcriptional space (Arendt, 2008; Arendt et al., 2019, 2016). This can be achieved by extracting the evolutionarily conserved features of a cell type (Musser et al., 2021), including conserved combinations of modules. As gene expression modules relate to discrete functions, their existence might be evolutionarily stable and observable (Zemke et al., 2023). Thus, the module, as compared to a single gene, might appear as a more evolutionarily conserved phenomenon, which can be better detected in evolutionarily distant organisms, especially in cases when direct homologs of individual genes are not clearly identifiable. The concept of evolutionary conserved gene expression modules might provide a great proxy and new metrics for evolutionary comparisons of cell types (Zemke et al., 2023).

Cells from different organisms might have different modules present or missing in the digital pattern, which can be at the base of a new metric of evolutionary comparisons between cell types, where the numbers of conserved modules and their unique combinations are converted into clear-cut distances. Furthermore, the gene composition of every module might drift during evolution, which can be studied as well and used for building better evolutionary connections within the spectrum of homologous cell phenotypes.

A similar view has been extensively outlined in the review article "The origin and evolution of cell types" by Detlev Arendt and coauthors, where the authors propose a definition of cell type, rooted in evolutionary theory rather than traditional phenotypic or developmental criteria (Arendt et al., 2016). Their central argument is that cell types should be understood as evolutionary units defined by their capacity for independent evolutionary change. This independence, they argue, stems from the presence of a unique gene regulatory mechanism – what they term the core regulatory complex (CoRC) – that underlies and maintains each cell type's distinct gene expression program. According to this framework, the CoRC consists of a specific, cooperative set of transcription factors that act as the molecular basis of cell identity. The formation of a new cell type, therefore, is marked by the establishment of a novel CoRC, which allows the ancestral regulatory network to split into two or more daughter networks, producing what the authors call sister cell types (Arendt et al., 2016). This notion of "genetic individuation" allows a cell type to diverge from its relatives over evolutionary time, even if its morphology, function, or developmental origin appears similar to others. The authors emphasize that cell type homology should be traced not by outward characteristics but by shared regulatory architecture. Thus, two cells that perform similar roles or look alike may not be homologous if their CoRCs are unrelated, whereas cells that appear divergent may in fact descend from a common regulatory ancestor. This evolutionary definition offers several key advantages. It provides clarity for tracing cell type homology across species, especially when phenotypic traits are misleading due to convergence or loss. By grounding identity in specific molecular mechanisms, it lends analytical precision and opens the door to systematic cross-species comparisons. It also aligns with the current direction of single-cell and multiomics research, which increasingly relies on transcriptional and regulatory signatures to define and compare cell states. In any case, by shifting the focus from what cells look like or where they come from to how their identities are encoded and inherited, this perspective has the potential to transform how we understand cellular diversity, both within species and across the tree of life.

However, this constructive view of cell types as minimal evolutionary units still stands in contrast to an alternative, and quite influential, perspective: gene expression modules (responsible for individual and discrete functions and features), rather than entire cells or their full GRNs, constitute the true minimal units of evolutionary change, selection, and conservation. In this modular view, evolution acts not primarily at the level of whole cell types, but through the reuse, recombination, and modification of smaller regulatory subcircuits – each responsible for specific cellular functions, such as secretion, motility, adhesion, or metabolism. These functional modules can be deployed in different contexts and cell types, allowing evolutionary innovation through the flexible reorganization of existing building blocks.

Having said that, I would like to note that there must be more of hierarchically upstream units of evolution at different levels and scales: cell types, tissues or organs, individual organisms, populations, and, finally, ecosystems.

It is plausible, and indeed likely, that full cell type-specific regulatory networks, including CoRCs, are hierarchically organized, incorporating multiple semi-autonomous gene regulatory subcircuits that control the expression of distinct functional modules. Within this architecture, the CoRC can be seen not as a monolithic determinant of identity, but as a higher-order integrator or coordinator of modular elements. This hierarchical organization allows for both stability and plasticity: the overarching identity of a cell type can be maintained, while individual modules can be added, removed, or repurposed in response to selective pressures or developmental constraints.

In this light, the evolution of new cell types may often proceed not through the invention of entirely novel regulatory systems but through modular recombination and co-option (Figure 3.10) (McQueen and Rebeiz, 2020).

A module responsible for, say, pigment production or neurotransmitter synthesis might be activated in a new cellular context through relatively minor regulatory changes. Such modular flexibility provides a powerful mechanism for evolutionary

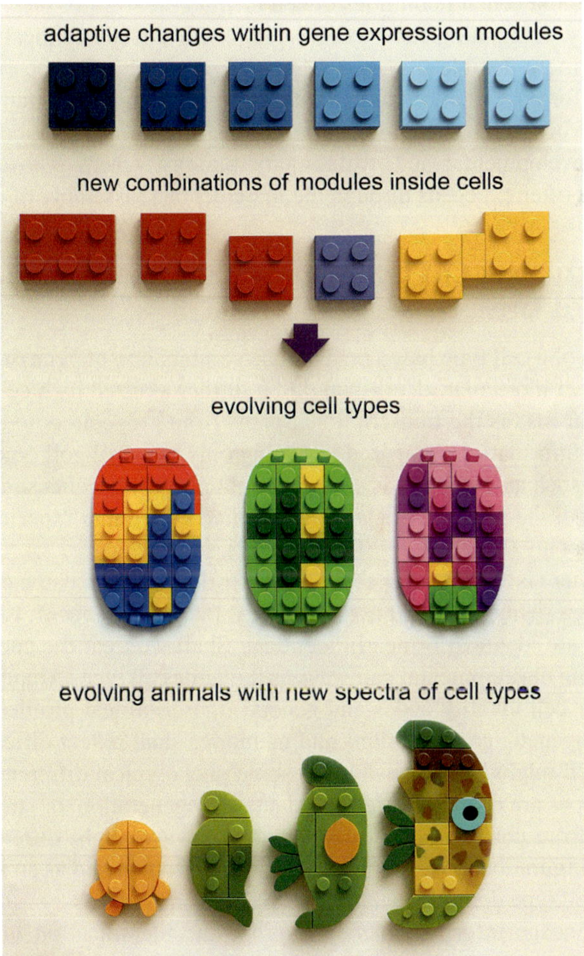

FIGURE 3.10 Evolution of cell identities and new functions. The evolutionary process may tackle module-level organization by diversifying homologous modules – shown as variations of the same basic color of Lego blocks. Also, the formation of new module sets in cells via co-options helps to achieve new functional combinations and repurposing of programs to new locations and cell identities.

innovation, enabling the emergence of new cell types through the shuffling and redeployment of existing genetic circuits rather than through wholesale rewiring.

This perspective does not necessarily contradict the CoRC-based model of cell type identity (Arendt et al., 2016), but rather complements it by emphasizing the combinatorial and layered nature of gene regulatory architecture. It suggests that the evolutionary history of a cell type is not only recorded in the lineage of its core regulatory program but also in the mosaic of functional modules it integrates. Understanding this interplay between modular subcircuits and higher-order regulatory structures may be key to reconstructing the pathways through which cellular diversity evolved, and to resolving apparent contradictions between morphological, functional, and molecular definitions of cell identity.

In summary, integrating the modular view with the hierarchical CoRC model might offer a more flexible framework for understanding cell type evolution – one that accounts for both the deep conservation of core identity programs and the creative potential of modular recombination. Such a framework may better reflect the realities of developmental and evolutionary biology, where innovation is rarely wholesale, but often proceeds through the incremental reassembly of existing parts.

3.6.2 Genealogical Cell Lineage Information Improves the Taxonomy of Cell Types

The definition of a cell type based on a unique combination of gene expression modules and distinct molecular identity might be further assisted by a cell lineage transcriptional analysis, as the transcriptional paths from progenitors to definitive cells provide more information compared to having only terminal cell types in comparison. Invoking such an ontogenesis concept might provide benefits somewhat similar to the benefits of the phylogenetic analysis applied to defining "species" in classical zoology and botany (Figure 3.11) (Doyle, 2022).

However, there is always an exception to the rule. In our case, the comparisons of terminal lineage segments might reveal cases of fate convergences, when the resulting cell types are identical or nearly identical. Shall we treat the cell types arising from converging developmental pathways as the same cell type (Arendt et al., 2016)? Although their cell identity codes and general transcriptional profiles might be the same, their epigenetic configurations and memories shall reflect differences in their cell history, and might be responsible for some phenotypical differences when convergent cell types are challenged (during damage, regeneration, or stress). Thus, cell types arising from converging developmental pathways might differ only by their epigenetic configurations, which, potentially, can be considered as an important confounder to a cell type definition.

Although lineage information may improve the identification of terminal cell types, there is a significant issue with defining developmental cell types via a "discrete molecular identity" approach that I sketched in the previous section. This issue is based on the fact that developmental lineage trajectories, when analyzed in a transcriptional space, appear uninterrupted (Kastriti et al., 2022; Wagner et al., 2018). A zygote is connected to a spectrum of terminal cell types by a smooth transition of

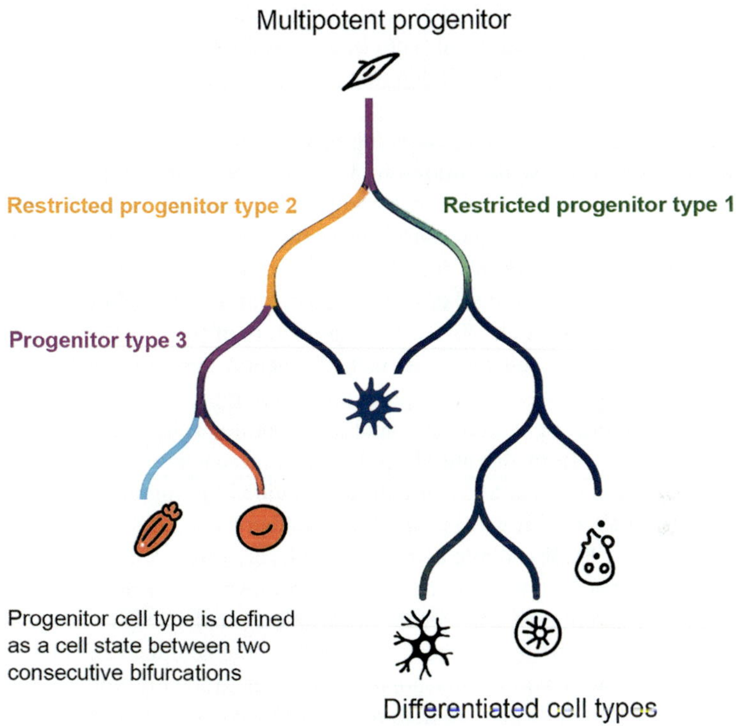

FIGURE 3.11 In development, the uninterrupted cell states can be discretized by fate split points, with intersegments between such points being referenced as developmental cell types. Such lineage-dependent approach also helps to define and resolve adult terminal cell types – via their connections to specific branches. This might highlight the formal cases of terminal fates converging from independent and different lineages. For instance, in the case of developing vertebrate face, the neural crest and mesoderm convergently produce skeletogenic cell types (Noden and Trainor, 2005).

transcriptional states with numerous branching points leading to specific families of cell types (Farrell et al., 2018). This picture is somewhat similar to a phylogenetic tree, where the extant species are not truly discrete, but all are connected via their evolutionary history, where the extinct intermediate forms shaped the uninterrupted continuum of adapting and changing animals. Therefore, both in evolution and in ontogeny, the terminal points on a tree appear discrete only because the intermediate forms are lost, and the continuity of a tree is not perceived at the single slice of evolutionary time or at the end of embryonic development. However, progenitor cells would then somehow have to be conceived as branches of this tree, raising the question: can we distinguish and define progenitor cell types using the same approach as for the terminal cell types? In the case of progenitor cells, molecular codes evolve smoothly, as switching between sets of transcriptional factors and their downstream programs takes time, which depends on RNA and protein synthesis/degradation curve (Kastriti et al., 2022; Soldatov et al., 2019). Thus, placing a separation line between consecutive progenitor cell types in a lineage becomes an arbitrary and non-objective task. Thus, the problem of defining progenitor cell types requires a different approach.

Let's dive first into the history of developmental biology and classic definitions of progenitor cell types. Those definitions are largely based on the anatomical position, a number of molecular markers, and the role in the lineage hierarchy (including potency). For instance, embryonic neural crest cells are defined as multipotent migratory cells delaminating from the dorsal neural tube and expressing Sox10, Foxd3, and a few other well-known marker genes (Kastriti et al., 2022; Simões-Costa and Bronner, 2015; Soldatov et al., 2019). The real picture of the neural crest cells is far from seeing them as "a cell type." In fact, the neural crest cannot be accurately described as a single, well-defined progenitor cell type. Rather, it represents a dynamic and evolving population that undergoes continuous transcriptional and phenotypic transitions – from the initial epithelial-to-mesenchymal transformation to differentiation into a diverse array of cell fates, guided by migratory pathways and local environmental cues (Kastriti et al., 2022; Soldatov et al., 2019; Tatarakis et al., 2021; Thiery et al., 2023b). While neural crest cells are often defined by their multipotency, in a manner analogous to canonical germ layers, they nonetheless experience progressive phenotype change, fate restrictions, and lineage bias throughout development, all while being collectively labeled as "the neural crest." In reality, the transcriptional changes within this population occur gradually and without clear demarcations, making it difficult to determine when a cell ceases to be a neural crest cell and becomes something else.

This ambiguity underscores a broader conceptual issue: like mesoderm, ectoderm, and endoderm, the neural crest encompasses a continuum of molecular and phenotypic states, yet is often treated as a uniform entity. In fact, the neural crest is more accurately described as a lineage – a trajectory through developmental space – rather than a single progenitor cell type. This example highlights the pressing need for a revised framework to define progenitor cell types and their boundaries, one that reflects the biological reality of smooth, continuous transitions rather than relying on traditional or historically convenient classifications.

Here, single-cell multiomics holds particular promise. By integrating transcriptomic, epigenomic, and proteomic data at single-cell resolution, this approach offers the possibility of identifying new, biologically grounded and universal criteria for defining progenitor states. Such insights could help reshape our understanding of cell identity, moving away from static, categorical definitions toward a more nuanced, dynamic, and data-driven view of cellular development.

Although the trajectories obtained with single-cell transcriptomics are smooth, they are naturally discretized by fate splits (Kastriti et al., 2022; Soldatov et al., 2019). After the fate split, new gene modules become activated, rendering the next lineage intersegment different from the previous one. For instance, in the case of the neural crest, the first fate split results in a separation of the trajectory segment leading to somatosensory neurons (Soldatov et al., 2019). In that segment, a cascade of specific neurogenic genes becomes activated after the bifurcation point demarcating the new segment, whereas the neural crest transcriptional factors from previous intersegments persist even after the bifurcation point for some time before they become downregulated. Therefore, the total combination of the modules in this intersegment (old plus new) appears unique for the entire neural crest development.

Overall, a concept of defining a progenitor cell type as an intersegment between fate splits makes biological sense (Figure 3.11). This perspective aligns with an evolutionary framework: intermediate states that are evolutionarily conserved and stable across species might be considered stable and homologous progenitor cell types (Arendt et al., 2016). Transient states unique to a single species, especially those not preserved through evolutionary time, are better viewed as species-specific cell states rather than stable progenitor cell types. However, this is highly debatable, as the degree of evolutionary conservation and comparative scale will inevitable influence such assessments.

Intuitively, such a concept stipulates that a progenitor cell type is biologically equivalent to a short period in developmental progression, which prepares the cell for a choice of the next-in-line fate-related option. During such a period, a unique set of modules is upregulated to mediate future choices and to perform the progenitor's functions. In agreement with this logic, previous studies revealed that the competing gene expression programs are co-activated in the intersegment prior to a bifurcation. When the cells approach the bifurcation, these opposing modules undergo competition and cross-repression to ultimately bias a population of progenitor cells toward different post-bifurcation outcomes implemented in the next intersegments. Thus, the nature of a trajectory intersegment is a preparation for the next upcoming bifurcation, and during such an "intersegment period," the cell positioning, signaling interactions, and epigenetic changes provide for an effective biasing or even commitment to a specific future fate. Although the commitment can be achieved before the bifurcation for some of the biased progenitors, the actual fate transition cannot be implemented before all differentially biased progenitors of this intersegment reach the bifurcation area or "approximated point" – independently of a particular choice, the intersegment needs to be completed.

The consequences of adopting this new scheme for redefining the progenitor cell types will be quite dramatic, as it will split the previously known progenitors into finer subgroups and also into new progenitor subtypes with "shifted borders" as compared to classical morphological/positional definitions. In principle, this process can be harmonized by keeping both systems in place or by adopting nested hierarchical categories, such as cell type families and orders, for better fitting to classical definitions.

For example, it is possible to use the cell lineage concept to organize terminal cell types into families of the first order that originate from the last fate split and belong to terminal intersegments connected by the lineage tree as closest neighbors via one hub. Then, a wider group of terminal cell types can be connected into a family of the second order that are connected via two or more fate splits down the lineage tree, and so on. Thus, we might end up with a system of cell type families (Arendt et al., 2019), classes, and orders, similar to how the animal species are organized in a systematic way. Implementing such a new "single-cell-based" system into practical use will necessarily require exceptionally high-resolution single-cell multiomics atlases instead of the currently available atlases, which are far from being saturated in terms of cell lineage resolution and sequencing depth. Also, despite being useful, this approach will have to take into the account multiple exceptions, especially when distant and evolutionary poorly related cell types become co-opted into the same developmental lineages (for example, cartilage and neurons arise from the neural crest).

Overall, looking at the current progress in single-cell multiomics (Lim et al., 2024), it is getting clear that the wealth of new knowledge will challenge the old definitions and taxonomy of cell types. This problem is not semantic in nature, and rather brings up a practical necessity to define cell types for evolutionary comparisons and for biomedical applications. The modular gene expression and hierarchy of transcriptional control are expected to assist future refinement of a cell type concept based on single-cell omics studies. Evolutionary comparisons will help to identify conserved cell type-specific modules and molecular cell identity codes, and those in return will provide new metrics for comparing cell types from different organisms. The pathological cell types might enter the same classification scheme via identification of discrete pathology-related modules. The concept of comparing gene expression modules will help to define cell subtypes based on persistent spatial variation, whereas cell states might be identified via modules with unstable temporal pattern of expression. Finally, the continuity of transcriptional changes along the lineage tree can be discretized based on intersegments separated by fate splits, and those intersegments might be rendered as progenitor cell types with unique combinations of gene expression modules (altogether being locked between two fate selection episodes on a temporal scale). Overall, the cell type taxonomy might embrace a hierarchical structure with cell type families of the first, second, and other orders based on the proximity of cell types on a lineage tree measures as separation via one, two, or more fate splits.

3.7 SUMMARY OF RECENT TRENDS

Overall, the current trends render the process of cell fate selection as a dynamic, probabilistic, and context-sensitive computation embedded within the fabric of GRNs. Emerging knowledge – ranging from competing genetic circuits and multistable transcriptional dynamics to probabilistic sampling and noise-mediated transitions – highlights the richness and flexibility of intracellular decision-making. These mechanisms collectively equip cells with the ability to explore multiple trajectories, respond to fluctuating environments, and achieve robust outcomes despite inherent biological noise. Such findings challenge traditional frameworks that view cell fate decisions as binary switches triggered by singular cues, replacing them with a paradigm of iterative, integrative processing akin to neural computation or Bayesian inference. This shift has profound implications for how we define and classify cell types and states. The classic view of progenitor cell types as discrete, functionally specialized entities is strained by the continuous and reversible nature of many developmental transitions, particularly in progenitor populations that display overlapping transcriptional programs and gradual bifurcations rather than sharp boundaries. Likewise, the notion of a fixed cell identity becomes increasingly ambiguous in the context of developmental plasticity and reprogramming, all of which underscore the modular and dynamic organization of gene expression programs. The challenge now lies in developing frameworks that can accommodate this fluidity – integrating molecular, functional, and temporal dimensions to define cell identity in a way that is both precise and biologically meaningful.

The future of studying and modeling cell fate decisions during embryonic development is poised for a conceptual and technological renaissance. One conceptual shift already underway is the move from static Waddington-like landscapes to dynamic, adaptive developmental topologies. Classic models treat fate decisions as valleys and ridges shaped by GRNs, but newer insights suggest that these landscapes are not fixed. Instead, they are dynamically reshaped by internal feedback loops, external signaling, metabolic status, noise and variation, and even mechanical cues. Modeling fate decisions as dynamically modifiable noisy manifolds – potentially using tools from dynamical systems theory and AI – could provide deeper, predictive insights into cell behavior. Another conceptual shift will be the deepening analogy between biological and artificial computation. Rather than seeing cells as simple "switches," many of us are starting to treat them as information-processing agents capable of integrating past and present inputs (memory), evaluating probabilities (inference), and adapting decisions through feedback (learning). Concepts from neural networks, Bayesian decision theory, and even reinforcement learning could reshape how we think about fate choices and information processing, especially in the context of noisy and changing environments.

While single-cell models have been incredibly useful, there is growing appreciation that fate decisions are often emergent properties of multicellular collectives. Cells influence each other through lateral inhibition, morphogen gradients, and mechanical coupling. Understanding how collective multipotency operates – where groups of cells collectively bias and refine each other's fate outcomes – may require new models inspired by swarm intelligence or distributed computing. This emerging perspective naturally integrates time as a core dimension. Fate selection is increasingly understood not as an instantaneous event but as a process that unfolds over time through co-activation of competing transcriptional programs, gradual biasing, and eventual commitment.

To truly understand these processes, new experimental capacities will be required. One promising direction is the development of *in vivo* molecular typewriters – biological "flight data recorders" that write a cell's transcriptional or signaling history into DNA or RNA barcodes (Choi et al., 2022). These could be read later to reconstruct the sequence of inputs and decisions a cell made throughout development. Advances in biosensors will likely enable the parallel medium-throughput real-time tracking of chromatin states, protein interactions, or enhancer activity inside living embryos, without the need to disrupt or destroy the tissue. In a distant future, these rich datasets will open the door to constructing "digital twin" embryos – simulated developmental systems that integrate empirical data with predictive models. Much like digital twins in engineering, these systems could be used to forecast developmental outcomes, identify failure points in morphogenesis, or test genetic or environmental perturbations before they are applied in vivo.

However, the increasing richness and complexity of descriptive data pose a significant challenge: paradoxically, the more data we accumulate, the harder it becomes to model the system's behavior in a meaningful way. Building useful models requires more than just data – it demands insight into the underlying causal mechanisms and principles of organization. In order to simplify effectively, we need to identify components or processes that can be treated as functional units, such as modeling the input–output relationships of larger subsystems without needing to account for every molecular detail within them. Achieving this depends on a deeper understanding of the system's modular and hierarchical architecture – specifically, which processes or networks can be abstracted and treated as coherent modules. This issue is discussed in detail in a chapter by Gerhard Schlosser, who directly addresses the importance of recognizing organizational and evolutionary-conserved principles in complex biological systems.

A particularly transformative trajectory lies in the integration of synthetic biology with developmental systems. Rather than passively observing fate decisions, we will increasingly be able to write new regulatory logic into cells. Synthetic transcriptional circuits, programmable enhancers, and modular signaling scaffolds could be inserted into progenitor cells to test how different computational architectures alter fate decisions. For example, we may want to create artificial bistable switches or multi-input logic gates that mimic natural decision nodes. These tools could be used not only to test the logic of endogenous gene networks but to engineer novel differentiation pathways or rescue defective ones. Synthetic oscillators – circuits designed to mimic the fluctuating expression of endogenous fate regulators – could allow precise control over the timing of biasing and commitment steps. Even more radically, synthetic biology may enable the construction of entirely novel developmental programs: organoids that develop new morphologies, synthetic embryos with programmable axis formation, or tissues that regenerate on demand. As we learn to harness the principles of endogenous computation, synthetic biology may evolve from a toolkit for gene expression control into a platform for programmable morphogenesis. Synthetic biology, with its capacity to rewrite the codes of life, could become the ultimate tool for probing this logic. By designing novel fate decision circuits, embedding synthetic feedback systems, and reconstructing early embryogenesis from engineered components, we will not only understand how life builds itself – we may begin to build it back, better.

In this context, organoids become not just models of development, but living computational systems in which hypotheses about fate decisions can be encoded, tested, and revised in real time. Researchers may build developmental "testbeds" – miniaturized embryos where multiple fate pathways are stimulated or suppressed under tightly controlled conditions, allowing us to interrogate the entire decision landscape of a progenitor population.

Finally, looking further ahead, the conceptual boundary between natural and artificial computation may blur entirely. Just as early discoveries in electromagnetism paved the way for understanding excitable tissues, our growing fluency in machine learning, neural networks, and systems design may unlock new interpretations of cell behavior. If we begin to think of genomes and cells as agents in a distributed computation – each integrating multiple signals, drawing on memory (epigenetics), sensing uncertainty, and making context-sensitive choices – then we must also reconsider how we define a "program," a "state," or even a "cell type." In this view, development is no longer the unfolding of a deterministic script written in the genome, but the emergent outcome of a complex, adaptive process – part computation, part self-organization, and part stochastic inference – deeply embedded in space and time.

ACKNOWLEDGMENTS

I would like to thank Prof. Michael Levin and Prof. Gerhard Schlosser for the insightful discussions that provided the foundation for this book chapter.

REFERENCES

Abdel Fattah, A.R., Grebenyuk, S., De Rooij, L.P.M.H., Salmon, I., Ranga, A., 2023. Neuroepithelial organoid patterning is mediated by a neighborhood watch mechanism. *Cell Rep.* 42, 113334. https://doi.org/10.1016/j.celrep.2023.113334

Aboelsoud, R., Kurtz, J., 2019. An HSP90-regulated *reduced-eye* phenotype in *Tribolium* shows fitness benefits and thus provides evidence for evolutionary capacitance. https://doi.org/10.1101/690727

Achille, A., Paolini, G., Soatto, S., 2019. Where is the Information in a Deep Neural Network? https://doi.org/10.48550/ARXIV.1905.12213

Adameyko, I., Bakken, T., Bhaduri, A., Chhatbar, C., Filbin, M.G., Gate, D., Hochgerner, H., Kim, C.N., Krull, J., La Manno, G., Li, Q., Linnarsson, S., Ma, Q., Mayer, C., Menon, V., Nano, P., Prinz, M., Quake, S., Walsh, C.A., Yang, J., Bayraktar, O.A., Gokce, O., Habib, N., Konopka, G., Liddelow, S.A., Nowakowski, T.J., 2024. Applying single-cell and single-nucleus genomics to studies of cellular heterogeneity and cell fate transitions in the nervous system. *Nat. Neurosci.* 27, 2278–2291. https://doi.org/10.1038/s41593-024-01827-9

Arani, E., Triantafillou, S., Kording, K.P., 2019. Reverse engineering neural networks from many partial recordings. https://doi.org/10.48550/ARXIV.1907.01588

Arenas, E.G.S. y G.F., 2025. *The Impact of Triangular-Toothed Gears on the Functionality of the Antikythera Mechanism*. https://doi.org/10.48550/ARXIV.2504.00327

Arendt, D., 2008. The evolution of cell types in animals: emerging principles from molecular studies. *Nat. Rev. Genet.* 9, 868–882. https://doi.org/10.1038/nrg2416

Arendt, D., Bertucci, P.Y., Achim, K., Musser, J.M., 2019. Evolution of neuronal types and families. *Curr. Opin. Neurobiol.* 56, 144–152. https://doi.org/10.1016/j.conb.2019.01.022

Arendt, D., Musser, J.M., Baker, C.V.H., Bergman, A., Cepko, C., Erwin, D.H., Pavlicev, M., Schlosser, G., Widder, S., Laubichler, M.D., Wagner, G.P., 2016. The origin and evolution of cell types. *Nat. Rev. Genet.* 17, 744–757. https://doi.org/10.1038/nrg.2016.127

Arumugam, K., Shin, W., Schiavone, V., Vlahos, L., Tu, X., Carnevali, D., Kesner, J., Paull, E.O., Romo, N., Subramaniam, P., Worley, J., Tan, X., Califano, A., Cosma, M.P., 2020. The master regulator protein BAZ2B can reprogram human hematopoietic lineage-committed progenitors into a multipotent state. *Cell Rep.* 33, 108474. https://doi.org/10.1016/j.celrep.2020.108474

Badia-i-Mompel, P., Wessels, L., Müller-Dott, S., Trimbour, R., Ramirez Flores, R.O., Argelaguet, R., Saez-Rodriguez, J., 2023. Gene regulatory network inference in the era of single-cell multi-omics. *Nat. Rev. Genet.* 24, 739–754. https://doi.org/10.1038/s41576-023-00618-5

Baek, D.D., Tegmark, M., 2025. *Towards Understanding Distilled Reasoning Models: A Representational Approach*. https://doi.org/10.48550/ARXIV.2503.03730

Baggiolini, A., Varum, S., Mateos, J.M., Bettosini, D., John, N., Bonalli, M., Ziegler, U., Dimou, L., Clevers, H., Furrer, R., Sommer, L., 2015. Premigratory and migratory neural crest cells are multipotent in vivo. *Cell Stem Cell* 16, 314–322. https://doi.org/10.1016/j.stem.2015.02.017

Bakken, T., Cowell, L., Aevermann, B.D., Novotny, M., Hodge, R., Miller, J.A., Lee, A., Chang, I., McCorrison, J., Pulendran, B., Qian, Y., Schork, N.J., Lasken, R.S., Lein, E.S., Scheuermann, R.H., 2017. Cell type discovery and representation in the era of high-content single cell phenotyping. *BMC Bioinform.* 18, 559. https://doi.org/10.1186/s12859-017-1977-1

Balaban, N.Q., Merrin, J., Chait, R., Kowalik, L., Leibler, S., 2004. Bacterial persistence as a phenotypic switch. *Science* 305, 1622–1625. https://doi.org/10.1126/science.1099390

Balázsi, G., van Oudenaarden, A., Collins, J.J., 2011. Cellular decision making and biological noise: from microbes to mammals. *Cell* 144, 910–925. https://doi.org/10.1016/j.cell.2011.01.030

Barua, A., Hatzikirou, H., 2023. Cell decision making through the lens of bayesian learning. *Entropy* 25, 609. https://doi.org/10.3390/e25040609

Blount, Z.D., Borland, C.Z., Lenski, R.E., 2008. Historical contingency and the evolution of a key innovation in an experimental population of *Escherichia coli*. *Proc. Natl. Acad. Sci.* 105, 7899–7906. https://doi.org/10.1073/pnas.0803151105

Boije, H., Rulands, S., Dudczig, S., Simons, B.D., Harris, W.A., 2015. The independent probabilistic firing of transcription factors: a paradigm for clonal variability in the Zebrafish Retina. *Dev. Cell* 34, 532–543. https://doi.org/10.1016/j.devcel.2015.08.011

Bravo González-Blas, C., De Winter, S., Hulselmans, G., Hecker, N., Matetovici, I., Christiaens, V., Poovathingal, S., Wouters, J., Aibar, S., Aerts, S., 2023. SCENIC+: single-cell multiomic inference of enhancers and gene regulatory networks. *Nat. Methods* 20, 1355–1367. https://doi.org/10.1038/s41592-023-01938-4

Cai, W., Zhou, W., Han, Z., Lei, J., Zhuang, J., Zhu, P., Wu, X., Yuan, W., 2020. Master regulator genes and their impact on major diseases. *PeerJ* 8, e9952. https://doi.org/10.7717/peerj.9952

Cao, S., 2023. A review of decision theory and methods, *Front. Soc. Sci. Technol.* 5. https://doi.org/10.25236/FSST.2023.050311

Cardelli, L., 2005. Abstract machines of systems biology, in: Priami, C., Merelli, E., Gonzalez, P., Omicini, A. (Eds.), *Transactions on Computational Systems Biology III*, Lecture Notes in Computer Science. Springer Berlin Heidelberg, Berlin, Heidelberg, pp. 145–168. https://doi.org/10.1007/11599128_10

Carter, B., Zhao, K., 2021. The epigenetic basis of cellular heterogeneity. *Nat. Rev. Genet.* 22, 235–250. https://doi.org/10.1038/s41576-020-00300-0

Chen, R.X.F., 2016. A brief introduction to Shannon's information theory. https://doi.org/10.48550/ARXIV.1612.09316

Choi, J., Chen, W., Minkina, A., Chardon, F.M., Suiter, C.C., Regalado, S.G., Domcke, S., Hamazaki, N., Lee, C., Martin, B., Daza, R.M., Shendure, J., 2022. A time-resolved, multi-symbol molecular recorder via sequential genome editing. *Nature* 608, 98–107. https://doi.org/10.1038/s41586-022-04922-8

Clevers, H., What is your conceptual definition of "cell type" in the context of a mature organism?, 2017. *Cell Syst.* 4, 255–259. https://doi.org/10.1016/j.cels.2017.03.006

Crabbé, J., van der Schaar, M., 2022. Concept activation regions: a generalized framework for concept-based explanations. https://doi.org/10.48550/ARXIV.2209.11222

Crispatzu, G., Rehimi, R., Pachano, T., Bleckwehl, T., Cruz-Molina, S., Xiao, C., Mahabir, E., Bazzi, H., Rada-Iglesias, A., 2021. The chromatin, topological and regulatory properties of pluripotency-associated poised enhancers are conserved in vivo. *Nat. Commun.* 12, 4344. https://doi.org/10.1038/s41467-021-24641-4

Crow, J.F., Kimura, M., 1970. *An Introduction to Population Genetics Theory*. Burgess Pub. Co., Minneapolis.

Cuvier, G., Thomas, H., Laurent, G., 1985. *Discours sur les révolutions de la surface du globe et sur les changements qu'elles ont produits dans le règne animal, Collection "Épistémè."* Bourgois, Paris.

Davis, E., 2021. Deep Learning and Mathematical Intuition: A Review of (Davies et al. 2021). https://doi.org/10.48550/ARXIV.2112.04324

Davis, T.L., Rebay, I., 2017. Master regulators in development: Views from the *Drosophila* retinal determination and mammalian pluripotency gene networks. *Dev. Biol.* 421, 93–107. https://doi.org/10.1016/j.ydbio.2016.12.005

De Robertis, E.M., 2006. Spemann's organizer and self-regulation in amphibian embryos. *Nat. Rev. Mol. Cell Biol.* 7, 296–302. https://doi.org/10.1038/nrm1855

Doyle, J.J., 2022. Cell types as species: Exploring a metaphor. *Front. Plant Sci.* 13, 868565. https://doi.org/10.3389/fpls.2022.868565

Efstathiou, K., Efstathiou, M., Basiakoulis, A., Kokkinos, N., 2021. The Antikythera mechanism: The prove of the accuracy of the astronomical calculations based on it. *Heritage* 4, 3848–3878. https://doi.org/10.3390/heritage4040211

Elowitz, M.B., Levine, A.J., Siggia, E.D., Swain, P.S., 2002. Stochastic gene expression in a single cell. *Science* 297, 1183–1186. https://doi.org/10.1126/science.1070919

Emmert-Streib, F., Dehmer, M., Haibe-Kains, B., 2014. Gene regulatory networks and their applications: Understanding biological and medical problems in terms of networks. *Front. Cell Dev. Biol.* 2. https://doi.org/10.3389/fcell.2014.00038

Erickson, A.G., Isaev, S., Artemov, A., He, J., Semsch, B., Murtazina, A., Sun, J., Mangold, K., Chalou, A., Frisen, J., Ratz, M., Andersson, E., Kharchenko, P.V., Adameyko, I., 2024. Unbiased profiling of multipotency landscapes reveals spatial modulators of clonal fate biases. https://doi.org/10.1101/2024.11.15.623687

Erickson, A.G., Kameneva, P., Adameyko, I., 2023. The transcriptional portraits of the neural crest at the individual cell level. *Semin. Cell Dev. Biol.* 138, 68–80. https://doi.org/10.1016/j.semcdb.2022.02.017

Fares, M.A., Ruiz-González, M.X., Moya, A., Elena, S.F., Barrio, E., 2002. GroEL buffers against deleterious mutations. *Nature* 417, 398–398. https://doi.org/10.1038/417398a

Farrell, J.A., Wang, Y., Riesenfeld, S.J., Shekhar, K., Regev, A., Schier, A.F., 2018. Single-cell reconstruction of developmental trajectories during zebrafish embryogenesis. *Science* 360, eaar3131. https://doi.org/10.1126/science.aar3131

Fatieieva, Y., Galimullina, R., Isaev, S., Klimovich, A., Lemaire, L.A., Adameyko, I., 2025. Melanocytes and photosensory organs share a common ancestry that illuminates the origins of the neural crest. *Commun. Biol.* 8, 1092. https://doi.org/10.1038/s42003-025-08502-0

Faunes, M., Francisco Botelho, J., Ahumada Galleguillos, P., Mpodozis, J., 2015. On the hodological criterion for homology. *Front. Neurosci.* 9. https://doi.org/10.3389/fnins.2015.00223

Faure, L., Soldatov, R., Kharchenko, P.V., Adameyko, I., 2023. scFates: a scalable python package for advanced pseudotime and bifurcation analysis from single-cell data. *Bioinformatics* 39, btac746. https://doi.org/10.1093/bioinformatics/btac746

Feil, R., Wagner, J., Metzger, D., Chambon, P., 1997. Regulation of Cre recombinase activity by mutated estrogen receptor ligand-binding domains. *Biochem. Biophys. Res. Commun.* 237, 752–757. https://doi.org/10.1006/bbrc.1997.7124

Galbraith, M., Bocci, F., Onuchic, J.N., 2022. Stochastic fluctuations promote ordered pattern formation of cells in the Notch-Delta signaling pathway. *PLOS Comput. Biol.* 18, e1010306. https://doi.org/10.1371/journal.pcbi.1010306

Gangaraju, V.K., Yin, H., Weiner, M.M., Wang, J., Huang, X.A., Lin, H., 2011. *Drosophila* Piwi functions in Hsp90-mediated suppression of phenotypic variation. *Nat. Genet.* 43, 153–158. https://doi.org/10.1038/ng.743

Garcia-Alonso, L., Holland, C.H., Ibrahim, M.M., Turei, D., Saez-Rodriguez, J., 2019. Benchmark and integration of resources for the estimation of human transcription factor activities. *Genome Res.* 29, 1363–1375. https://doi.org/10.1101/gr.240663.118

Gonzalez-Ferrer, J., Lehrer, J., O'Farrell, A., Paten, B., Teodorescu, M., Haussler, D., Jonsson, V.D., Mostajo-Radji, M.A., 2024. SIMS: A deep-learning label transfer tool for single-cell RNA sequencing analysis. *Cell Genomics* 4, 100581. https://doi.org/10.1016/j.xgen.2024.100581

Grabowski, F., Czyż, P., Kochańczyk, M., Lipniacki, T., 2019. Limits to the rate of information transmission through the MAPK pathway. *J. R. Soc. Interface* 16, 20180792. https://doi.org/10.1098/rsif.2018.0792

Gregor, T., Tank, D.W., Wieschaus, E.F., Bialek, W., 2007. Probing the limits to positional information. *Cell* 130, 153–164. https://doi.org/10.1016/j.cell.2007.05.025

Griswold, C.K., Masel, J., 2009. Complex adaptations can drive the evolution of the capacitor [PSI+], even with realistic rates of yeast sex. *PLoS Genet.* 5, e1000517. https://doi.org/10.1371/journal.pgen.1000517

Grossmann, N., Gröller, E., Waldner, M., 2022. Concept splatters: Exploration of latent spaces based on human interpretable concepts. *Comput. Graph.* 105, 73–84. https://doi.org/10.1016/j.cag.2022.04.013

Grozinger, L., Miró-Bueno, J., Goñi-Moreno, Á., 2025. Genetic designs for stochastic and probabilistic biocomputing. *Phys. Rev. E* 111, 054412. https://doi.org/10.1103/PhysRevE.111.054412

Hartl, B., Levin, M., 2025. What does evolution make? Learning in living lineages and machines. *Trends Genet.* 41, 480–496. https://doi.org/10.1016/j.tig.2025.04.002

Hartree, D.R., 1940. The bush differential analyser and its applications *. *Nature* 146, 319–323. https://doi.org/10.1038/146319a0

Haviv, D., Remšík, J., Gatie, M., Snopkowski, C., Takizawa, M., Pereira, N., Bashkin, J., Jovanovich, S., Nawy, T., Chaligne, R., Boire, A., Hadjantonakis, A.-K., Pe'er, D., 2025. The covariance environment defines cellular niches for spatial inference. *Nat. Biotechnol.* 43, 269–280. https://doi.org/10.1038/s41587-024-02193-4

He, J., Zhang, G., Almeida, A.D., Cayouette, M., Simons, B.D., Harris, W.A., 2012. How variable clones build an invariant retina. *Neuron* 75, 786–798. https://doi.org/10.1016/j.neuron.2012.06.033

Hilscher, M.M., Langseth, C.M., Kukanja, P., Yokota, C., Nilsson, M., Castelo-Branco, G., 2022. Spatial and temporal heterogeneity in the lineage progression of fine oligodendrocyte subtypes. *BMC Biol.* 20, 122. https://doi.org/10.1186/s12915-022-01325-z

Hornischer, L., 2021. *Dynamical Systems via Domains: toward a Unified Foundation of Symbolic and Non-symbolic Computation*. Institute for Logic, Language and Computation, Amsterdam.

Huang, S., 2009. Non-genetic heterogeneity of cells in development: more than just noise. *Development* 136, 3853–3862. https://doi.org/10.1242/dev.035139

Huang, S., Guo, Y.-P., May, G., Enver, T., 2007. Bifurcation dynamics in lineage-commitment in bipotent progenitor cells. *Dev. Biol.* 305, 695–713. https://doi.org/10.1016/j.ydbio.2007.02.036

Huneman, P., 2013. Developmental biology, classical Sea Urchin experiments, in: Dubitzky, W., Wolkenhauer, O., Cho, K.-H., Yokota, H. (Eds.), *Encyclopedia of Systems Biology*. Springer New York, NY, pp. 566–567. https://doi.org/10.1007/978-1-4419-9863-7_929

Jonas, E., Kording, K.P., 2017. Could a neuroscientist understand a microprocessor? *PLOS Comput. Biol.* 13, e1005268. https://doi.org/10.1371/journal.pcbi.1005268

Joyner, A.L., Zervas, M., 2006. Genetic inducible fate mapping in mouse: Establishing genetic lineages and defining genetic neuroanatomy in the nervous system. *Dev. Dyn.* 235, 2376–2385. https://doi.org/10.1002/dvdy.20884

Kalcheim, C., 2000. Mechanisms of early neural crest development: From cell specification to migration, in: *International Review of Cytology*. Elsevier, pp. 143–196.

Kamenev, D., Kameneva, P., Adameyko, I., 2025. The role of microheterogeneity in cell fate decisions in neural progenitors and neural crest. *Curr. Opin. Neurobiol.* 92, 103031. https://doi.org/10.1016/j.conb.2025.103031

Kamimoto, K., Stringa, B., Hoffmann, C.M., Jindal, K., Solnica-Krezel, L., Morris, S.A., 2023. Dissecting cell identity via network inference and in silico gene perturbation. *Nature* 614, 742–751. https://doi.org/10.1038/s41586-022-05688-9

Karlebach, G., Shamir, R., 2008. Modelling and analysis of gene regulatory networks. *Nat. Rev. Mol. Cell Biol.* 9, 770–780. https://doi.org/10.1038/nrm2503

Kastriti, M.E., Faure, L., Von Ahsen, D., Bouderlique, T.G., Boström, J., Solovieva, T., Jackson, C., Bronner, M., Meijer, D., Hadjab, S., Lallemend, F., Erickson, A., Kaucka, M., Dyachuk, V., Perlmann, T., Lahti, L., Krivanek, J., Brunet, J., Fried, K., Adameyko, I., 2022. Schwann cell precursors represent a neural crest-like state with biased multipotency. *EMBO J.* 41, e108780. https://doi.org/10.15252/embj.2021108780

Kelsh, R.N., Camargo Sosa, K., Farjami, S., Makeev, V., Dawes, J.H.P., Rocco, A., 2021. Cyclical fate restriction: a new view of neural crest cell fate specification. *Development* 148, dev176057. https://doi.org/10.1242/dev.176057

Kimura, M., 1962. On the probability of fixation of mutant genes in a population. *Genetics* 47, 713–719. https://doi.org/10.1093/genetics/47.6.713

Kingma, D.P., Welling, M., 2019. An introduction to variational autoencoders. Found. *Trends® Mach. Learn.* 12, 307–392. https://doi.org/10.1561/2200000056

Knill, D.C., Pouget, A., 2004. The Bayesian brain: the role of uncertainty in neural coding and computation. *Trends Neurosci.* 27, 712–719. https://doi.org/10.1016/j.tins.2004.10.007

Kramer, B.A., Sarabia Del Castillo, J., Pelkmans, L., 2022. Multimodal perception links cellular state to decision-making in single cells. *Science* 377, 642–648. https://doi.org/10.1126/science.abf4062

Krammer, T., Stuart, H.T., Gromberg, E., Ishihara, K., Cislo, D., Melchionda, M., Becerril Perez, F., Wang, J., Costantini, E., Lehr, S., Arbanas, L., Hörmann, A., Neumüller, R.A., Elvassore, N., Siggia, E., Briscoe, J., Kicheva, A., Tanaka, E.M., 2024. Mouse neural tube organoids self-organize floorplate through BMP-mediated cluster competition. *Dev. Cell* 59, 1940–1953.e10. https://doi.org/10.1016/j.devcel.2024.04.021

La Manno, G., Soldatov, R., Zeisel, A., Braun, E., Hochgerner, H., Petukhov, V., Lidschreiber, K., Kastriti, M.E., Lönnerberg, P., Furlan, A., Fan, J., Borm, L.E., Liu, Z., Van Bruggen, D., Guo, J., He, X., Barker, R., Sundström, E., Castelo-Branco, G., Cramer, P., Adameyko, I., Linnarsson, S., Kharchenko, P.V., 2018. RNA velocity of single cells. *Nature* 560, 494–498. https://doi.org/10.1038/s41586-018-0414-6

Le Rouzic, A., Álvarez-Castro, J.M., Hansen, T.F., 2013. The evolution of canalization and evolvability in stable and fluctuating environments. *Evol. Biol.* 40, 317–340. https://doi.org/10.1007/s11692-012-9218-z

Lehr, S., Brückner, D.B., Greunz-Schindler, M., Minchington, T., Merrin, J., Hannezo, E., Kicheva, A., 2023. Self-organised pattern formation in the developing neural tube by a temporal relay of BMP signalling. https://doi.org/10.1101/2023.11.15.567070

Levin, M., 2023a. Collective Intelligence of Morphogenesis as a Teleonomic Process, in: Corning, P.A., Kauffman, S.A., Noble, D., Shapiro, J.A., Vane-Wright, R.I., Pross, A. (Eds.), *Evolution "On Purpose."* The MIT Press, pp. 175–198. https://doi.org/10.7551/mitpress/14642.003.0013

Levin, M., 2023b. Darwin's agential materials: evolutionary implications of multiscale competency in developmental biology. *Cell. Mol. Life Sci.* 80, 142. https://doi.org/10.1007/s00018-023-04790-z

Levy, S.F., Ziv, N., Siegal, M.L., 2012. Bet hedging in yeast by heterogeneous, age-correlated expression of a stress protectant. *PLoS Biol.* 10, e1001325. https://doi.org/10.1371/journal.pbio.1001325

Leyes Porello, E.A., Trudeau, R.T., Lim, B., 2023. Transcriptional bursting: stochasticity in deterministic development. *Development* 150, dev201546. https://doi.org/10.1242/dev.201546

Li, L., Bowling, S., Lin, H., Chen, D., Wang, S.-W., Camargo, F.D., 2025. DARLIN mouse for in vivo lineage tracing at high efficiency and clonal diversity. *Nat. Protoc.* https://doi.org/10.1038/s41596-025-01141-z

Li, Y., Guo, P., 2018. Training neural networks by marginalizing out hidden layer noise. *Neural Comput. Appl.* 29, 401–412. https://doi.org/10.1007/s00521-017-2864-4

Liberali, P., Schier, A.F., 2024. The evolution of developmental biology through conceptual and technological revolutions. *Cell* 187, 3461–3495. https://doi.org/10.1016/j.cell.2024.05.053

Lim, J., Park, C., Kim, M., Kim, H., Kim, J., Lee, D.-S., 2024. Advances in single-cell omics and multiomics for high-resolution molecular profiling. *Exp. Mol. Med.* 56, 515–526. https://doi.org/10.1038/s12276-024-01186-2

Linden, W.V.D., Dose, V., Toussaint, U.V., 2014. *Bayesian Probability Theory: Applications in the Physical Sciences*, 1st ed. Cambridge University Press. https://doi.org/10.1017/CBO9781139565608

Losick, R., Desplan, C., 2008. Stochasticity and cell fate. *Science* 320, 65–68. https://doi.org/10.1126/science.1147888

Manhaeve, R., Marra, G., De Raedt, L., 2021. Approximate Inference for Neural Probabilistic Logic Programming, in: Proceedings of the Eighteenth International Conference on Principles of Knowledge Representation and Reasoning. Presented at the *18th International Conference on Principles of Knowledge Representation and Reasoning {KR-2021}, International Joint Conferences on Artificial Intelligence Organization*, Hanoii, Vietnam, pp. 475–486. https://doi.org/10.24963/kr.2021/45

Mansisidor, A.R., Risca, V.I., 2022. Chromatin accessibility: methods, mechanisms, and biological insights. *Nucleus* 13, 238–278. https://doi.org/10.1080/19491034.2022.2143106

Marblestone, A.H., Wayne, G., Kording, K.P., 2016. Toward an integration of deep learning and neuroscience. *Front. Comput. Neurosci.* 10. https://doi.org/10.3389/fncom.2016.00094

Masel, J., 2013. Q&A: Evolutionary capacitance. *BMC Biol.* 11, 103. https://doi.org/10.1186/1741-7007-11-103

Mason, J.E., Nie, X., Coca, D., Andrews, P.W., 2025. Stem cell fate decisions: Substates and attractors. *Stem Cell Rep.* 20, 102532. https://doi.org/10.1016/j.stemcr.2025.102532

McKenna, A., Findlay, G.M., Gagnon, J.A., Horwitz, M.S., Schier, A.F., Shendure, J., 2016. Whole-organism lineage tracing by combinatorial and cumulative genome editing. *Science* 353, aaf7907. https://doi.org/10.1126/science.aaf7907

McMillen, P., Levin, M., 2024. Collective intelligence: A unifying concept for integrating biology across scales and substrates. *Commun. Biol.* 7, 378. https://doi.org/10.1038/s42003-024-06037-4

McQueen, E., Rebeiz, M., 2020. On the specificity of gene regulatory networks: How does network co-option affect subsequent evolution?, in: *Current Topics in Developmental Biology*. Elsevier, pp. 375–405. https://doi.org/10.1016/bs.ctdb.2020.03.002

Meijer, W.H.M., Sonnen, K.F., 2024. From signalling oscillations to somite formation. *Curr. Opin. Syst. Biol.* 39, 100520. https://doi.org/10.1016/j.coisb.2024.100520

Melo, D., Pallares, L.F., Ayroles, J.F., 2024. Reassessing the modularity of gene co-expression networks using the Stochastic Block Model. *PLOS Comput. Biol.* 20, e1012300. https://doi.org/10.1371/journal.pcbi.1012300

Minoux, M., Holwerda, S., Vitobello, A., Kitazawa, T., Kohler, H., Stadler, M.B., Rijli, F.M., 2017. Gene bivalency at Polycomb domains regulates cranial neural crest positional identity. *Science* 355, eaal2913. https://doi.org/10.1126/science.aal2913

Mitchell, K.J., Cheney, N., 2025. The Genomic Code: the genome instantiates a generative model of the organism. *Trends Genet.* 41, 462–479. https://doi.org/10.1016/j.tig.2025.01.008

Mojtahedi, M., Skupin, A., Zhou, J., Castaño, I.G., Leong-Quong, R.Y.Y., Chang, H., Trachana, K., Giuliani, A., Huang, S., 2016. Cell fate decision as high-dimensional critical state transition. *PLOS Biol.* 14, e2000640. https://doi.org/10.1371/journal.pbio.2000640

Montesinos López, O.A., Montesinos López, A., Crossa, J., 2022. Fundamentals of Artificial Neural Networks and Deep Learning, in: *Multivariate Statistical Machine Learning Methods for Genomic Prediction*. Springer International Publishing, Cham, pp. 379–425. https://doi.org/10.1007/978-3-030-89010-0_10

Moris, N., Pina, C., Arias, A.M., 2016. Transition states and cell fate decisions in epigenetic landscapes. *Nat. Rev. Genet.* 17, 693–703. https://doi.org/10.1038/nrg.2016.98

Morris, S.A., 2019. The evolving concept of cell identity in the single cell era. *Development* 146, dev169748. https://doi.org/10.1242/dev.169748

Müller-Dott, S., Tsirvouli, E., Vázquez, M., Ramirez Flores, R.O., Badia-i-Mompel, P., Fallegger, R., Lægreid, A., Saez-Rodriguez, J., 2023. Expanding the coverage of regulons from high-confidence prior knowledge for accurate estimation of transcription factor activities. https://doi.org/10.1101/2023.03.30.534849

Musser, J.M., Schippers, K.J., Nickel, M., Mizzon, G., Kohn, A.B., Pape, C., Ronchi, P., Papadopoulos, N., Tarashansky, A.J., Hammel, J.U., Wolf, F., Liang, C., Hernández-Plaza, A., Cantalapiedra, C.P., Achim, K., Schieber, N.L., Pan, L., Ruperti, F., Francis, W.R., Vargas, S., Kling, S., Renkert, M., Polikarpov, M., Bourenkov, G., Feuda, R., Gaspar, I., Burkhardt, P., Wang, B., Bork, P., Beck, M., Schneider, T.R., Kreshuk, A., Wörheide, G., Huerta-Cepas, J., Schwab, Y., Moroz, L.L., Arendt, D., 2021. Profiling cellular diversity in sponges informs animal cell type and nervous system evolution. *Science* 374, 717–723. https://doi.org/10.1126/science.abj2949

Noden, D.M., Trainor, P.A., 2005. Relations and interactions between cranial mesoderm and neural crest populations. *J. Anat.* 207, 575–601. https://doi.org/10.1111/j.1469-7580.2005.00473.x

Orhan, A.E., Ma, W.J., 2017. Efficient probabilistic inference in generic neural networks trained with non-probabilistic feedback. *Nat. Commun.* 8, 138. https://doi.org/10.1038/s41467-017-00181-8

Parmigiani, G., 2001. Decision theory: Bayesian, in: *International Encyclopedia of the Social & Behavioral Sciences*. Elsevier, pp. 3327–3334. https://doi.org/10.1016/B0-08-043076-7/00403-4

Pigliucci, M., 2008. Is evolvability evolvable? *Nat. Rev. Genet.* 9, 75–82. https://doi.org/10.1038/nrg2278

Quandt, E.M., Gollihar, J., Blount, Z.D., Ellington, A.D., Georgiou, G., Barrick, J.E., 2015. Fine-tuning citrate synthase flux potentiates and refines metabolic innovation in the Lenski evolution experiment. *eLife* 4, e09696. https://doi.org/10.7554/eLife.09696

Queitsch, C., Sangster, T.A., Lindquist, S., 2002. Hsp90 as a capacitor of phenotypic variation. *Nature* 417, 618–624. https://doi.org/10.1038/nature749

Rackauckas, C., Schilling, T., Nie, Q., 2018. Mean-Independent Noise Control of Cell Fates via Intermediate States. *iScience* 3, 11–20. https://doi.org/10.1016/j.isci.2018.04.002

Raj, A., Rifkin, S.A., Andersen, E., Van Oudenaarden, A., 2010. Variability in gene expression underlies incomplete penetrance. *Nature* 463, 913–918. https://doi.org/10.1038/nature08781

Raj, A., Van Oudenaarden, A., 2008. Nature, nurture, or chance: Stochastic gene expression and its consequences. *Cell* 135, 216–226. https://doi.org/10.1016/j.cell.2008.09.050

Raj, B., Wagner, D.E., McKenna, A., Pandey, S., Klein, A.M., Shendure, J., Gagnon, J.A., Schier, A.F., 2018. Simultaneous single-cell profiling of lineages and cell types in the vertebrate brain. *Nat. Biotechnol.* 36, 442–450. https://doi.org/10.1038/nbt.4103

Regot, S., Macia, J., Conde, N., Furukawa, K., Kjellén, J., Peeters, T., Hohmann, S., De Nadal, E., Posas, F., Solé, R., 2011. Distributed biological computation with multicellular engineered networks. *Nature* 469, 207–211. https://doi.org/10.1038/nature09679

Richardson, M.K., Hanken, J., Gooneratne, M.L., Pieau, C., Raynaud, A., Selwood, L., Wright, G.M., 1997. There is no highly conserved embryonic stage in the vertebrates: implications for current theories of evolution and development. *Anat. Embryol. (Berl.)* 196, 91–106. https://doi.org/10.1007/s004290050082

Riedel-Kruse, I.H., Müller, C., Oates, A.C., 2007. Synchrony dynamics during initiation, failure, and rescue of the segmentation clock. *Science* 317, 1911–1915. https://doi.org/10.1126/science.1142538

Rodriguez-Fraticelli, A.E., Wolock, S.L., Weinreb, C.S., Panero, R., Patel, S.H., Jankovic, M., Sun, J., Calogero, R.A., Klein, A.M., Camargo, F.D., 2018. Clonal analysis of lineage fate in native haematopoiesis. *Nature* 553, 212–216. https://doi.org/10.1038/nature25168

Roeder, I., Glauche, I., 2006. Towards an understanding of lineage specification in hematopoietic stem cells: A mathematical model for the interaction of transcription factors GATA-1 and PU.1. *J. Theor. Biol.* 241, 852–865. https://doi.org/10.1016/j.jtbi.2006.01.021

Rood, J.E., Wynne, S., Robson, L., Hupalowska, A., Randell, J., Teichmann, S.A., Regev, A., 2025. The Human Cell Atlas from a cell census to a unified foundation model. *Nature* 637, 1065–1071. https://doi.org/10.1038/s41586-024-08338-4

Rossi, R., Torelli, S., Ala, P., Weston, W., Morgan, J., Malhotra, J., Muntoni, F., 2023. MyoD-induced reprogramming of human fibroblasts and urinary stem cells in vitro: protocols and their applications. *Front. Physiol.* 14, 1145047. https://doi.org/10.3389/fphys.2023.1145047

Rusin, L.Y., 2023. Evolution of homology: From archetype towards a holistic concept of cell type. *J. Morphol.* 284, e21569. https://doi.org/10.1002/jmor.21569

Rutherford, S.L., Lindquist, S., 1998. Hsp90 as a capacitor for morphological evolution. *Nature* 396, 336–342. https://doi.org/10.1038/24550

Sáez, M., Blassberg, R., Camacho-Aguilar, E., Siggia, E.D., Rand, D.A., Briscoe, J., 2022. Statistically derived geometrical landscapes capture principles of decision-making dynamics during cell fate transitions. *Cell Syst.* 13, 12–28.e3. https://doi.org/10.1016/j.cels.2021.08.013

Sander, K., 1992. Shaking a concept: Hans Driesch and the varied fates of sea urchin blastomeres. *Rouxs Arch. Dev. Biol.* 201, 265–265. https://doi.org/10.1007/BF00592106

Sander, K., 1997. "Mosaic work" and "assimilating effects" in embryogenesis: Wilhelm Roux's conclusions after disabling frog blastomeres, in: *Landmarks in Developmental Biology 1883–1924*. Springer Berlin Heidelberg, Berlin, Heidelberg, pp. 13–15. https://doi.org/10.1007/978-3-642-60492-8_5

Sangster, T.A., Salathia, N., Lee, H.N., Watanabe, E., Schellenberg, K., Morneau, K., Wang, H., Undurraga, S., Queitsch, C., Lindquist, S., 2008. HSP90-buffered genetic variation is common in *Arabidopsis thaliana*. *Proc. Natl. Acad. Sci.* 105, 2969–2974. https://doi.org/10.1073/pnas.0712210105

Schlosser, G., 2021. *Evolutionary Origin of Sensory and Neurosecretory Cell Types*, 1st edition. ed, Vertebrate cranial placodes. CRC Press, Taylor & Francis Group, Boca Raton London New York.

Shao, W.-W., Shao, Q., Xu, H.-H., Qiao, G.-J., Wang, R.-X., Ma, Z.-Y., Meng, W.-W., Yang, Z.-B., Zang, Y.-L., Li, X.-H., 2025. Repetitive training enhances the pattern recognition capability of cultured neural networks. *PLOS Comput. Biol.* 21, e1013043. https://doi.org/10.1371/journal.pcbi.1013043

Shreyas D. K., Srivatsa N. Joshi, S., Kumar, V.H., Venkataramanan, V., Kaliprasad C. S., 2023. A Review on Neural Networks and Its Applications. *J. Comput. Technol. Appl.* https://doi.org/10.37591/jocta.v14i2.1062

Simões-Costa, M., Bronner, M.E., 2015. Establishing neural crest identity: a gene regulatory recipe. *Development* 142, 242–257. https://doi.org/10.1242/dev.105445

Smith, J.J., Antonacci, F., Eichler, E.E., Amemiya, C.T., 2009. Programmed loss of millions of base pairs from a vertebrate genome. *Proc. Natl. Acad. Sci.* 106, 11212–11217. https://doi.org/10.1073/pnas.0902358106

Soldatov, R., Kaucka, M., Kastriti, M.E., Petersen, J., Chontorotzea, T., Englmaier, L., Akkuratova, N., Yang, Y., Häring, M., Dyachuk, V., Bock, C., Farlik, M., Piacentino, M.L., Boismoreau, F., Hilscher, M.M., Yokota, C., Qian, X., Nilsson, M., Bronner, M.E., Croci,L., Hsiao, W.-Y., Guertin, D.A., Brunet, J.-F., Consalez, G.G., Ernfors, P., Fried, K., Kharchenko, P.V., Adameyko, I., 2019. Spatiotemporal structure of cell fate decisions in murine neural crest. *Science* 364, eaas9536. https://doi.org/10.1126/science.aas9536

Steib, N., Büchter, T., Eichler, A., Binder, K., Krauss, S., Böcherer-Linder, K., Vogel, M., Hilbert, S., 2025. How to teach Bayesian reasoning: An empirical study comparing four different probability training courses. *Learn. Instr.* 95, 102032. https://doi.org/10.1016/j.learninstruc.2024.102032

Swift, M., Horns, F., Quake, S.R., 2023. Lineage tracing reveals fate bias and transcriptional memory in human B cells. *Life Sci. Alliance* 6, e202201792. https://doi.org/10.26508/lsa.202201792

Symmons, O., Raj, A., 2016. What's luck got to do with it: Single cells, multiple fates, and biological nondeterminism. *Mol. Cell* 62, 788–802. https://doi.org/10.1016/j.molcel.2016.05.023

Takahashi, K., Yamanaka, S., 2006. Induction of pluripotent stem cells from mouse embryonic and adult fibroblast cultures by defined factors. *Cell* 126, 663–676. https://doi.org/10.1016/j.cell.2006.07.024

Tassinari, R., Olivi, E., Cavallini, C., Taglioli, V., Zannini, C., Marcuzzi, M., Fedchenko, O., Ventura, C., 2023. Mechanobiology: A landscape for reinterpreting stem cell heterogeneity and regenerative potential in diseased tissues. *iScience* 26, 105875. https://doi.org/10.1016/j.isci.2022.105875

Tatarakis, D., Cang, Z., Wu, X., Sharma, P.P., Karikomi, M., MacLean, A.L., Nie, Q., Schilling, T.F., 2021. Single-cell transcriptomic analysis of zebrafish cranial neural crest reveals spatiotemporal regulation of lineage decisions during development. *Cell Rep.* 37, 110140. https://doi.org/10.1016/j.celrep.2021.110140

The Tabula Muris Consortium, Overall coordination, Logistical coordination, Organ collection and processing, Library preparation and sequencing, Computational data analysis, Cell type annotation, Writing group, Supplemental text writing group, Principal investigators, 2018. Single-cell transcriptomics of 20 mouse organs creates a Tabula Muris. *Nature* 562, 367–372. https://doi.org/10.1038/s41586-018-0590-4

Thiagalingam, S., 2020. Epigenetic memory in development and disease: Unraveling the mechanism. *Biochim. Biophys. Acta BBA Rev. Cancer* 1873, 188349. https://doi.org/10.1016/j.bbcan.2020.188349

Thiery, A.P., Buzzi, A.L., Hamrud, E., Cheshire, C., Luscombe, N.M., Briscoe, J., Streit, A., 2023a. scRNA-sequencing in chick suggests a probabilistic model for cell fate allocation at the neural plate border. *eLife* 12, e82717. https://doi.org/10.7554/eLife.82717

Thiery, A.P., Buzzi, A.L., Hamrud, E., Cheshire, C., Luscombe, N.M., Briscoe, J., Streit, A., 2023b. scRNA-sequencing in chick suggests a probabilistic model for cell fate allocation at the neural plate border. *eLife* 12, e82717. https://doi.org/10.7554/eLife.82717

Thomson, J., III. 1876. On an integrating machine having a new kinematic principle, 1876. *Proc. R. Soc. Lond.* 24, 262–265. https://doi.org/10.1098/rspl.1875.0033

Tkačik, G., Gregor, T., 2021. The many bits of positional information. *Development* 148, dev176065. https://doi.org/10.1242/dev.176065

Tkačik, G., Walczak, A.M., 2011. Information transmission in genetic regulatory networks: A review. *J. Phys. Condens. Matter* 23, 153102. https://doi.org/10.1088/0953-8984/23/15/153102

Topolewski, P., Komorowski, M., 2021. Information-theoretic analyses of cellular strategies for achieving high signaling capacity—dynamics, cross-wiring, and heterogeneity of cellular states. *Curr. Opin. Syst. Biol.* 27, 100352. https://doi.org/10.1016/j.coisb.2021.06.003

Veening, J.-W., Smits, W.K., Kuipers, O.P., 2008. Bistability, Epigenetics, and Bet-Hedging in Bacteria. *Annu. Rev. Microbiol.* 62, 193–210. https://doi.org/10.1146/annurev.micro.62.081307.163002

Velik, R., Bruckner, D., 2008. Neuro-symbolic networks: introduction to a new information processing principle, in: 2008 6th IEEE International Conference on Industrial Informatics. Presented at the *2008 6th IEEE International Conference on Industrial Informatics (INDIN), IEEE,* Daejeon, South Korea, pp. 1042–1047. https://doi.org/10.1109/INDIN.2008.4618256

Verd, B., Monk, N.A., Jaeger, J., 2019. Modularity, criticality, and evolvability of a developmental gene regulatory network. *eLife* 8, e42832. https://doi.org/10.7554/eLife.42832

Waddington, C.H., 1942. Canalization of development and the inheritance of acquired characters. *Nature* 150, 563–565. https://doi.org/10.1038/150563a0

Waddington, C.H., 1957 and 2014 (republished). *The Strategy of the Genes*, 0 ed. Routledge. https://doi.org/10.4324/9781315765471

Wagner, D.E., Weinreb, C., Collins, Z.M., Briggs, J.A., Megason, S.G., Klein, A.M., 2018. Single-cell mapping of gene expression landscapes and lineage in the zebrafish embryo. *Science* 360, 981–987. https://doi.org/10.1126/science.aar4362

Wagner, G.P., 2021. Devo-Evo of cell types, in: Nuño De La Rosa, L., Müller, G.B. (Eds.), *Evolutionary Developmental Biology*. Springer International Publishing, Cham, pp. 511–528. https://doi.org/10.1007/978-3-319-32979-6_153

Wagner, G.P., Pavlicev, M., Cheverud, J.M., 2007. The road to modularity. *Nat. Rev. Genet.* 8, 921–931. https://doi.org/10.1038/nrg2267

Walczak, A.M., Mugler, A., Wiggins, C.H., 2012. Analytic methods for modeling stochastic regulatory networks, in: Liu, X., Betterton, M.D. (Eds.), *Computational Modeling of Signaling Networks, Methods in Molecular Biology*. Humana Press, Totowa, NJ, pp. 273–322. https://doi.org/10.1007/978-1-61779-833-7_13

Wang, H., Yang, Y., Liu, J., Qian, L., 2021. Direct cell reprogramming: approaches, mechanisms and progress. *Nat. Rev. Mol. Cell Biol.* 22, 410–424. https://doi.org/10.1038/s41580-021-00335-z

Wang, J., Wang, L., Feng, G., Wang, Y., Li, Y., Li, X., Liu, C., Jiao, G., Huang, C., Shi, J., Zhou, T., Chen, Q., Liu, Z., Li, W., Zhou, Q., 2018. Asymmetric expression of LincGET biases cell fate in two-cell mouse embryos. *Cell* 175, 1887–1901.e18. https://doi.org/10.1016/j.cell.2018.11.039

Wang, Y., Papayova, M., Warren, E., Pears, C.J., 2024. mTORC1 pathway activity biases cell fate choice. *Sci. Rep.* 14, 20832. https://doi.org/10.1038/s41598-024-71298-2

Weinreb, C., Klein, A.M., 2020. Lineage reconstruction from clonal correlations. *Proc. Natl. Acad. Sci.* 117, 17041–17048. https://doi.org/10.1073/pnas.2000238117

Wellmann, J., 2022. History of embryology: Visualizations through series and animation, in: McCallum, D. (Ed.), *The Palgrave Handbook of the History of Human Sciences*. Springer Nature Singapore, Singapore, pp. 259–290. https://doi.org/10.1007/978-981-16-7255-2_22

Welper, G., 2024. Approximation and gradient descent training with neural networks. https://doi.org/10.48550/ARXIV.2405.11696

Wilkins, J.S., 2011. Philosophically speaking, how many species concepts are there? *Zootaxa* 2765. https://doi.org/10.11646/zootaxa.2765.1.5

Wilkinson, D.J., 2006. Bayesian methods in bioinformatics and computational systems biology. *Brief. Bioinform.* 8, 109–116. https://doi.org/10.1093/bib/bbm007

Zemke, N.R., Armand, E.J., Wang, W., Lee, S., Zhou, J., Li, Y.E., Liu, H., Tian, W., Nery, J.R., Castanon, R.G., Bartlett, A., Osteen, J.K., Li, D., Zhuo, X., Xu, V., Chang, L., Dong, K., Indralingam, H.S., Rink, J.A., Xie, Y., Miller, M., Krienen, F.M., Zhang, Q., Taskin, N., Ting, J., Feng, G., McCarroll, S.A., Callaway, E.M., Wang, T., Lein, E.S., Behrens, M.M., Ecker, J.R., Ren, B., 2023. Conserved and divergent gene regulatory programs of the mammalian neocortex. *Nature* 624, 390–402. https://doi.org/10.1038/s41586-023-06819-6

Zeng, H., 2022. What is a cell type and how to define it? *Cell* 185, 2739–2755. https://doi.org/10.1016/j.cell.2022.06.031

Zhang, X., Sheng, V.S., 2024. *Neuro-Symbolic AI: Explainability, Challenges, and Future Trends*. https://doi.org/10.48550/ARXIV.2411.04383

Zhou, J.X., Aliyu, M.D.S., Aurell, E., Huang, S., 2012. Quasi-potential landscape in complex multi-stable systems. https://doi.org/10.48550/ARXIV.1206.2311

4 The Nodes and Edges of Developmental Gene Regulation

Jake Cornwall-Scoones and James Briscoe

4.1 INTRODUCTION

Cells exist in fluctuating surroundings, from changing environmental conditions to signals from neighbouring cells. To survive and function, cells must interpret, process and respond to these external stimuli, and do so in a robust manner. The primary solution to this challenge is for cells to modulate gene expression, allowing cells to adjust concentrations and combinations of expressed genes in response to specific external cues. This positions gene regulation as the fundamental information-processing system that enables cellular decision-making and adaptive responses.

Jacob and Monod's seminal work revealed that gene regulation operates not through independent control of individual genes, but via interconnected regulatory networks (Jacob and Monod, 1961). They proposed that specialised "regulator genes" form complex feedback circuits that respond to external cues by binding to "operator genes" to control transcription, creating regulatory circuits that enable cellular information processing. Deciphering these circuits promised to solve "the fundamental problem of chemical physiology and of embryology… to understand why tissue cells do not all express, all the time, all the potentialities inherent in their genome" (Jacob and Monod, 1961).

Early insights into gene regulation came from studying the lac operon, and it is remarkable how well these early insights have held up over time. Regulators – now called transcription factors (TFs) – control the expression of other genes, including other TFs, by binding "operator genes", now termed cis-regulatory elements (CREs). These regulatory elements, exemplified by promoters and enhancers, encompass any DNA sequence with regulatory capacity that functions through protein binding rather than transcription. Together, TFs and CREs form gene regulatory networks (GRNs). This network perspective transforms gene regulation from a collection of independent switches into an integrated information-processing system.

4.2 GENE REGULATORY NETWORKS: NODES AND EDGES WITH INFORMATION-PROCESSING CAPABILITIES

GRNs have been proposed to encode how eukaryotic cells acquire and maintain specific states (Levine and Davidson, 2005). In this view, a cell type or state is defined

DOI: 10.1201/9781003560784-6

by its gene expression programme – the set of genes a cell expresses – which is determined by the set of TFs that regulate their transcription. Gene expression programmes have internal feedback, with TFs regulating the expression of one another. Together, this codifies the concept of a GRN: TFs are the nodes that modulate the transcription of each other through differential binding at their corresponding CREs, which act as edges in the network (Figure 4.1A).

More generally, regulatory networks can be thought of as dynamical systems, where the state of the system is described by the vector of TF concentrations, evolving in time through transcription, translation and degradation (Figure 4.1B,C). Such systems will have a set of stable states (so-called attractors), representing cell types in terms of defined combinations and concentrations of TFs (Goodwin, 1963; Jaeger et al., 2004) (Figure 4.1D). To transition between states, extrinsic signals controlling downstream effector TFs feed into a GRN by modulating the transcription rate of the TFs in the network. Such stimuli can change the dynamical system by moving attractors and changing their number (bifurcations). Therefore, external signals can influence the path of a cell through state-space, providing a description of a differentiation process. Moreover, if external factors are supplied in a spatially polarised manner, state transitions may diverge spatially. Feedback within the GRN therefore enables the processing of externally supplied information. Knowing the structure and dynamics of GRNs can reveal mechanisms by which cell identity is established and maintained (Briscoe and Small, 2015; Davidson and Erwin, 2006; Delás and Briscoe, 2020; Karlebach and Shamir, 2008; Levine and Davidson, 2005).

TFs are continuously reused for different cellular functions in diverse cell types and time points during development (S. A. Lambert et al., 2018). This means that cell identity depends not only on which TFs are present but on how they are connected

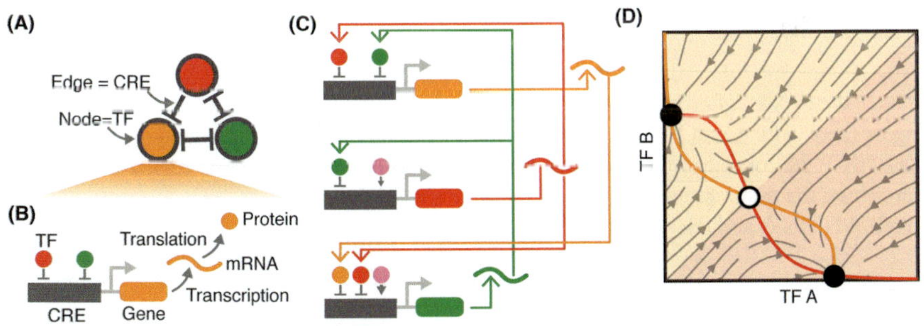

FIGURE 4.1 Quantitative and structural representations of GRNs. (A) A GRN is denoted as a wiring diagram. Nodes represent TFs and the edges represent regulatory interactions, encoded by CREs. (B) GRN models in principle need to describe all steps of gene regulation, from differential binding of TFs, their modulation of transcription rate and the subsequent translation of mRNA. (C) The simpler wiring diagram in A implicitly describes hypotheses for the binding of TFs to regulatory elements. (D) Together, GRNs define dynamical systems, with attractors, saddles and repellers. Stable states and their cognate basins of attraction (shaded regions) provide a framework to describe cell fates or states. Figure adapted from Zhu et al. (2022).

(Continued)

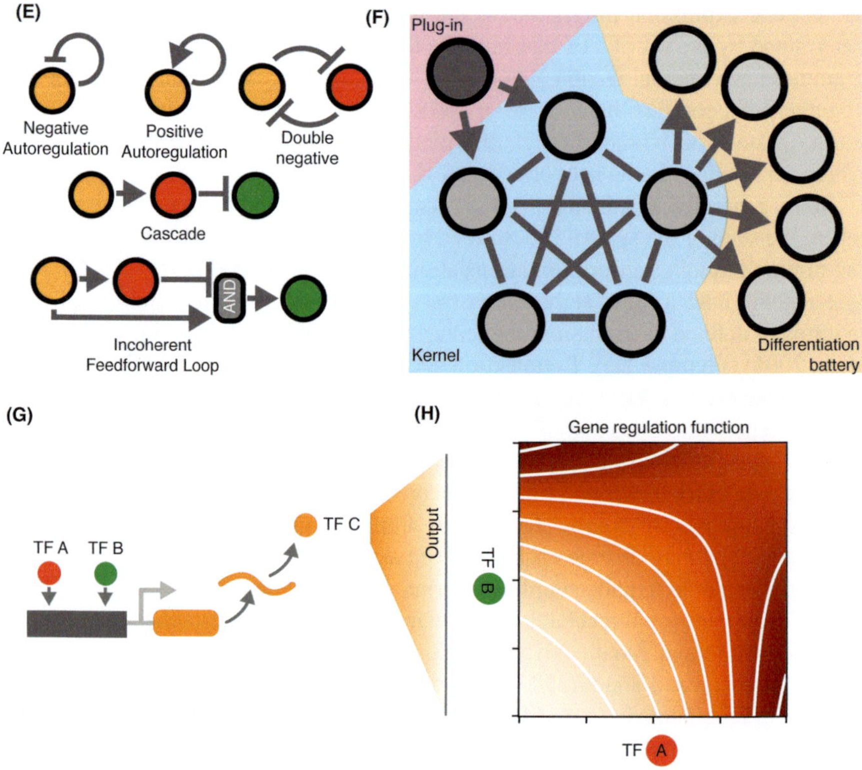

FIGURE 4.1 (Continued) (E) Distinct network motifs display characteristic dynamics (Alon, 2007). Such motifs highlight how even simple motifs often contain multiple regulatory inputs to a single TF. (F) An alternative view of the structural classification of GRNs, inspired by Davidson and Erwin (2006), where kernels are dense networks of interconnected TFs, plugins modulate the dynamics and selection of kernels and differentiation batteries encode the necessary proteins of terminally-fated cell types. (G) When a CRE has multiple TF inputs (here, TFs A and B), one must specify how these inputs are integrated to yield a single transcriptional response. (H) This can be understood in terms of a gene regulation function, describing the quantitative mapping of output expression level (here, TF C), across the input manifold (TFs A and B).

through regulatory networks (Davidson et al., 2002; Davidson and Erwin, 2006). Thus, the key to a GRN lies in the CREs that wire the TFs together (Levine and Davidson, 2005). An understanding of a GRN requires two components: a wiring diagram that maps which TFs regulate each CRE and quantitative descriptions of how gene expression rates respond to different TF concentrations. Signal-dependent GRNs encode and integrate information about how signalling pathways modulate TF activity, which CREs receive these signals and how multiple inputs quantitatively combine.

4.3 GRAPHICAL, MODULAR AND FUNCTIONAL DESCRIPTIONS OF GRNS

GRNs involve many TFs connected through complex, many-to-many interactions. Understanding how these connections influence cell identity and fate transitions represents a major challenge. Theoretical analyses of small network modules have revealed fundamental principles of GRN behaviour, particularly how network structure enables information processing of external signals. Identifying small networks with defined behaviours within GRNs – the so-called "motifs" (Alon, 2007; Milo et al., 2002) – can help inform how the network as a whole behaves (Figure 4.1E). These regulatory modules or "motifs" serve as building blocks that help predict network-wide behaviour (Alon, 2007; Milo et al., 2002). But even simple regulatory circuits can produce complex dynamics. For example, negative auto-regulation, where a TF reduces its own transcription rate, can accelerate the approach to steady-state levels, but it can also result in sustained oscillations depending on regulatory parameters (Alon, 2007; Maeda and Sano, 2006; Rosenfeld et al., 2005). Conversely, positive autoregulation slows steady-state establishment and can create bistable switches. More complex behaviours emerge in three-node networks such as feedforward loops, where one TF regulates both a second factor and their common target. The eight possible combinations of activating and repressing interactions in this simple motif each produce distinct dynamic responses and signal processing capabilities (Alon, 2007). This complexity illustrates why understanding GRNs requires more than mapping regulatory connections – it demands quantitative characterisation of how TFs combine to control gene expression. Regulatory architecture alone cannot predict dynamic behaviour – the quantitative parameters of each regulatory interaction determine whether a circuit reaches a stable steady-state, oscillates, or exhibits other complex temporal patterns (Strogatz, 2007).

Analysis of detailed and well-validated GRNs of animal development (predominantly sea urchin (Davidson et al., 2002)) has also provided a set of definitions of sub-networks that are rooted in developmental ontogeny (Figure 4.1F). "Kernels" are proposed to be groups of TFs that display extensive regulatory interconnection, typically implicated in core developmental processes such as patterning (Davidson and Erwin, 2006). A key feature of kernels is the mutual interdependence of their nodes and edges: disruption of any part has major gene expression and phenotypic effects, meaning kernels are predicted to be evolutionarily entrenched and highly conserved. In addition, "plug-ins" are smaller sub-networks that feed signalling information into kernels, whereas "switches" connect together kernels, enabling cells to flip between distinct gene expression programmes. This decision-making machinery in turn feeds into "differentiation batteries", which define the physiology and functioning of differentiated cell types.

Despite the conceptual insight offered by the GRN paradigm, even simplified GRN models require quantitative descriptions of how multiple TF inputs combine to regulate target gene expression. This relationship – the "cis-regulatory input function" or "gene regulation function" – remains the critical missing component for predictive network models (Rosenfeld et al., 2005) (Figure 4.1G,H). A diagram with multiple (activating or repressing) arrows pointing to a single TF is insufficient to describe how that TF will be expressed, and in turn how the network will behave. Moreover, we generally lack foundational rules or null hypotheses to provide a best first guess of how transcription depends on multiple TFs, with biology seemingly choosing different gene regulation functions in different contexts. Even for the best studied, canonical, "simple" CRE – the lac operon – transcription rate depends on the concentration of its two input factors in a complex relationship between an "AND" and an "OR" gate (Setty et al., 2003). Such pictures are further complicated by networks of dense overlapping regulons with multiple simultaneous inputs, typically found in developing tissues.

Two fundamental challenges emerge from this complexity. First, network wiring diagrams alone cannot predict even qualitative network behaviours. Second, understanding GRNs requires quantitative characterisation of how CREs integrate multiple TF inputs. Without these quantitative input–output relationships, GRN models remain largely descriptive rather than predictive.

For GRNs to provide explanations of the developmental ontogeny and stability of cell types, and their spatiotemporal allocation downstream of morphogens, we need to define the ways in which regulatory elements integrate information of TF levels, especially those controlled by signalling, to regulate the transcription of genes. Consequently, a major step forward in defining and interpreting GRNs would be to provide a quantitative null expectation for the structure of gene regulation functions. One possibility is that regulatory inputs multiply with some level of saturation unless additional synergies are at play. This hypothesis is the heuristic used in "connectionist" continuous modelling frameworks of GRNs (Jaeger et al., 2004; Mjolsness et al., 1991). Yet this proposition – at least in mammalian systems – lacks a comprehensive empirical grounding at present, and leaves open prospects for quantitatively or qualitatively distinct null expectations across CREs or cell states.

4.4 GRNS AS POSITIONAL INFORMATION DECODERS

Tissue formation during development relies on the precise allocation of distinct sets of cell types along embryonic axes. One solution to this problem involves diffusible signalling molecules, termed morphogens, which are produced from a source and spread to form a gradient (Kicheva and Briscoe, 2023; Rogers and Schier, 2011); the receiving field of cells responds to different concentrations to elicit discrete cell-fate responses (Briscoe and Small, 2015). Together, this defines the concept of "positional information" (Wolpert, 1969), where cells are able to measure their location with respect to a signalling gradient.

How morphogen signalling creates spatial patterns of gene expression has been subject to debate (Briscoe and Small, 2015). A core feature of all models is the requirement for CREs to show some form of signal-dependent activity, as demonstrated by dose-dependent gene expression to the anterior morphogen of *Drosophila* blastoderm patterning, bicoid (Struhl et al., 1989), and to the ventral morphogen of vertebrate neural tube patterning, Sonic Hedgehog (Shh) (Ericson et al., 1997). This behaviour could be achieved by individual CREs autonomously interpreting the morphogen. Some CREs exhibit sharp, threshold-like responses to morphogen concentration, suggesting they can generate discrete expression boundaries without input from other regulatory factors (Burz et al., 1998; Gregor et al., 2007). In this autonomous model, spatial gene expression boundaries reflect the intrinsic sensitivity of each regulatory element to morphogen signalling. Experimental support comes from studies where altering the strength of morphogen-responsive binding sites shifts expression boundaries (Driever et al., 1989; Struhl et al., 1989). However, such manipulations do not always produce expected results, indicating additional complexity (Ochoa-Espinosa et al., 2005).

An alternative mechanism involves signal-decoding GRNs converting the graded morphogen inputs into sharp gene expression boundaries (Figure 4.2A). The morphogen effector TF serves as an input into the network by binding CREs of key patterning genes. The network's internal feedback architecture transforms the analogue morphogen signal into discrete, digital gene expression outputs. Secondary feedback between signal-responsive TFs and other TFs in the GRN in turn sharpens boundaries. This relaxes the constraint of CRE-autonomous thresholded expression. A network-based mechanism often shows a repeated motif of mutual repression among TFs marking alternative fates. The relative strengths of morphogen-dependent activation and cross-repressive inhibition determine where expression boundaries form

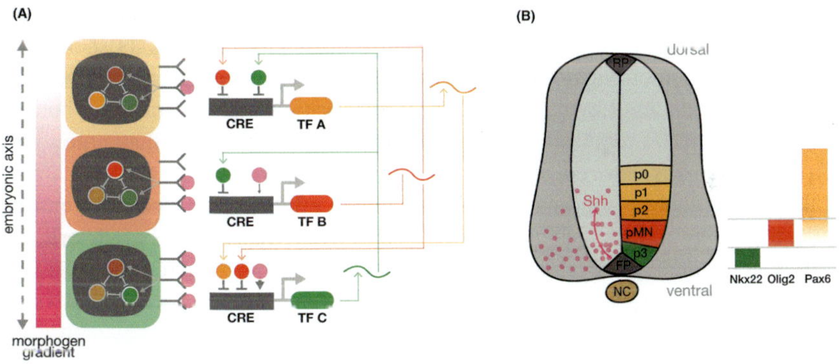

FIGURE 4.2 Morphogen decoding GRNs. (A) Morphogen interpretation involves the conversion of an analogue graded input (pink) into discrete, digital cell fates (yellow, red, green) at distinct threshold concentrations. The GRN hypothesis of morphogen interpretation proposes that regulatory interactions (typically cross-repressing) among a set of TFs establish distinct stable states, where signalling inputs to some of these TFs break spatial symmetry to define domain boundaries. Consequently, patterning is a dynamical read-out of signalling (history), with the GRN functioning to process and integrate signalling information. This hypothesis is distinct from the binding affinity hypothesis, which proposes that CREs only receive regulatory inputs from signalling TFs. (B) Schematic of vertebrate ventral neural tube patterning. A gradient of the morphogen Shh emanating from the floor plate (FP) and notochord (NC) promotes the expression of distinct combinations of TFs at threshold concentrations. These establish a set of progenitor domains: p3, pMN (motor neuron), p2, p1 and p0. These domains are marked by combinatorial expression of key TFs, notably Nkx2-2 in the p3 domain, Olig2 in the pMN domain and Pax6, at low levels in the pMN domain and higher more dorsally.

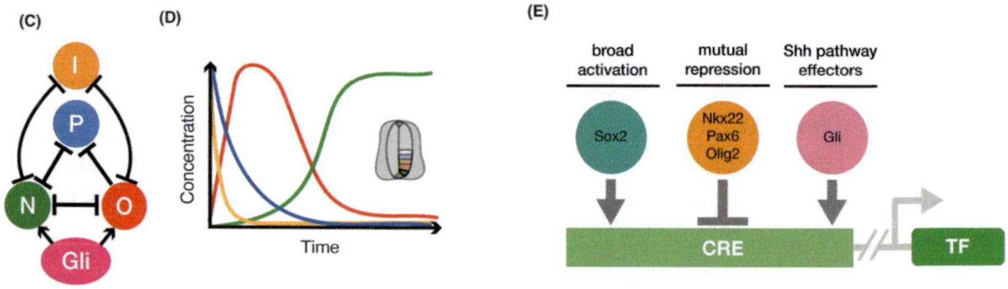

FIGURE 4.2 (Continued) (C) Wiring diagram of ventral neural tube patterning GRN, where O = Olig2, N = Nkx2-2, P = Pax6 and I = Irx3. (D) Dynamics of gene expression of these markers (schematic of simulation from Cohen et al. (2014)) for a ventral, high Shh coordinate. Note that Nkx2-2 expression occurs only after transient Olig2 expression. (E) Summary of regulatory inputs of CREs, as inferred through developmental genetics (Delás & Briscoe, 2020).

(Briscoe and Small, 2015; Cohen et al., 2014; Delás and Briscoe, 2020). Cross-repression sharpens boundaries beyond what individual regulatory elements could achieve autonomously, relaxing the requirement for inherently threshold-responsive elements. This hypothesis is supported by cases in which gene expression boundaries shift despite temporally uniform morphogen supply (Clark and Akam, 2016; Jaeger et al., 2004; Verd et al., 2019), and functional tests where the deletion of certain TFs results in the expansion of the spatial domains of other TFs, typically those otherwise expressed in neighbouring domains (Balaskas et al., 2012; Briscoe et al., 2000; Clyde et al., 2003; Kraut and Levine, 1991). These observations demonstrate that boundaries depend on active repression between competing factors rather than autonomous morphogen sensing. Indeed, it is hard to conceive of mechanisms where a single regulatory element responding to one morphogen effector could generate multiple sharp stripes or complex spatial patterns (Cotterell and Sharpe, 2010; Schaerli et al., 2014). Understanding morphogen-mediated patterning therefore requires characterising how CREs integrate multiple TF inputs to create signal-responsive GRNs.

4.5 THE SIGNAL DECODING GRN CONTROLLING NEURAL TUBE PATTERNING

The dorsoventral patterning of progenitors in the developing vertebrate neural tube is among the best studied examples of morphogen interpretation (Briscoe and Small, 2015). Here, progenitor cells express distinct combinations of TFs depending on their dorsoventral position, forming domains of progenitor subtypes (Delás and Briscoe, 2020). These progenitors differentiate into distinct classes of neurons that comprise the adult spinal cord (Sagner and Briscoe, 2019). The dorsoventral symmetry of the neural tube is broken by two sets of signalling centres at opposite poles to establish antiparallel morphogen gradients: Shh from the notochord and later floor plate (Echelard et al., 1993; Krauss et al., 1993; Yamada et al., 1993), and bone Bone Morphogentic Proteins (BMPs) and Wingless-integrated proteins (Wnts) from the surface ectoderm and roof plate (Liem et al., 1997; Muroyama et al., 2002). These together establish eleven discrete progenitor domains (Frith et al., 2024). The most ventral cells respond predominantly to differences in Hedgehog signalling, and the most dorsal cells to BMPs and Wnts, with intermediate progenitor domains decoding positional information using both gradients (Zagorski et al., 2017) (Figure 4.2B).

How is positional information encoded? Experiments with tissue explanted from the developing neural tube showed that neural progenitors are able to decode different concentrations of both Shh and BMP (Liem et al., 1997; Yamada et al., 1993), consistent with a classical French Flag model (Wolpert, 1969). However, patterning also depends on the duration of signalling input – the same morphogen concentration can specify different fates depending on how long cells are exposed (Dessaud et al., 2007; Tozer et al., 2013). Adding to the complexity, the signalling landscape itself changes over time in the neural tube. Shh expression increases during development (Cohen et al., 2015; Zagorski et al., 2017), while the neural tube expands through cell proliferation (Kicheva et al., 2014). Consequently, the concentration of morphogen to which a given cell is exposed will vary over time (Kicheva and Briscoe, 2023; Plum and Serra, 2025).

Shh signals through the Hedgehog pathway to modulate the transcriptional activity of its downstream effectors: the Gli family of TFs (Hui and Angers, 2011). Hedgehog signalling blocks the proteolytic processing of Gli2 and Gli3 into their cognate shorter-length repressive forms, yielding transcriptional activators (Pan et al., 2006; Wang et al., 2000). Both repressive (GliR) and activating (GliA) isoforms bind the same binding site, meaning the transcriptional response to Hedgehog signalling is potentially complex, integrating the stoichiometry and relative avidities of these two isoforms. Consequently, it is not trivial to predict changes in expression upon mutating or changing the affinity of a Gli transcription factor binding site (TFBS.) Moreover, Hedgehog signalling has multiple inbuilt feedback loops (Cohen et al., 2015; Hui and Angers, 2011), with Hedgehog signalling regulating the expression of both Glis (Lee et al., 1997) and Ptch1 (Cohen et al., 2015) – a negatively regulating receptor.

Neural progenitors decode dynamic morphogen gradients through a GRN that converts graded Hedgehog signalling into discrete progenitor domains along the dorsoventral axis (Figure 4.2C). This network consists of mutually repressing TFs, with the most ventral factors – Olig2 and Nkx2-2 – receiving direct positive input from Hedgehog signalling (Balaskas et al., 2012; Briscoe et al., 2000; Novitch et al., 2001). The cross-repressive architecture creates a dynamical system where each progenitor subtype represents a distinct attractor state (Balaskas et al., 2012; Cohen et al., 2015; Exelby et al., 2021; Pezzotta and Briscoe, 2023; Rayon et al., 2020). Evidence for mutual repression came from genetic perturbation experiments. Deletion of individual TFs resulted in the neighbouring factors expanding their expression domains to fill the vacant space (Balaskas et al., 2012; Briscoe et al., 1999, 2000; Ericson et al., 1997). For example, loss of Olig2 or Pax6 allows the Nkx2-2-expressing p3 domain to expand dorsally (Balaskas et al., 2012; Ericson et al., 1997; Q. Zhou and Anderson, 2002). Conversely, forced expression of key TFs cause the downregulation of the other TFs in the GRN (Briscoe et al., 2000), both those associated with adjacent domains as well as those that are expressed non-adjacently (Kutejova et al., 2016; Nishi et al., 2015).

These findings have led to the development of dynamical mathematical models to capture the emergent logic of this Hedgehog-decoding GRN (Figure 4.2D). An ordinary differential equation model codifying a GRN composed of dynamically varying Olig2, Nkx2-2, Irx3 and Pax6 supplied with spatially varying Hedgehog inputs was sufficient to reproduce the pattern of TF expression in both wildtype and a range of mutants (Cohen et al., 2014). Moreover, this model also captured the dynamics of the system: starting from an initial, intermediate state – marked by Pax6 and Irx3 expression – Hedgehog signalling stimulates the sequential transition towards increasingly ventral fates – first pMN marked by Olig2 expression and then p3 marked by Nkx2-2 expression – instead of instantaneously committing to specific dorsoventral identities (Cohen et al., 2014). These observations also explain the hysteresis observed in the network: ventral cell types resist reverting to more dorsal fates if Hedgehog signalling is decreased (Balaskas et al., 2012; Cohen et al., 2014; Perez-Carrasco et al., 2018). Dynamical systems theory provides a framework for understanding these behaviours: Hedgehog concentration shapes the underlying dynamical landscape by determining the number and position of attractor states. Higher Hedgehog levels create bifurcations that establish new attractors corresponding to ventral cell fates, making these states increasingly stable and resistant to reversal (Perez-Carrasco et al., 2016, 2018).

Such dynamical models implicitly describe hypotheses for the molecular mechanism of gene regulation. Specifically, TF dynamics are defined by non-linear activation and linear decay terms, the former being the gene regulation function of the CREs regulating each of the TFs, formalised via a thermodynamic model of CRE activity (Ackers et al., 1982; Bintu et al., 2005; Cohen et al., 2014; Shea and Ackers, 1985; Wong and Gunawardena, 2020) (Figure 4.2E). In this model, CREs receive: (1) uniform positive inputs; (2) Hedgehog-dependent inputs, encoded as the ratio of positively and negatively effecting Gli isoforms; and (3) repressive inputs from other TFs in the network. However, cooperativities and/or post-binding synergies were not varied in fitting nor constrained to molecular data. Given there are many potential gene regulatory functions that are likely sufficient to describe the same data, it is therefore challenging to relate the quantitative logic of activity to the structure and sequence of an individual CRE. Consequently, it remains an open question how structurally and mechanistically the CREs of the neural tube GRN encode the regulatory architecture and Hedgehog-decoding capacity of the ventral neural tube GRN.

A related line of enquiry has been the molecular characterisation of CREs controlling dorsoventrally regulated genes. By analysing sequence conservation (Oosterveen et al., 2012; Peterson et al., 2012), as well as TF binding and chromatin accessibility (Delás et al., 2023; Nishi et al., 2015; I. Zhang et al., 2024), a set of putative CREs regulating each of the key patterning TFs has been identified that are sufficient to drive activity in in vivo reporter assays (Oosterveen et al., 2012; Peterson et al., 2012). Among these CREs, the CRE positioned 33kb upstream of Olig2 is the best studied. A knock-out of this CRE results in a substantial decrease in Olig2 expression and an expansion of neighbouring domains (Exelby et al., 2021). Such molecular characterisation has revealed that SoxB1 TF is the likely candidate of uniform positive input (Peterson et al., 2012) and that Shh signalling regulates the ratios of GliA/R occupancies to their corresponding binding sites to break dorsoventral symmetry

(Oosterveen et al., 2012; Peterson et al., 2012). This work also suggests that the positive inputs from either Gli or SoxB1 are individually insufficient to promote expression, arguing that the gene regulation functions of ventral patterning CREs may be intrinsically combinatorial. Moreover, key patterning TFs appear to bind and repress regulatory elements corresponding to alternative TFs in the network, corresponding to both adjacent and distant domains (Kutejova et al., 2016; Nishi et al., 2015).

4.6 THE MOLECULAR ANATOMY OF REGULATORY LOCI

The behaviour of a GRN depends on how the transcription rate of one TF depends on the levels of other TFs in the network. Consequently, the structure of a GRN is defined by its "edges", the CREs. An integrative understanding of GRNs must therefore identify the important CREs in the network, and how information about TF binding at CREs is transmitted and transduced into TF transcription rates.

CREs are generally divided into promoters and enhancers (Figure 4.3A). The latter were first discovered as sequence elements from the SV40 viral genome that could increase the expression of a beta-globin transgene in a manner that was independent of its position or orientation (Banerji et al., 1981). Similar CREs were soon discovered in metazoans (Hammer et al., 1987; Johnson et al., 1989; Pinkert et al.,

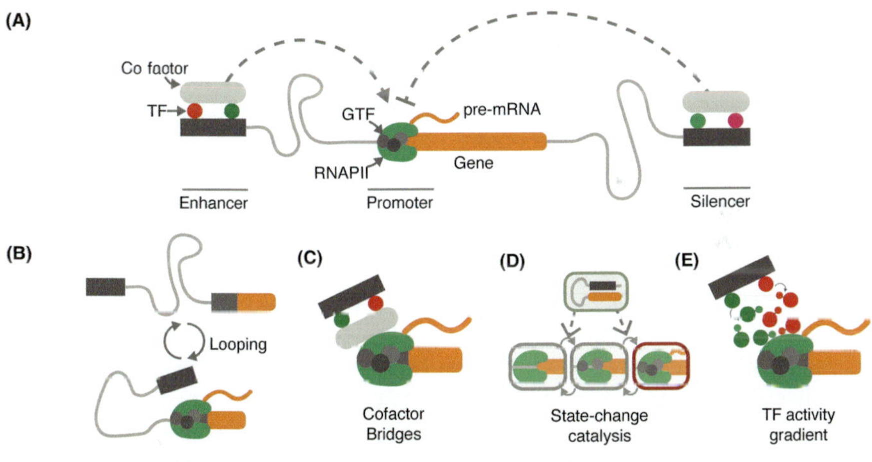

FIGURE 4.3 The mechanism of cis-regulation. (A) Distal cis-regulatory elements (enhancers, silencers) are found up to ~1 Mb from their cognate genes, either upstream or downstream. TF binding regulates the transcription rate typically through bound co-factors. Alternatively, proximal cis-regulatory elements (i.e. promoters) are found around the transcription start site of the gene, binding a distinct set of TFs (general TFs), which help recruit RNA polymerase II (RNAPII) to engage in transcription. Promoters enable transcription, whereas distal cis-regulatory elements modulate its rate. (B) CREs and promoters are thought to interact in three dimensions through DNA looping. (C) Information about TF binding at CREs may be communicated through direct contact, for example mediated by physical bridges of cofactors. (D) CRE-promoter contact may catalyse state changes at the promoter, conferring "memory" of regulatory inputs. (E) TF binding information may alternatively be communicated at a distance, via gradients of post-transcriptionally modified TFs.

1987; Swift et al., 1984), regulating the expression of genes as far as 1 Mb away (Lettice et al., 2003; Pennacchio et al., 2013). Distal CREs have also been argued to actively repress transcription (Figure 4.3A). These are termed silencers (S. Kim and Wysocka, 2023; Segert et al., 2021). Indeed CREs can flip their functionality between silencing and enhancing in a context-dependent manner (Delás et al., 2023; Delás and Briscoe, 2020; Gisselbrecht et al., 2020; Kehayova et al., 2011).

Across the taxonomy of CREs, function is thought to be confined to accessible chromatin, profiled by Assay for Transposase-Accessible Chromatin with sequencing (ATAC-seq) or DNase-seq (Buenrostro et al., 2013; Domcke et al., 2020; Li et al., 2023; Thurman et al., 2012; K. Zhang et al., 2021). The pattern of histone marks and co-factor association within these domains is indicative of its state, profiled by Chromatin ImmunoPrecipitation with sequencing (ChIP-seq) or CUTandRUN/CUTandTAG (Kaya-Okur et al., 2019; Skene and Henikoff, 2017). While the histone code appears reasonable at discriminating regulatory sequence properties, whether they play mechanistic roles in gene regulation remains less clear (for discussion, see Catarino and Stark (2018)). Enhancer RNAs (eRNAs) also provide a correlate of distal CREs, based on initial reports of transcription at the enhancer itself of the beta-globin gene (Tuan et al., 1992), with 70% of sequence fragments with divergent transcription displaying enhancing properties in reporter assays (Andersson et al., 2014) (for discussion, see Andersson and Sandelin (2020) and Catarino and Stark (2018)).

In metazoa, distal CREs can be kilo- to mega-bases away from the promoters and transcription start sites (TSSs), posing the question: how is TF binding information communicated from these distal sites to modulate transcriptional initiation? Well-documented examples, such as the Zone of polarising activity Regulatory Sequence (ZRS) (Lettice et al., 2003), as well as high-throughput functional ablation approaches (Fulco et al., 2019; Gasperini et al., 2019), suggest information can be transmitted over long distances, and associated promoter–CRE pairs are separated by active genes as frequently as 30%–50% of the time (Gasperini et al., 2019; Yang and Hansen, 2024). Additionally, multiple enhancers can influence a single promoter (Allahyar et al., 2018; Hong et al., 2008; Oudelaar et al., 2018), and multiple promoters can be influenced by a single enhancer (Fukaya et al., 2016; Oh et al., 2021).

A diversity of models has been proposed to explain how information is transmitted from distal regulatory elements to promoters to modulate their transcription (reviewed in Yang and Hansen (2024)). Originally, the linear arrangement of regulatory elements on DNA was proposed to encode the mapping of promoters to their regulatory elements (Moreau et al., 1981). However, to explain the long-distance, promoter-skipping action of regulatory elements, three-dimensional (3D) models are now favoured, suggesting CREs and promoters communicate through physical proximity via DNA looping (Figure 4.3B).

A subset of 3D models proposes that direct physical contact between promoters and CREs is required for information transmission (Figure 4.3C). Evidence from chromatin conformation capture technologies has shown that active promoters and enhancers have enhanced proximity (Fulco et al., 2019; Levo et al., 2022; Montavon et al., 2011; Oudelaar et al., 2018; Rekaik et al., 2023). While initially conceived of as stable contacts, live imaging studies have shown that interactions are dynamic, and contact, if any, is likely transient (Alexander et al., 2019; Brückner et al., 2023; H. Chen et al., 2018). Loops engaged in ultra-long contacts (e.g. ZRS) have been proposed to be stabilised by the broader loop extrusion machinery of the nucleus by CCCTC-binding factor (CTCF) (Paliou et al., 2019) or cohesin (Kane et al., 2022). Alternative dynamic contact-based information transfer models have been proposed, with the "hit-and-run" (Schaffner, 1988) and "kiss-and-kick" models (Pownall et al., 2023) arguing that contact is required to transfer regulatory information, but not for continued transcription.

What is the physical form of information transmission from regulatory elements to TSSs? Contact-mediated transmission mechanisms have been proposed to occur through a range of mechanisms. Co-factor complexes such as Mediator may act as physical bridges between transcriptional effectors and RNA polymerase (Abdella et al., 2021; X. Chen et al., 2021; Rengachari et al., 2021), although contacts appear to be maintained upon Mediator depletion (El Khattabi et al., 2019). Alternatively, enhancers may catalyse state changes at the promoter (e.g. histone modifications or RNA polymerase II (RNAPII) phosphorylation), such that contact is not required for sustained transcription and allows promoters to sustain a "memory" of the state of their cognate promoter (Soutourina, 2018) (Figure 4.3D).

An alternative set of hypotheses for how information is transmitted without sustained contact implicate transcriptional "condensates". Given TFs, co-factors and polymerases have been shown to phase-separate in the nucleus (Furlong and Levine, 2018), one possibility is that information transmission is enabled by pockets of enhanced concentration that allow for the direct transmission of regulatory information over a longer range (Boija et al., 2018; Sabari et al., 2018). Theoretical frameworks for how gene expression maps to combinatorial TF binding at regulatory elements are, however, currently in their infancy (Bialek et al., 2019; Xiao et al., 2021). Moreover, some have argued that the relatively large radii of such phase-separated droplets, wide enough to envelope around 40 genes, make it challenging to rationalise promoter–CRE specificity (Yang and Hansen, 2024).

Contact-mediated information transmission models have been tested in several ways. As a sufficiency argument, gene expression of otherwise silent genes can be driven by forced chromatin looping using synthetic zinc-fingers (Deng et al., 2012, 2014). Moreover, the changes in gene expression levels upon high-throughput inhibition of CREs are consistent with an "activity by contact" model, where transcription correlates with enhancer status convolved with the frequency of promoter–CRE contacts (Fulco et al., 2019). However, while some live-imaging studies have suggested that sustained physical contact is required for transcriptional induction (H. Chen et al., 2018), in a steady-state context, correlations between contact and transcriptional activation have been weak (Mateo et al., 2019) or absent (Alexander et al., 2019). Moreover, some CREs involved in regulating the Shh gene have been shown to reduce their contact frequencies upon gene induction, and synthetic recruitment of a diversity of transcriptional activators to these regulatory elements simultaneously boosts Shh transcription and reduces contact frequency (Benabdallah et al., 2019). Alternative models invoke diffusive mechanisms, such as the "TF activity gradient" model (Karr et al., 2022) (Figure 4.3E). This suggests that activated regulatory elements catalyse the post-translational modification (e.g. acetylation) of bound transcriptional effectors, which form sharp 3D concentration gradients by diffusion and fast deacetylation, to stimulate nearby promoters. While mechanistic details of this have been speculated (Karr et al., 2022), conclusive tests of this fundamentally alternative paradigm have yet to be explored experimentally.

4.7 MEASURING CIS-REGULATORY ELEMENT ACTIVITY

In general, the activity of CREs is measured in one of two ways. Either the CRE is functionally modulated at its endogenous locus, through knock-out or epigenetic modulation, or its activity is measured via its capacity to drive the expression of a reporter gene with a "minimal promoter" in an ectopic genomic context.

Reporter assays historically have been low throughput, studying the spatiotemporal pattern of luciferase, lacZ, or fluorescent protein production (Armendariz et al., 2023) (Figure 4.4A). Such approaches have established atlases of spatiotemporal CRE activity in developmental model systems (Kvon et al., 2014; Visel et al., 2009; Zinzen et al., 2009). Modifications to improve the temporal resolution of the assay include the use of destabilised fluorophores or methods to tag and measure mRNA molecules directly (Garcia et al., 2013).

More recently, reporter assays have been multiplexed in "massively parallel reporter assays" (MPRAs) (Figure 4.4B), where libraries of reporter constructs are supplied to cells or tissues, deconvolving the activities of each construct from the pool by high-throughput sequencing (La Fleur et al., 2024). Strategies to achieve this vary in experimental design (Klein et al., 2020). The sequences assayed by an MPRA may either be fragmented genomes or, more frequently, synthesised regions. For the latter, these may be synthesised genomic sequences selected based on enhancer-like features (Agarwal et al., 2025; Gosai et al., 2024; Inoue et al., 2019), synthetic regulatory elements (Frömel et al., 2024; Sahu et al., 2022; R. P. Smith, Taher, et al., 2013), or random N-mers (Sahu et al., 2022; R. P. Smith, Riesenfeld, et al., 2013). In some strategies, cells are transduced or transfected with a pool of vectors such that each cell receives a single reporter construct, and relative activities are determined by sequencing identifiers of regulatory elements in groups of cells stratified by reporter levels (Kinney et al., 2010; Sharon et al., 2012). Alternatively, sequence identifiers are transcribed, meaning activities can be determined by the ratio of RNA to DNA counts (Agarwal et al., 2025; Inoue et al., 2019). The identifier can be a short "barcode" sequence or, in the case of Self-Transcribing Active Regulatory Region sequencing (STARR-seq), the regulatory element placed in the reporter gene so that it is transcribed along with the reporter (Arnold et al., 2013). Finally, vectors may either be integrated into the genome using lentivirus (Agarwal et al., 2025; Inoue et al., 2017, 2019) or can be supplied episomally through plasmid transfection, electroporation or transduction with non-integrating viruses (integration-deficient lentiviruses or adeno-associated viruses) (Arnold et al., 2013; Inoue et al., 2017; Klein et al., 2020; J. T. Lambert et al., 2021; R. P. Smith, Taher, et al., 2013). MPRA approaches have recently been combined with single-cell technologies to measure regulatory element activity as a function of transcriptional states and cell fates (Lalanne et al., 2024; Zhao et al., 2023).

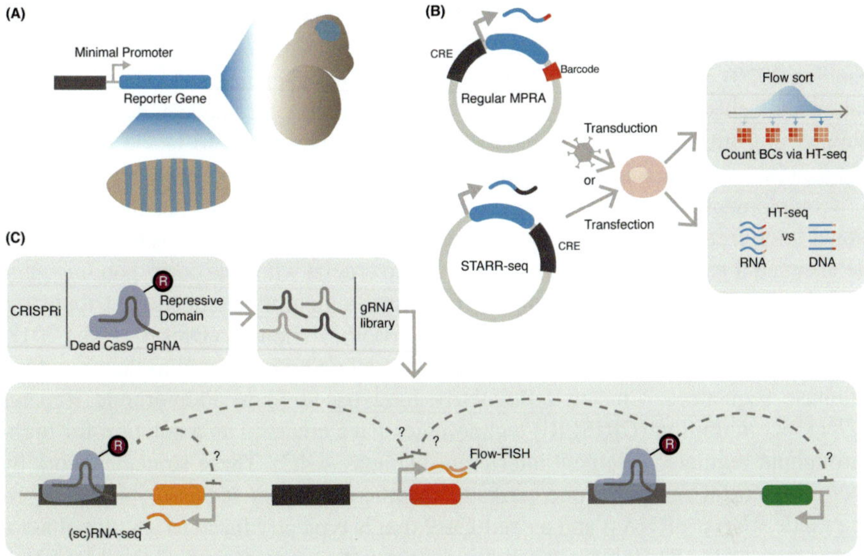

FIGURE 4.4 Experimental strategies to measure CRE activity. (A) Traditional reporter assays, where a CRE is placed adjacent (typically upstream) of a minimal promoter and a reporter cassette (fluorescent protein, lacZ, luciferase). CRE activity is measured by quantifying the magnitude or spatial expression of the reporter construct. Reporters may be transiently transfected or stably integrated depending on the experimental context. (B) Massively parallel reporter assays (MPRAs). Schematic illustrating the multiple experimental design choices in MPRAs, including using barcodes or the CRE sequence itself as the molecular identifier. Constructs are introduced into cells in a pool, either through transfection or via lentiviral transduction. Activities of CREs across the library are deconvolved either through fluorescence-gated flow sorting and sequencing or a comparison of barcode abundance in cellular RNA, normalised by abundance in DNA. (C) Strategies for CRISPRi-mediated CRE perturbations. CRISPRi relies on guide RNAs (gRNAs) targeting a dead Cas9 (dCas9) to a locus of interest. dCas9 typically harbours a repressive domain (e.g. KRAB), which modulates the expression of the genes corresponding to the targeted locus. Here, the read-out can be either variants of in situ hybridisation (Fulco et al., 2019; Reilly et al., 2021) or single-cell RNA sequencing (Gasperini et al., 2019).

MPRAs and reporter assays, in general, must be considered in the context of several conceptual and methodological challenges. One important consideration is the choice of the minimal promoter, with different promoters having different levels of background expression (Kvon et al., 2020) and potentially different CRE compatibilities (Zabidi et al., 2015). In terms of trade-offs in MPRA technology, episomal assays typically enable higher throughput, but this must be balanced against potentially less faithful chromatin context (Inoue et al., 2017; Klein et al., 2020). A second issue is that many developmentally interesting CREs are not enhancers, or at least are not enhancers all the time, but function as silencers. Consequently, profiling silencers with MPRAs presents methodological challenges, with a lower signal-to-noise ratio when using a "minimal" promoter given already low background expression, or requiring additional enhancing elements or less "minimal" promoters, which may interact with the silencers in confounding ways (Doni Jayavelu et al., 2020; Pang and Snyder, 2020). Strategies to combine combinatorially silencing and enhancing regulatory elements or fragments thereof – to probe trade-offs between activating and suppressing transcription – are now starting to be developed to tackle these challenges (Cornwall-Scoones et al., 2025).

Complementing reporter assays, CRE activity can be measured through functional interference. The necessity of a regulatory element for gene expression can be determined through knock-outs. These tests are crucial when the corresponding gene for a regulatory element is not known, if multiple regulatory elements contribute to a gene, or if the element acts as a (context-specific) silencer (Dickel et al., 2018; Osterwalder et al., 2018). Given challenges of scaling up homozygous knock-outs of kilobase sequences, Clustered Regularly Interspaced Short Palindromic Repeats (CRISPR) inhibition (CRISPRi) technologies have emerged as a solution for high-throughput regulatory element interference (Figure 4.4C). These strategies work by recruiting co-factors or histone marks to specific regulatory sequences using targeting guide RNAs (gRNAs) and a dead-Cas9 that is typically fused to a Krüppel associated box (KRAB) domain, promoting epigenetic suppression of the regulatory element (Larson et al., 2013). CRISPRi has been used for pooled screening, where libraries of gRNAs are introduced to cells, and the corresponding response on the target gene is measured through single-cell RNA sequencing (Gasperini et al., 2019) or through flow-sorting bins of a target gene's mRNA levels by fluorescence in-situ hybridisation (FISH) (Fulco et al., 2019) or hybridisation chain reaction fluorescence (HCR) (Reilly et al., 2021).

4.8 MORPHOGENETIC FIELDS AS CHROMATIN LANDSCAPES

The clustering of TFBS on CREs creates distinct chromatin configurations when bound by TFs that are experimentally detectable with techniques such as ATAC-seq. Different tissues display characteristic chromatin accessibility profiles that reflect their unique TF repertoires and gene expression programmes (Zaret and Mango, 2016). This tissue-specific chromatin architecture provides a molecular foundation for the morphogenetic field concept. Originally proposed by embryologists to describe discrete modular units of embryonic development, in which cell organisation is spatially coordinated, a morphogenetic field could now be understood as a domain of cells sharing a common chromatin accessibility profile.

In this view, the establishment of a morphogenetic field follows a characteristic developmental sequence. Pioneer TFs initially open chromatin at regulatory elements that will later participate in tissue-specific GRNs (Zaret and Carroll, 2011). This pioneering activity creates a shared epigenetic foundation across the developing tissue, delimiting the subset of CREs and genes that participate in a GRN. Once established, this common chromatin landscape enables cells throughout the field to respond in a coordinated manner to patterning signals.

In this model, cells adopt distinct identities based on their position within the field, due to the differential binding of a set of specific TFs. The combination of TFs active in a cell determines the transcriptional output and cell fate. Positive and negative transcriptional inputs are integrated at the commonly accessible regulatory elements. Cell-fate choice, therefore, depends on the balance of activating and repressing TFs bound to these common, pre-allocated elements. The CREs remain open and available for TF binding regardless of whether their target genes are actively transcribed. So together, the combination and configuration of TFs bound to the common set of CREs determines cell identity. This strategy enables cell-fate transitions without requiring extensive chromatin remodelling, as the regulatory infrastructure is already established and available for occupancy by different TF combinations.

The morphogenetic field concept bridges classical developmental biology (Barresi and Gilbert, 2023) with modern chromatin biology. Fields represent both spatial domains of patterned gene expression and genomic domains defined by shared regulatory potential. This dual nature explains how tissues can maintain coherent responses to developmental signals while generating the cellular diversity necessary for organ function. The morphogenetic field concept, when understood through the lens of chromatin accessibility, provides a framework for understanding how developmental patterning achieves both coordination and specificity. The shared chromatin landscape of a morphogenetic field ensures coordinated responses to patterning signals, while cell-type-specific modifications to this landscape enable the generation of distinct cell identities within the field. Pioneer TFs serve as the architects of morphogenetic fields, establishing the initial chromatin accessibility patterns that define tissue identity. Their binding creates the regulatory infrastructure upon which subsequent gene expression programmes are elaborated.

The neural tube provides an example of this mechanism (Delás et al., 2023). Most ventral neural progenitor domains share a common chromatin accessibility landscape established during neural induction, creating a shared regulatory foundation. These elements are subsequently bound by different combinations of TFs depending on the local levels of morphogen signalling. The same regulatory element that binds Sox2 in all neural progenitors additionally binds Pax6 in more dorsal progenitors, which have only been exposed to low levels of Shh signalling, while it binds Nkx6-1 and Olig2 in ventral progenitors that receive high levels of Shh signalling and have thus induced expression of these TFs.

This strategy of differential TF binding contrasts with cell-fate specification mechanisms that involve chromatin remodelling to establish unique regulatory landscapes to distinguish one cell type from another within the same tissue. Establishing cell-type-specific CRE accessibility requires cell-type-specific pioneer factors capable of binding to closed chromatin and opening new regulatory regions, fundamentally rewiring the GRN. Moreover, transitioning between morphogenetic fields through pioneering provides a hypothesis for how epigenetic memory may be achieved, where transient expression of the pioneer factor establishes lasting changes in chromatin configuration that persist even after the downregulation of the pioneer. This provides a molecular explanation for how brief developmental signals can have lasting consequences on cell-fate specification.

It is likely that a combination of both differential TF binding and differential CRE accessibility operates in many tissues. For example, although most ventral progenitors in the neural tube share a common chromatin landscape, one group of cells – p3 progenitors – possess a distinct set of accessible CREs not found in neighbouring progenitors. The pioneer TF FoxA2, expressed transiently in response to high Shh signalling, opens these p3-specific regulatory elements during early neural tube development. Once opened by FoxA2, these elements remain accessible and are subsequently bound by the p3-specific TF Nkx2-2, establishing and maintaining p3 cell identity.

Within established morphogenetic fields, graded morphogen signals control gene expression by modulating the activity and binding of TFs to the pre-established regulatory landscape. The morphogen gradient acts primarily by determining which combinations of TFs are active at each position within the field. This explains how a single morphogen can specify multiple distinct cell fates: the same signal activates different transcriptional programmes within the context of the field's chromatin landscape. In this way, the chromatin landscape serves as both the memory of developmental history and the platform for interpreting current signalling inputs, linking past developmental decisions with present cell-fate choices.

4.9 THE EVOLUTION OF NEW CRES

The establishment of new regulatory elements and the modification of existing elements have been proposed to be a major driver of phenotypic innovation (Rebeiz and Tsiantis, 2017). Unlike TFs, which are reused in different developmental contexts, CREs are generally thought to be tissue-specific, allowing the same TF to perform different gene regulatory functions in different cellular contexts. Indeed, while CREs are variable across genomes, TFs show remarkable conservation in their preferred binding sites (He et al., 2011; Zheng et al., 2010).

Given that TFBSs are small, it is conceivable that random mutagenesis of existing "neutral" sequence can lead to the evolution of a particular gene regulatory behaviour, provided selection pressures exist. Indeed, high-throughput screens of large oligonucleotide libraries have shown that a subset of random sequences can function as regulatory elements, sometimes driving robust and tissue-specific activity (Sahu et al., 2022; R. P. Smith, Riesenfeld, et al., 2013). Moreover, given that CREs can, in principle, act long-range, the genomic space of mutagenic effect in principle is large, making it plausible that a given TFBS appears within the window of influence purely by chance. Using such approximate calculations, it was estimated that it would take between 0.5 and 10 million years to evolve new anterior–posterior *Drosophila* blastoderm patterning CREs from neutral DNA (Duque and Sinha, 2015). Indeed, comparative sequence analysis suggests that between ~50% and 70% of CREs arose de novo (Villar et al., 2015).

New CREs may also arise from transposable elements (Britten and Davidson, 1969, 1971; Fueyo et al., 2022; Trizzino et al., 2017) and are the likely origins for the majority of primate-specific regulatory elements (Jacques et al., 2013), as well as uterine tissue innovation revealed through cross-species transcriptomic comparisons (Lynch et al., 2015). Further, regulatory changes spurred by transposable elements have been linked to phenotypic innovations, for example, in Gdf6 expression, promoting increases in body armour size in sticklebacks during their marine to freshwater transition (Indjeian et al., 2016). Transposable elements may also give de novo evolution a head-start by providing a sequence substrate that is "closer" to functionality; for example, enhancer-like properties can be gained from Alu repetitive elements, which populate 10% of our genomes, by just a few mutations (Su et al., 2014).

A third source of regulatory innovation has been proposed to occur through "co-option", where novel regulatory behaviours are added to or modified from existing CREs. Regulatory elements may acquire a small number of mutations to gain or lose binding sites, or modify cooperative interactions, to facilitate phenotypic innovation, as has been reported for innovations in *Drosophila biarmipes* wing pigmentation patterning (Gompel et al., 2005), or Nep1 gene expression in the optic lobe of *Drosophila santomea* (Rebeiz et al., 2011). Such co-option events might be enabled by the intrinsic modularity of regulatory sequences, meaning different aspects of their information integration can be tuned somewhat independently without catastrophic phenotypic implications. Indeed, in many instances, there is deep functional conservation of regulatory elements with minimal or absent sequence conservation; TFBS composition is approximately conserved, but the overall structure of the sequence among species has diverged substantially. Instances of deeply conserved enhancer gene expression patterns even across major morphological transitions of homologous body parts, such as spatially organised Hox gene expression in the fin to limb transition (Gehrke et al., 2015), indicate that regulatory elements may have more ancient origins than the developmental processes they regulate. Viewing CREs not in isolation but embedded within GRNs, such as in sub-networks – Davidson's "kernels" – may impose strict functional conservation of regulatory connections, where evolutionary innovations occur through subtle tinkering of core regulatory interactions or through swapping out less-connected "plug-ins" or "switches" (Davidson and Erwin, 2006).

4.10 PRINCIPLED ENGINEERING OF SYNTHETIC REGULATORY ELEMENTS

The constraints and composability rules of regulatory element information integration could be used to inform principled engineering of synthetic regulatory elements. Studies concatenating regulatory elements stress the combinatorial requirement of TFBSs and their relative positioning, with optimal sequences only identified in retrospect of massively parallel screening (Fiore and Cohen, 2016; Georgakopoulos-Soares et al., 2023; Sahu et al., 2022). One solution to this problem has been to exploit the large and growing amount of genomics data on natural CRE functionality and train neural network-based deep-learning models of regulatory element activity (Alipanahi et al., 2015; Avsec, Agarwal, et al., 2021; Avsec, Weilert, et al., 2021; H. Janssens et al., 2006; Kelley et al., 2016, 2018; D. S. Kim et al., 2021; Minnoye et al., 2020; Oh et al., 2021; Pampari et al., 2025; J. Zhou and Troyanskaya, 2015).

Using these predictive models, various optimisation strategies have been implemented to design synthetic regulatory sequences with specific functionalities. A simple solution is through random sequence filtering and prioritisation, which showed reasonable performance using genome-wide STARR-seq data on *Drosophila* S2 cells to train the model (de Almeida et al., 2022). This strategy could be further adapted to establish tissue-specific CRE patterns by performing transfer learning. Here, the weights of a pre-trained *Drosophila* S2 cell CNN model were tuned by secondary fitting using a curated set of spatiotemporal expression patterns of defined natural regulatory elements, producing synthetic regulatory elements for which ~70% were expressed in their target tissue (De Almeida et al., 2024). However, random sequence search is unlikely to yield optimal results with finite computational power.

Consequently, alternative approaches have been developed to guide the optimisation of synthetic regulatory elements. Specifically, an implicit assumption is that the "fitness" landscape of regulatory sequences is sufficiently smooth that an optimal solution can be found through incremental optimisation. This has been used to optimise yeast promoters using models trained on expression data (Vaishnav et al., 2022) and cell-type-specific synthetic *Drosophila* enhancers using models trained on single-nuclear ATAC-seq data (i.e. chromatin accessibility) (Taskiran et al., 2024). In the latter study, functional cell-type-specific enhancers could be "evolved", with a ~75% validation rate, within 15 mutation steps from a random sequence. Analysis of the mutational paths indicated that inactive random sequences seemingly showed a predominance of repressive TFBS, which had to be eroded before activating TFBSs were added. Optimal sequences contained approximately four to six TFBSs that contributed to the predicted "activity" (strictly speaking, accessibility). In the same study, an orthogonal approach that embedded TFBS in a random sequence yielded comparable results, where the model-selected sequences appeared to optimise TFBS composition and relative spacing. While validation datasets were limited to the 10s, validations of predictions of machine-learning-guided cell-type-specific synthetic regulatory element designs were recently scaled up to ~10^4, using MPRAs both to train the model and to validate predictions (Gosai et al., 2024).

While such approaches are exciting for engineering purposes, they face a series of technical and conceptual challenges if they are going to prove useful to understand natural regulatory elements. Firstly, the choice of training data can fundamentally influence the predictions of any model. Virtually all deep-learning models are trained on naturally occurring sequences, meaning data contamination is unavoidable (de Boer and Taipale, 2024). Moreover, the data modality used to train the model is typically not rooted in CRE (reporter) activity but TF binding or chromatin accessibility. These are not equivalent. We should therefore expect such technologies to perform less well in distinguishing a more similar set of cell types – for example, progenitor cells – especially those that exist within a single morphogenetic field. When models are trained on activity, this is exclusively through MPRA approaches, which generally have input sequence size limits that are far smaller than average regulatory elements. Yet parts of a regulatory sequence interact to make the whole, meaning the space of regulatory possibilities is feasibly limited or biased in these approaches, and insights may not always generalise to larger, more natural sequences.

More fundamentally, much explanatory power is relinquished when deferring to black-box models typically produced by deep-learning approaches. Such models are not currently structured to infer general principles of how the parts of a regulatory sequence combine to make the whole. This is problematic even if the goal is purely prediction, as each new application requires an accompanying large-scale dataset. Nonetheless, the remarkable predictive power of these deep learning approaches reveals some hidden correlative structure in active regulatory DNA. This then poses perhaps a deeper question: to what extent is the correlative structure intuitive to humans? For example, does the activity of a regulatory element depend on a set of high-order interactions between its many base-pairs, or alternatively can the activity of a sequence be rationalised in terms of its building-blocks which combine together with simple first-order or pairwise composability rules?

As a complementary perspective, we have recently shown that Hedgehog-dependent synthetic CREs (synCREs) with a diversity of behaviours can be readily generated by fusing fragments of endogenous neural patterning CREs (Cornwall-Scoones et al., 2025). Using a high-throughput combinatorial screening strategy, termed NeMECiS, to generate and profile the exhaustive combinatorial space of three-way fragment fusions, we discovered a surprising flexibility and compatibility among combinations. Statistical modelling established quantitative rules governing the signal-dependent CRE activity: regulatory sequence fragments exhibited characteristic activities that multiply to achieve strong, signal-dependent CREs; these activities are tuned by synergies among pairs of fragments, and these synergies often depend on spacing. That synCRE activity can be explained by purely modular and pairwise contributions aids the interpretability of explanations and activity to be predicted for arbitrary fragment combinations.

We used these rules to determine whether the natural ordering of fragments in an Olig2 CRE was "optimal". This revealed that alternative orderings can both diminish and boost the signal-dependent activity of this CRE. These out-of-distribution predictions both demonstrate the generalisability of the rules and emphasise the role of order and spacing of fragments for tuning natural CRE activity. We then explored the consequences for patterning when rewiring this GRN with full-length synCREs: minimising and maximising the syntactical (order and spacing) contributions in a natural CRE altered the expression level of Olig2 and domain size of motor neuron progenitors in neural tube organoids in the predicted manner.

4.11 UNDERSTANDING THE PARTS TO BUILD THE WHOLE: RECONSTRUCTING GRNs

How do the rules of information processing in regulatory elements shape GRNs? To answer this, it is necessary to examine how insights have been both applied and overlooked when building GRNs from data. One of the grand challenges in biology, especially developmental biology, is to devise predictive models of (the dynamics of) gene expression programmes. This would provide insights into how cells make decisions, given a diverse set of temporally stratified cues, and how natural or engineered perturbations in the signalling or the underlying genetics influence such decision-making. In turn, this has applications in devising better tissue engineering approaches or understanding how disease-associated genetic variation becomes manifest in pathologies through the generative processes of GRNs (Badia-I-Mompel et al., 2023; Briscoe and Marín, 2020; Levine and Davidson, 2005).

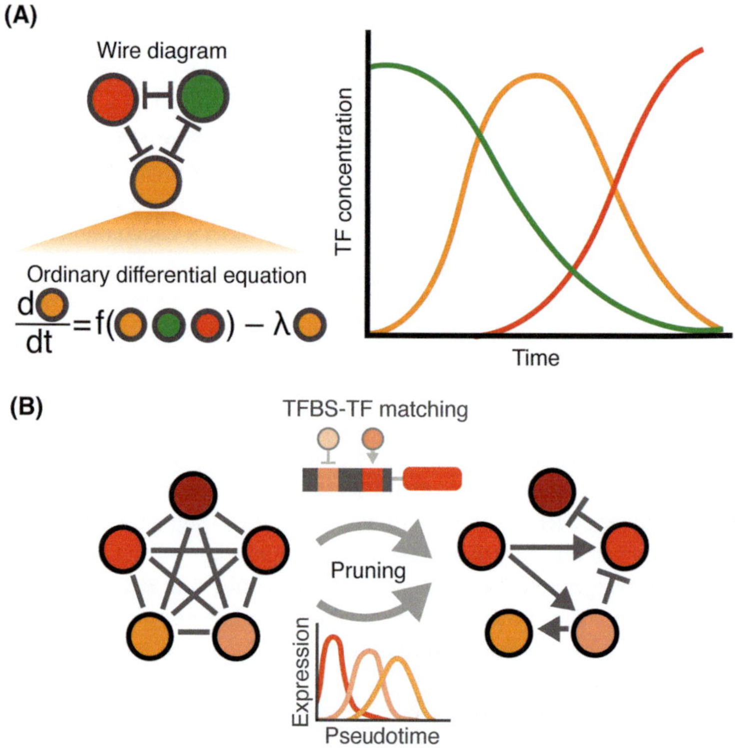

FIGURE 4.5 Mechanistic and statistical models of gene regulatory networks. (A) Mechanistic models. GRNs are often represented as wiring diagrams that can be cast as systems of ordinary differential equations, where the expression of a TF depends on the concentrations of the other TFs in the system. The function "f" here is the gene regulation function. This yields a dynamical system with stable states, allowing for dynamic simulations from initial conditions to one of multiple steady states. Figure adapted from Zhu et al. (2022). (B) Statistical models. Gene expression co-variance can help produce networks of putatively interacting TFs. Then, through a series of pruning approaches, including using the dynamics of TF expression in (pseudo)time and through assigning CREs to TFs, and TFBSs to CREs, causal arrows can be inferred.

One class of models can broadly be described as "mechanistic" (Figure 4.5A). These seek to describe the rules by which the expression of TFs depends on each other, then simulate and analyse behaviours of such models to gain an understanding of the operation of a GRN. Broadly, these can be subdivided into logical models, where the expression state of each TF is Boolean, temporal progression is discrete and interaction rules between TFs follow logical operations (e.g. AND/OR) (Glass and Kauffman, 1973; Thomas, 1973) or more generically decision trees (Segal et al., 2003). While this is a simplification of the quantitative nature of gene expression control, where information integration is continuous, not discrete, such approaches have had remarkable success, most notably in sea urchin development (Davidson et al., 2002; J. Smith et al., 2007). Such approaches are also more computationally tractable, with algorithms to infer feasible networks, given sparsely validated interactions (Lähdesmäki et al., 2003) (for discussion, see Karlebach and Shamir (2008)).

Alternatively, mechanistic models may be continuous in both state-space (i.e. TF concentrations) and time. Such approaches typically cast the temporal evolution of GRNs as a set of differential equations, an approach dating back to Goodwin (Goodwin, 1963), that describe transcription rate in terms of the levels of each TF protein through parameterised gene regulation functions. Choices of regulation function present a trade-off between computational/analytical tractability and the potentially complex biological realities, with at one extreme strictly linear models capable of analytical solutions and at the other fully parameterised sets of thermodynamic-formalism-inspired gene regulation functions, which can take potentially thousands of parameters and are likely under-constrained by data. The latter approaches have been used successfully to model the gap-gene network of *Drosophila* anterior–posterior patterning (Jaeger et al., 2004) and dorsoventral patterning of neural progenitors by Shh in the vertebrate neural tube (Balaskas et al., 2012; Cohen et al., 2014; Exelby et al., 2021). Both case studies are highly constrained by a large amount of genetic data, which dramatically reduce the parameter search space both in terms of TFs and modes of interaction.

For GRNs where such a wealth of functional data is lacking, regulatory interactions can be alternatively constrained through parameterising gene regulation functions based on sequence, typically TFBS composition (Fiore and Cohen, 2016; Segal et al., 2008), although at present, we lack well-defined predictors of which TFBSs are likely to be bound, functionally important or involved in cooperative interactions. Consequently, continuous models have an even larger challenge of under-determination, where empirical heuristics regarding how sequence maps to regulatory activity could prove of great use.

Continuous mechanistic models of GRNs can provide a deeper level of understanding of developmental ontogeny. For example, models of the neural tube patterning GRN have indicated the basis of hysteresis in cell-fate acquisition in the mutually inhibiting architecture of the network (Balaskas et al., 2012; Cohen et al., 2014), identified how GRNs can process information from dynamically changing gradients (Richards and Saunders, 2015) and tested roles of protein degradation in the scaling of developmental tempo (Rayon et al., 2020). Additionally, analysis of the gap-gene network in *Drosophila* revealed three overlapping gene regulatory motifs, differentially regulated by maternal gradients, involving the same TFs in qualitatively different dynamical sub-systems in different portions of the anterior–posterior axis (Verd et al., 2019). This study underscores how predictions and intuitions about GRN dynamics are only possible with a quantitative description of the gene regulation function.

"Statistical" or data-driven GRNs constitute the other class of models (Figure 4.5B). These aim to extrapolate the successes of mechanistic models to situations in which the underlying regulatory interactions are unknown by exploiting correlative data afforded by technologies such as single-cell RNA- and ATAC-seq (Badia-I-Mompel et al., 2023). Generally, the output of these models is a prospective wiring diagram of TFs, constructed from correlation metrics among gene-sets (Huynh-Thu et al., 2010) (Moerman et al., 2019). Given the dramatic co-variance of TFs in our genome, attributing causality to correlative weights is a fundamental challenge, addressed through "chronological" arguments (e.g. through pseudotemporal ordering; Papili Gao et al., 2018; Specht and Li, 2017) or by constraining feasible interactions using TFBS contents of nearby CREs (Aibar et al., 2017; Bravo González-Blas et al., 2023; J. Janssens et al., 2022; Kamimoto et al., 2023). In order to cast such networks as dynamical models with potential for prediction, most approaches assume TFs modulate transcription in a *linear* and *independent* manner (see Table 1 in Badia-I-Mompel et al. (2023)), with some studies proposing capacities of such models to predict fate responses upon perturbation (Bravo González-Blas et al., 2023; Kamimoto et al., 2023).

Statistical GRN models are currently plagued with methodological issues. At present, approaches are highly divergent in their predictions, are sensitive to parameter choices, cannot recover some key regulatory interactions and are not able to reproduce causal TF interactions from perturbation experiments (Badia-I-Mompel et al., 2024). The significance of these setbacks depends on the use-case of such methodologies. Methods simplifying the complex web of correlations in gene expression data suggest possible regulatory connections between TFs. However, without experimental validation, it is unclear whether this simplification reveals the actual regulatory connections or just makes random guesses. Moreover, given the interdependence of TF binding in driving gene expression, it is often unclear how meaningful a single regulatory arrow in such wiring diagrams is. If, instead, we would like to build dynamical, causally motivated, predictive models of both cell-fate stability and cell-fate transitions, we need dynamical descriptions. This is at odds with the current modelling paradigms in several important regards. Firstly, inference packages typically build one GRN per cell type, yet dynamical systems theory tells us that cell types are dynamical attractors manifest within a single GRN; consequently, there is no coherent description of a cell-fate transition. Secondly, such models do not represent gene regulation functions in biophysically meaningful ways, yet we know that the quantitative parameterisation of such functions, even with identical inputs, can yield qualitatively different behaviours. Hence, thirdly, such approaches over-emphasise the value of a perfect wiring diagram.

Is statistical GRN inference totally in vain? With the birth of these technologies, it is worth heeding the recommendation from Levine and Davidson some 20 years prior: "We argue that there is only one way to authenticate a GRN, and that is by means of direct experimental manipulation of individual cis-regulatory modules" (Levine and Davidson, 2005). Can this piecemeal approach be scaled up? An exciting interface in the years to come will be a combination of data-driven models of CRE activity with their broader connection into lineage- or embryo-wide dynamic GRNs, holding promise to parameterise gene regulation functions without needing extensive perturbation in all of the ~10^5 regulatory elements in our genome. Nonetheless, to constrain models for computational tractability and to provide a layer of interpretability about how regulatory inputs to a particular node are integrated and affect global expression dynamics, we need some empirical grounding on how the parts of the regulatory sequence come together to define gene expression as a whole and how these parts differentially respond to changing stimuli.

4.12 PROGRAMMING DEVELOPMENTAL DYNAMICS WITH SYNTHETIC GRNs

One goal of developmental biology and of mammalian bioengineering is to re-engineer GRNs controlling cell-fate decisions. This goes beyond simply generating regulatory sequences that are active at a particular place or tissue. The ambition is to engineer CREs and TFs to alter tissue ontogenesis or cellular physiology itself. The driving principle is that if all the nodes and edges in a GRN are known, we should be able to genetically intervene to reshape the dynamical system and thus redirect cell-fate decisions. Achieving these capabilities, and the endeavour of trying, should yield an understanding of fundamental questions in developmental biology, helping test hypotheses about the genetic and causal structure of natural GRNs, and derive, through experimentation, principles about how GRN construction relates to phenotype, for example, determining trade-offs between the pace, robustness and sensitivity of cell-fate decisions.

This emerging experimental paradigm pushes developmental biology towards convergence with synthetic biology – termed "generative biology" – where design principles of natural systems are derived by building synthetic alternatives (Church et al., 2014), separating functional principles from a system's evolutionary history. Specifically, one ambition of the field is to establish suites of synthetic GRNs to map how genotype and phenotype inter-relate, making the connection between meta-networks of genotypes and morphospaces of phenotypes (Babtie et al., 2014; Ma et al., 2009). This is particularly exciting in studying questions of evolutionary developmental biology. It would distinguish ecological, historical and ontogenic contingencies and constraints in the construction of GRNs (Currie, 2012; McGhee, 2011). Moreover, with the greater control afforded by synthetic approaches, synthetic GRNs have the potential to act as therapeutic devices, decoding cellular context to retune physiology (Allen et al., 2022; Bragdon et al., 2023).

The promise of synthetic GRNs is, however, met with the challenge of constructing them. Generating synthetic GRNs from scratch has proved challenging. The assembly of synthetic GRNs started the field of synthetic biology with the advent of bioengineered repressillators and toggle switches in bacteria (Elowitz and Leibler, 2000; Gardner et al., 2000). Work in bacteria has continued over the past quarter-century to establish synthetic morphogen decoding circuits, and has provided new insights into the functional and evolutionary trade-offs implicit in circuit design (Baier et al., 2023; Santos-Moreno et al., 2020; Schaerli et al., 2014). Yet despite major success stories, translating these principles into mammalian systems has proved challenging (Cabrera et al., 2022; Glykofrydis and Elfick, 2022; Lienert et al., 2014). Advances have been made in generating synthetic TFs (Khalil et al., 2012; F. Zhang et al., 2011), including through co-opting CRISPR-Cas9 (Benzinger et al., 2022; Perez-Pinera et al., 2013), using programmed oligomerisation (Zhu et al., 2022), or through tailored protease circuits (Z. Chen et al., 2024; Fink et al., 2019; Gao et al., 2018). But less attention has been given to the non-coding half of GRNs – the CREs – which act as the edges in these networks.

Establishing predictive models of the cis-regulatory logic underpinning gene regulation and cell-fate decisions has implications for our understanding of more macroscopic developmental processes. An understanding of how CREs decode concentrations of Hedgehog ligands helps rationalise how the neural tube is patterned to specify distinct bands of dorsoventral progenitors (Balaskas et al., 2012). It highlights the roles of a tissue's initial and boundary conditions in determining subsequent development by affecting the dynamics and inputs of the underlying GRN. Yet studying synCREs in natural mammalian tissues is technically challenging due to experimental bottlenecks for genetic engineering. Hence, an exciting emerging avenue is the use of in vitro stem cell-derived culture platforms, which recapitulate tissue-scale properties of natural embryos (J. Kim et al., 2020). Moreover, their combination with advanced 3D microscopy and image analysis techniques enables the construction and analysis of synthetic developmental patterning at scale.

The combination of programmable genetic circuits with the emerging field of "synthetic embryology" has the potential to reveal design principles of development (Cornwall-Scoones and Zernicka-Goetz, 2021; Martínez-Ara et al., 2022; Toda et al., 2019). An exciting prospect of this intersection is to explore more non-linear patterning systems, going beyond cells "simply" reading out concentrations of pre-ordained gradients, towards scenarios where cellular interactions mutually enable the coordination of developmental patterns (Collinet and Lecuit, 2021; Green and Sharpe, 2015). For example, this synthesis may enable the construction of synthetic Turing patterning circuits by coupling intracellular synGRNs to the expression of external signalling molecules (Sekine et al., 2018; Tica et al., 2024), where the combination of local activation and long-range inhibition can drive spontaneous symmetry breaking with a defined wavelength (Hiscock and Megason, 2015; Koch and Meinhardt, 1994; Turing, 1952). While Turing mechanisms have been posited to underpin several key developmental processes (Cornwall-Scoones and Hiscock, 2020; Economou et al., 2012; Ho et al., 2019; Raspopovic et al., 2014; Sick et al., 2006), our understanding of the quantitative interactions underpinning this symmetry breaking has remained confined to simulation (Marcon et al., 2016). Moreover, understanding and programming how CREs interface with signalling molecules – both in signal reception and signal production – has broader implications in revealing principles of tissue homeostasis, pattern scaling and growth control, which are likely underpinned by tissue-scale signalling feedback loops whose set points are encoded by this intracellular machinery (Kitadate et al., 2019; Rahimi et al., 2020).

4.13 PERSPECTIVE

Understanding GRNs as dynamical systems requires integrating molecular mechanisms with computational frameworks to decode how cells interpret complex developmental signals. Here, we have highlighted the critical gap between mapping regulatory connections and achieving predictive understanding of cell-fate decisions. The quantitative logic of CREs remains the missing component that transforms descriptive network diagrams into mechanistic models capable of explaining developmental ontogeny and engineering synthetic systems.

Several challenges remain. We need empirical frameworks that define how TFs combine at regulatory elements to control gene expression, moving beyond assumed regulatory functions towards experimentally validated input–output relationships.

The development of massively parallel approaches and synthetic regulatory element design offers promising avenues for establishing these quantitative rules. We also need to understand information transmission between distal regulatory elements and promoters. This requires resolving the mechanistic basis of enhancer–promoter communication, whether through direct contact, diffusive signals or alternative paradigms. Finally, integrating morphogenetic field concepts with chromatin landscapes provides a framework for understanding how pioneer factors establish regulatory infrastructure that enables coordinated tissue responses while maintaining cellular diversity.

The convergence of synthetic biology with developmental biology presents unprecedented opportunities to test fundamental principles of GRN construction and evolution. Programmable regulatory circuits can reveal design principles of natural networks while offering therapeutic applications for tissue engineering and regenerative medicine. Success in this endeavour depends on establishing predictive models that connect sequence to function, enabling rational design of regulatory elements that can reshape developmental dynamics. The ultimate goal is to understand how the molecular anatomy of regulatory DNA encodes the information-processing capabilities that transform single cells into complex multicellular organisms, providing both fundamental insights into development and practical tools for bioengineering applications.

ACKNOWLEDGEMENTS

We are grateful to Joaquina Delas and Sujay Raj for their constructive comments on the manuscript. Work in the group is supported by the Francis Crick Institute, which receives its core funding from Cancer Research UK (CC001051), the UK Medical Research Council (CC001051) and the Wellcome Trust (CC001051 and 220379/D/20/Z). JCS was funded by a PhD fellowship from the Boehringer Ingelheim Fonds.

REFERENCES

Abdella, R., Talyzina, A., Chen, S., Inouye, C.J., Tjian, R., He, Y., 2021. Structure of the human Mediator-bound transcription preinitiation complex. *Science* 372(6537), 52–56. https://doi.org/10.1126/science.abg3074

Ackers, G.K., Johnson, A.D., Shea, M.A., 1982. Quantitative model for gene regulation by lambda phage repressor. *Proceedings of the National Academy of Sciences*, 79(4) 1129–1133. https://doi.org/10.1073/pnas.79.4.1129

Agarwal, V., Inoue, F., Schubach, M., Penzar, D., Martin, B.K., Dash, P.M., Keukeleire, P., Zhang, Z., Sohota, A., Zhao, J., Georgakopoulos-Soares, I., Noble, W.S., Yardımcı, G.G., Kulakovskiy, I.V., Kircher, M., Shendure, J., Ahituv, N., 2025. Massively parallel characterization of transcriptional regulatory elements. *Nature* 1–10. https://doi.org/10.1038/s41586-024-08430-9

Aibar, S., González-Blas, C.B., Moerman, T., Huynh-Thu, V.A., Imrichova, H., Hulselmans, G., Rambow, F., Marine, J.-C., Geurts, P., Aerts, J., van den Oord, J., Atak, Z.K., Wouters, J., Aerts, S., 2017. SCENIC: Single-cell regulatory network inference and clustering. *Nature Methods* 14(11) 1083–1086. https://doi.org/10.1038/nmeth.4463

Alexander, J.M., Guan, J., Li, B., Maliskova, L., Song, M., Shen, Y., Huang, B., Lomvardas, S., Weiner, O.D., 2019. Live-cell imaging reveals enhancer-dependent Sox2 transcription in the absence of enhancer proximity. *eLife* 8, e41769.

Alipanahi, B., Delong, A., Weirauch, M.T., Frey, B.J., 2015. Predicting the sequence specificities of DNA- and RNA-binding proteins by deep learning. *Nature Biotechnology* 33(8), 831–838. https://doi.org/10.1038/nbt.3300

Allahyar, A., Vermeulen, C., Bouwman, B.A.M., Krijger, P.H.L., Verstegen, M.J.A.M., Geeven, G., van Kranenburg, M., Pieterse, M., Straver, R., Haarhuis, J.H.I., Jalink, K., Teunissen, H., Renkens, I.J., Kloosterman, W.P., Rowland, B.D., de Wit, E., de Ridder, J., de Laat, W., 2018. Enhancer hubs and loop collisions identified from single-allele topologies. *Nature Genetics* 50(8), 1151–1160. https://doi.org/10.1038/s41588-018-0161-5

Allen, G. M., Frankel, N.W., Reddy, N.R., Bhargava, H.K., Yoshida, M.A., Stark, S.R., Purl, M., Lee, J., Yee, J.L., Yu, W., Li, A.W., Garcia, K.C., El-Samad, H., Roybal, K.T., Spitzer, M.H., Lim, W.A., 2022. Synthetic cytokine circuits that drive T cells into immune-excluded tumors. *Science* 378(6625), eaba1624. https://doi.org/10.1126/science.aba1624

Alon, U., 2007. Network motifs: Theory and experimental approaches. *Nature Reviews. Genetics* 8(6), 450–461. https://doi.org/10.1038/nrg2102

Andersson, R., Gebhard, C., Miguel-Escalada, I., Hoof, I., Bornholdt, J., Boyd, M., Chen, Y., Zhao, X., Schmidl, C., Suzuki, T., Ntini, E., Arner, E., Valen, E., Li, K., Schwarzfischer, L., Glatz, D., Raithel, J., Lilje, B., Rapin, N., … Sandelin, A., 2014. An atlas of active enhancers across human cell types and tissues. *Nature* 507(7493), 455–461. https://doi.org/10.1038/nature12787

Andersson, R., Sandelin, A., 2020. Determinants of enhancer and promoter activities of regulatory elements. *Nature Reviews Genetics* 21(2), 71–87. https://doi.org/10.1038/s41576-019-0173-8

Armendariz, D.A., Sundarrajan, A., Hon, G.C., 2023. Breaking enhancers to gain insights into developmental defects. *eLife* 12, e88187.

Arnold, C. D., Gerlach, D., Stelzer, C., Boryń, Ł. M., Rath, M., Stark, A., 2013. Genome-Wide Quantitative Enhancer Activity Maps Identified by STARR-seq. *Science* 339(6123), 1074–1077. https://doi.org/10.1126/science.1232542

Avsec, Ž., Agarwal, V., Visentin, D., Ledsam, J.R., Grabska-Barwinska, A., Taylor, K.R., Assael, Y., Jumper, J., Kohli, P., Kelley, D.R., 2021. Effective gene expression prediction from sequence by integrating long-range interactions. *Nature Methods* 18(10), 1196–1203. https://doi.org/10.1038/s41592-021-01252-x

Avsec, Ž., Weilert, M., Shrikumar, A., Krueger, S., Alexandari, A., Dalal, K., Fropf, R., McAnany, C., Gagneur, J., Kundaje, A., Zeitlinger, J., 2021. Base-resolution models of transcription-factor binding reveal soft motif syntax. *Nature Genetics* 53(3), 354–366. https://doi.org/10.1038/s41588-021-00782-6

Babtie, A.C., Kirk, P., Stumpf, M.P.H., 2014. Topological sensitivity analysis for systems biology. *Proceedings of the National Academy of Sciences of the United States of America* 111(52), 18507–18512. https://doi.org/10.1073/pnas.1414026112

Badia-I-Mompel, P., Casals-Franch, R., Wessels, L., Müller-Dott, S., Trimbour, R., Yang, Y., Flores, R.O.R., Saez-Rodriguez, J., (2024). *Comparison and evaluation of methods to infer gene regulatory networks from multimodal single-cell data* (p. 2024.12.20.629764). bioRxiv. https://doi.org/10.1101/2024.12.20.629764

Badia-I-Mompel, P., Wessels, L., Müller-Dott, S., Trimbour, R., Ramirez Flores, R.O., Argelaguet, R., Saez-Rodriguez, J., 2023. Gene regulatory network inference in the era of single-cell multi-omics. *Nature Reviews. Genetics* 24(11), 739–754. https://doi.org/10.1038/s41576-023-00618-5

Baier, F., Gauye, F., Perez-Carrasco, R., Payne, J.L., Schaerli, Y., 2023. Environment-dependent epistasis increases phenotypic diversity in gene regulatory networks. *Science Advances* 9(21), eadf1773. https://doi.org/10.1126/sciadv.adf1773

Balaskas, N., Ribeiro, A., Panovska, J., Dessaud, E., Sasai, N., Page, K.M., Briscoe, J., Ribes, V., 2012. Gene regulatory logic for reading the Sonic Hedgehog signaling gradient in the vertebrate neural tube. *Cell* 148(1–2), 273–284. https://doi.org/10.1016/j.cell.2011.10.047

Banerji, J., Rusconi, S., Schaffner, W., 1981. Expression of a beta-globin gene is enhanced by remote SV40 DNA sequences. *Cell* 27(2 Pt 1), 299–308.

Barresi, M.J.F., Gilbert, S.F., 2023. *Developmental biology*, International thirteenth edition. Oxford University Press.

Benabdallah, N.S., Williamson, I., Illingworth, R.S., Kane, L., Boyle, S., Sengupta, D., Grimes, G.R., Therizols, P., Bickmore, W.A., 2019. Decreased Enhancer-Promoter Proximity Accompanying Enhancer Activation. *Molecular Cell* 76(3), 473–484.e7. https://doi.org/10.1016/j.molcel.2019.07.038

Benzinger, D., Ovinnikov, S., Khammash, M., 2022. Synthetic gene networks recapitulate dynamic signal decoding and differential gene expression. *Cell Systems* 13(5), 353–364.e6. https://doi.org/10.1016/j.cels.2022.02.004

Bialek, W., Gregor, T., Tkačik, G., 2019. Action at a distance in transcriptional regulation, No. arXiv:1912.08579. arXiv. https://doi.org/10.48550/arXiv.1912.08579

Bintu, L., Buchler, N.E., Garcia, H.G., Gerland, U., Hwa, T., Kondev, J., Phillips, R., 2005. Transcriptional regulation by the numbers: Models. *Current Opinion in Genetics and Development* 15(2), 116–124. https://doi.org/10.1016/j.gde.2005.02.007

Boija, A., Klein, I. A., Sabari, B. R., Dall'Agnese, A., Coffey, E.L., Zamudio, A.V., Li, C.H., Shrinivas, K., Manteiga, J.C., Hannett, N.M., Abraham, B.J., Afeyan, L.K., Guo, Y.E., Rimel, J.K., Fant, C.B., Schuijers, J., Lee, T.I., Taatjes, D.J., Young, R.A., 2018. Transcription Factors Activate Genes through the Phase-Separation Capacity of Their Activation Domains. *Cell* 175(7), 1842–1855.e16. https://doi.org/10.1016/j.cell.2018.10.042

Bragdon, M.D.J., Patel, N., Chuang, J., Levien, E., Bashor, C.J., Khalil, A.S., 2023. Cooperative assembly confers regulatory specificity and long-term genetic circuit stability. *Cell* 186(18), 3810–3825.e18. https://doi.org/10.1016/j.cell.2023.07.012

Bravo González-Blas, C., De Winter, S., Hulselmans, G., Hecker, N., Matetovici, I., Christiaens, V., Poovathingal, S., Wouters, J., Aibar, S., Aerts, S., 2023. SCENIC+: Single-cell multiomic inference of enhancers and gene regulatory networks. *Nature Methods* 20(9), 1355–1367. https://doi.org/10.1038/s41592-023-01938-4

Briscoe, J., Marín, O., 2020. Looking at neurodevelopment through a big data lens. *Science* 369(6510), eaaz8627. https://doi.org/10.1126/science.aaz8627

Briscoe, J., Pierani, A., Jessell, T.M., Ericson, J., 2000. A homeodomain protein code specifies progenitor cell identity and neuronal fate in the ventral neural tube. *Cell* 101(4), 435–445.

Briscoe, J., Small, S., 2015. Morphogen rules: Design principles of gradient-mediated embryo patterning. *Development* 142(23), 3996–4009. https://doi.org/10.1242/dev.129452

Briscoe, J., Sussel, L., Serup, P., Hartigan-O'Connor, D., Jessell, T.M., Rubenstein, J.L., Ericson, J., (1999). Homeobox gene Nkx2.2 and specification of neuronal identity by graded Sonic hedgehog signalling. *Nature*, *398*(6728), 622–627. https://doi.org/10.1038/19315

Britten, R.J., Davidson, E.H., 1969. Gene regulation for higher cells: A theory. *Science* 165(3891), 349–357. https://doi.org/10.1126/science.165.3891.349

Britten, R.J., Davidson, E.H., 1971. Repetitive and non-repetitive DNA sequences and a speculation on the origins of evolutionary novelty. *The Quarterly Review of Biology* 46(2), 111–138. https://doi.org/10.1086/406830

Brückner, D.B., Chen, H., Barinov, L., Zoller, B., Gregor, T., 2023. Stochastic motion and transcriptional dynamics of pairs of distal DNA loci on a compacted chromosome. *Science* 380(6652), 1357–1362. https://doi.org/10.1126/science.adf5568

Buenrostro, J.D., Giresi, P.G., Zaba, L.C., Chang, H.Y., Greenleaf, W.J., 2013. Transposition of native chromatin for fast and sensitive epigenomic profiling of open chromatin, DNA-binding proteins and nucleosome position. *Nature Methods* 10(12), 1213–1218. https://doi.org/10.1038/nmeth.2688

Burz, D.S., Rivera-Pomar, R., Jäckle, H., Hanes, S.D., 1998. Cooperative DNA-binding by Bicoid provides a mechanism for threshold-dependent gene activation in the *Drosophila* embryo. *The EMBO Journal* 17(20), 5998–6009. https://doi.org/10.1093/emboj/17.20.5998

Cabrera, A., Edelstein, H.I., Glykofrydis, F., Love, K.S., Palacios, S., Tycko, J., Zhang, M., Lensch, S., Shields, C.E., Livingston, M., Weiss, R., Zhao, H., Haynes, K.A., Morsut, L., Chen, Y.Y., Khalil, A.S., Wong, W.W., Collins, J.J., Rosser, S.J., … Deans, T.L., 2022. The sound of silence: Transgene silencing in mammalian cell engineering. *Cell Systems* 13(12), 950–973. https://doi.org/10.1016/j.cels.2022.11.005

Catarino, R.R., Stark, A., 2018. Assessing sufficiency and necessity of enhancer activities for gene expression and the mechanisms of transcription activation. *Genes and Development* 32(3–4), 202–223. https://doi.org/10.1101/gad.310367.117

Chen, H., Levo, M., Barinov, L., Fujioka, M., Jaynes, J.B., Gregor, T., 2018. Dynamic interplay between enhancer–promoter topology and gene activity. *Nature Genetics* 50(9), 1296–1303. https://doi.org/10.1038/s41588-018-0175-z

Chen, X., Yin, X., Li, J., Wu, Z., Qi, Y., Wang, X., Liu, W., Xu, Y., 2021. Structures of the human Mediator and Mediator-bound preinitiation complex. *Science* 372(6546), eabg0635. https://doi.org/10.1126/science.abg0635

Chen, Z., Linton, J.M., Xia, S., Fan, X., Yu, D., Wang, J., Zhu, R., Elowitz, M.B., 2024. A synthetic protein-level neural network in mammalian cells. *Science* 386(6727), 1243–1250. https://doi.org/10.1126/science.add8468

Church, G.M., Elowitz, M.B., Smolke, C.D., Voigt, C.A., Weiss, R., 2014. Realizing the potential of synthetic biology. *Nature Reviews Molecular Cell Biology* 15(4), 289–294. https://doi.org/10.1038/nrm3767

Clark, E., Akam, M., 2016. Odd-paired controls frequency doubling in *Drosophila* segmentation by altering the pair-rule gene regulatory network. *eLife* 5, e18215. https://doi.org/10.7554/eLife.18215

Clyde, D.E., Corado, M.S.G., Wu, X., Paré, A., Papatsenko, D., Small, S., 2003. A self-organizing system of repressor gradients establishes segmental complexity in *Drosophila*. *Nature* 426(6968), 849–853. https://doi.org/10.1038/nature02189

Cohen, M., Kicheva, A., Ribeiro, A., Blassberg, R., Page, K.M., Barnes, C.P., Briscoe, J., 2015. Ptch1 and Gli regulate Shh signalling dynamics via multiple mechanisms. *Nature Communications* 6(1), 6709. https://doi.org/10.1038/ncomms7709

Cohen, M., Page, K.M., Perez-Carrasco, R., Barnes, C.P., Briscoe, J., 2014. A theoretical framework for the regulation of Shh morphogen-controlled gene expression. *Development* 141(20), 3868–3878. https://doi.org/10.1242/dev.112573

Collinet, C., Lecuit, T., 2021. Programmed and self-organized flow of information during morphogenesis. *Nature Reviews Molecular Cell Biology* 22(4), 245–265. https://doi.org/10.1038/s41580-020-00318-6

Cornwall-Scoones, J., Benzinger, D., Yu, T., Pezzotta, A., Sagner, A., Gerontogianni, L., Bernadet, S., Finnie, E., Boezio, G.L.M., Stuart, H.T., Melchionda, M., Inge, O.C.K., Dumitrascu, B., Briscoe, J., Delás, M.J., 2025. *Predictable Engineering of Signal-Dependent Cis-Regulatory Elements*, p. 2025.03.07.642002. bioRxiv. https://doi.org/10.1101/2025.03.07.642002

Cornwall-Scoones, J., Hiscock, T.W., 2020. A dot-stripe Turing model of joint patterning in the tetrapod limb. *Development* 147(8), dev183699. https://doi.org/10.1242/dev.183699

Cornwall-Scoones, J., Zernicka-Goetz, M., 2021. Unifying synthetic embryology. *Developmental Biology* 474, 1–4. https://doi.org/10.1016/j.ydbio.2021.03.007

Currie, A.M., 2012. Convergence, contingency and morphospace. *Biology and Philosophy* 27(4), 583–593. https://doi.org/10.1007/s10539-012-9319-1

Davidson, E.H., Erwin, D.H., 2006. Gene regulatory networks and the evolution of animal body plans. *Science* 311(5762), 796–800. https://doi.org/10.1126/science.1113832

Davidson, E.H., Rast, J.P., Oliveri, P., Ransick, A., Calestani, C., Yuh, C.-H., Minokawa, T., Amore, G., Hinman, V., Arenas-Mena, C., Otim, O., Brown, C.T., Livi, C.B., Lee, P. Y., Revilla, R., Rust, A. G., Pan, Z., Schilstra, M.J., Clarke, P.J.C., … Bolouri, H., 2002. A genomic regulatory network for development. *Science* 295(5560), 1669–1678. https://doi.org/10.1126/science.1069883

de Almeida, B.P., Reiter, F., Pagani, M., Stark, A., 2022. DeepSTARR predicts enhancer activity from DNA sequence and enables the de novo design of synthetic enhancers. *Nature Genetics* 54(5), 613–624. https://doi.org/10.1038/s41588-022-01048-5

De Almeida, B.P., Schaub, C., Pagani, M., Secchia, S., Furlong, E.E.M., Stark, A., 2024. Targeted design of synthetic enhancers for selected tissues in the *Drosophila* embryo. *Nature* 626(7997), 207–211. https://doi.org/10.1038/s41586-023-06905-9

de Boer, C.G., Taipale, J., 2024. Hold out the genome: A roadmap to solving the cis-regulatory code. *Nature* 625(7993), 41–50. https://doi.org/10.1038/s41586-023-06661-w

Delás, M.J., Briscoe, J., 2020. Chapter Eight - Repressive interactions in gene regulatory networks: When you have no other choice. In: I. S. Peter, (Ed.), *Current Topics in Developmental Biology*, 139, pp. 239–266. Academic Press. https://doi.org/10.1016/bs.ctdb.2020.03.003

Delás, M.J., Kalaitzis, C.M., Fawzi, T., Demuth, M., Zhang, I., Stuart, H.T., Costantini, E., Ivanovitch, K., Tanaka, E.M., Briscoe, J., 2023. Developmental cell fate choice in neural tube progenitors employs two distinct cis-regulatory strategies. *Developmental Cell* 58(1), 3–17.e8. https://doi.org/10.1016/j.devcel.2022.11.016

Deng, W., Lee, J., Wang, H., Miller, J., Reik, A., Gregory, P.D., Dean, A., Blobel, G.A., 2012. Controlling long-range genomic interactions at a native locus by targeted tethering of a looping factor. *Cell* 149(6), 1233–1244. https://doi.org/10.1016/j.cell.2012.03.051

Deng, W., Rupon, J.W., Krivega, I., Breda, L., Motta, I., Jahn, K.S., Reik, A., Gregory, P.D., Rivella, S., Dean, A., Blobel, G., 2014. Reactivation of developmentally silenced globin genes by forced chromatin looping. *Cell* 158(4), 849–860. https://doi.org/10.1016/j.cell.2014.05.050

Dessaud, E., Yang, L.L., Hill, K., Cox, B., Ulloa, F., Ribeiro, A., Mynett, A., Novitch, B.G., Briscoe, J., 2007. Interpretation of the sonic hedgehog morphogen gradient by a temporal adaptation mechanism. *Nature* 450(7170), 717–720. https://doi.org/10.1038/nature06347

Dickel, D.E., Ypsilanti, A.R., Pla, R., Zhu, Y., Barozzi, I., Mannion, B.J., Khin, Y.S., Fukuda-Yuzawa, Y., Plajzer-Frick, I., Pickle, C.S., 2018. Ultraconserved enhancers are required for normal development. *Cell* 172(3), 491–499. e15.

Domcke, S., Hill, A.J., Daza, R.M., Cao, J., O'Day, D.R., Pliner, H.A., Aldinger, K.A., Pokholok, D., Zhang, F., Milbank, J.H., Zager, M.A., Glass, I.A., Steemers, F.J., Doherty, D., Trapnell, C., Cusanovich, D.A., Shendure, J., 2020. A human cell atlas of fetal chromatin accessibility. *Science* 370(6518), eaba7612. https://doi.org/10.1126/science.aba7612

Doni Jayavelu, N., Jajodia, A., Mishra, A., Hawkins, R.D., 2020. Candidate silencer elements for the human and mouse genomes. *Nature Communications* 11(1), 1061. https://doi.org/10.1038/s41467-020-14853-5

Driever, W., Thoma, G., Nüsslein-Volhard, C., 1989. Determination of spatial domains of zygotic gene expression in the *Drosophila* embryo by the affinity of binding sites for the bicoid morphogen. *Nature* 340(6232), 363–367. https://doi.org/10.1038/340363a0

Duque, T., Sinha, S., 2015. What does it take to evolve an enhancer? A simulation-based study of factors influencing the emergence of combinatorial regulation. *Genome Biology and Evolution* 7(6), 1415–1431. https://doi.org/10.1093/gbe/evv080

Echelard, Y., Epstein, D. J., St-Jacques, B., Shen, L., Mohler, J., McMahon, J.A., McMahon, A. P., 1993. Sonic hedgehog, a member of a family of putative signaling molecules, is implicated in the regulation of CNS polarity. *Cell* 75(7), 1417–1430.

Economou, A.D., Ohazama, A., Porntaveetus, T., Sharpe, P.T., Kondo, S., Basson, M.A., Gritli-Linde, A., Cobourne, M.T., Green, J.B.A., 2012. Periodic stripe formation by a Turing mechanism operating at growth zones in the mammalian palate. *Nature Genetics* 44(3), 348–351. https://doi.org/10.1038/ng.1090

El Khattabi, L., Zhao, H., Kalchschmidt, J., Young, N., Jung, S., Van Blerkom, P., Kieffer-Kwon, P., Kieffer-Kwon, K.-R., Park, S., Wang, X., Krebs, J., Tripathi, S., Sakabe, N., Sobreira, D.R., Huang, S.-C., Rao, S.S.P., Pruett, N., Chauss, D., Sadler, E., … Casellas, R., 2019. A Pliable Mediator Acts as a Functional Rather Than an Architectural Bridge between Promoters and Enhancers. *Cell* 178(5), 1145–1158.e20. https://doi.org/10.1016/j.cell.2019.07.011

Elowitz, M.B., Leibler, S., 2000. A synthetic oscillatory network of transcriptional regulators. *Nature* 403(6767), 335–338. https://doi.org/10.1038/35002125

Ericson, J., Rashbass, P., Schedl, A., Brenner-Morton, S., Kawakami, A., van Heyningen, V., Jessell, T. M., Briscoe, J., 1997. Pax6 controls progenitor cell identity and neuronal fate in response to graded Shh signaling. *Cell* 90(1), 169–180.

Exelby, K., Herrera-Delgado, E., Perez, L.G., Perez-Carrasco, R., Sagner, A., Metzis, V., Sollich, P., Briscoe, J., 2021. Precision of tissue patterning is controlled by dynamical properties of gene regulatory networks. *Development* 148(4), dev197566. https://doi.org/10.1242/dev.197566

Fink, T., Lonzarić, J., Praznik, A., Plaper, T., Merljak, E., Leben, K., Jerala, N., Lebar, T., Strmšek, Ž., Lapenta, F., Benčina, M., Jerala, R., 2019. Design of fast proteolysis-based signaling and logic circuits in mammalian cells. *Nature Chemical Biology* 15(2), 115–122. https://doi.org/10.1038/s41589-018-0181-6

Fiore, C., Cohen, B.A., 2016. Interactions between pluripotency factors specify cis-regulation in embryonic stem cells. *Genome Research* 26(6), 778–786. https://doi.org/10.1101/gr.200733.115

Frith, T.J.R., Briscoe, J., Boezio, G.L.M., 2024. From signalling to form: The coordination of neural tube patterning. *Current Topics in Developmental Biology* 159, 168–231. https://doi.org/10.1016/bs.ctdb.2023.11.004

Frömel, R., Rühle, J., Martinez, A.B., Szu-Tu, C., Pastor, F.P., Corral, R.M., Velten, L., 2024. Synthetic enhancers reveal design principles of cell state specific regulatory elements in hematopoiesis, p. 2024.08.26.609645. bioRxiv. https://doi.org/10.1101/2024.08.26.609645

Fueyo, R., Judd, J., Feschotte, C., Wysocka, J., 2022. Roles of transposable elements in the regulation of mammalian transcription. *Nature Reviews Molecular Cell Biology* 23(7), 481–497. https://doi.org/10.1038/s41580-022-00457-y

Fukaya, T., Lim, B., Levine, M., 2016. Enhancer Control of Transcriptional Bursting. *Cell* 166(2), 358–368. https://doi.org/10.1016/j.cell.2016.05.025

Fulco, C.P., Nasser, J., Jones, T.R., Munson, G., Bergman, D.T., Subramanian, V., Grossman, S.R., Anyoha, R., Doughty, B.R., Patwardhan, T.A., Nguyen, T.H., Kane, M., Perez, E.M., Durand, N.C., Lareau, C.A., Stamenova, E.K., Aiden, E.L., Lander, E.S., Engreitz, J.M., 2019. Activity-by-contact model of enhancer–promoter regulation from thousands of CRISPR perturbations. *Nature Genetics* 51(12), 1664–1669. https://doi.org/10.1038/s41588-019-0538-0

Furlong, E.E.M., Levine, M., 2018. Developmental enhancers and chromosome topology. *Science* 361(6409), 1341–1345. https://doi.org/10.1126/science.aau0320

Gao, X.J., Chong, L.S., Kim, M.S., Elowitz, M.B., 2018. Programmable protein circuits in living cells. *Science* 361(6408), 1252–1258. https://doi.org/10.1126/science.aat5062

Garcia, H.G., Tikhonov, M., Lin, A., Gregor, T., 2013. Quantitative Imaging of Transcription in Living *Drosophila* Embryos Links Polymerase Activity to Patterning. *Current Biology* 23(21), 2140–2145. https://doi.org/10.1016/j.cub.2013.08.054

Gardner, T.S., Cantor, C.R., Collins, J.J., 2000. Construction of a genetic toggle switch in *Escherichia coli*. *Nature* 403(6767), 339–342. https://doi.org/10.1038/35002131

Gasperini, M., Hill, A.J., McFaline-Figueroa, J.L., Martin, B., Kim, S., Zhang, M.D., Jackson, D., Leith, A., Schreiber, J., Noble, W.S., Trapnell, C., Ahituv, N., Shendure, J., 2019. A Genome-wide Framework for Mapping Gene Regulation via Cellular Genetic Screens. *Cell* 176(1), 377–390.e19. https://doi.org/10.1016/j.cell.2018.11.029

Gehrke, A.R., Schneider, I., de la Calle-Mustienes, E., Tena, J.J., Gomez-Marin, C., Chandran, M., Nakamura, T., Braasch, I., Postlethwait, J.H., Gómez-Skarmeta, J.L., Shubin, N.H., 2015. Deep conservation of wrist and digit enhancers in fish. *Proceedings of the National Academy of Sciences*, 112(3) 803–808. https://doi.org/10.1073/pnas.1420208112

Georgakopoulos-Soares, I., Deng, C., Agarwal, V., Chan, C.S.Y., Zhao, J., Inoue, F., Ahituv, N., 2023. Transcription factor binding site orientation and order are major drivers of gene regulatory activity. *Nature Communications* 14(1), 2333. https://doi.org/10.1038/s41467-023-37960-5

Gisselbrecht, S.S., Palagi, A., Kurland, J.V., Rogers, J.M., Ozadam, H., Zhan, Y., Dekker, J., Bulyk, M. L., 2020. Transcriptional Silencers in *Drosophila* Serve a Dual Role as Transcriptional Enhancers in Alternate Cellular Contexts. *Molecular Cell* 77(2), 324–337.e8. https://doi.org/10.1016/j.molcel.2019.10.004

Glass, L., Kauffman, S.A., 1973. The logical analysis of continuous, non-linear biochemical control networks. *Journal of Theoretical Biology*, 39(1) 103–129.

Glykofrydis, F., Elfick, A., 2022. Exploring standards for multicellular mammalian synthetic biology. *Trends in Biotechnology*, 40(11), 1299–1312. https://doi.org/10.1016/j.tibtech.2022.06.001

Gompel, N., Prud'homme, B., Wittkopp, P.J., Kassner, V.A., Carroll, S.B., 2005. Chance caught on the wing: Cis-regulatory evolution and the origin of pigment patterns in *Drosophila*. *Nature* 433(7025), 481–487. https://doi.org/10.1038/nature03235

Goodwin, B.C., 1963. *Temporal organization in cells; a dynamic theory of cellular control processes*. London, New York, Academic Press. http://archive.org/details/temporalorganiza00good

Gosai, S.J., Castro, R.I., Fuentes, N., Butts, J.C., Mouri, K., Alasoadura, M., Kales, S., Nguyen, T.T.L., Noche, R.R., Rao, A.S., Joy, M.T., Sabeti, P.C., Reilly, S.K., Tewhey, R., 2024. Machine-guided design of cell-type-targeting cis-regulatory elements. *Nature* 634(8036), 1211–1220. https://doi.org/10.1038/s41586-024-08070-z

Green, J.B.A., Sharpe, J., 2015. Positional information and reaction-diffusion: Two big ideas in developmental biology combine. *Development* 142(7), 1203–1211. https://doi.org/10.1242/dev.114991

Gregor, T., Tank, D.W., Wieschaus, E.F., Bialek, W., 2007. Probing the limits to positional information. *Cell* 130(1), 153–164. https://doi.org/10.1016/j.cell.2007.05.025

Hammer, R.E., Swift, G.H., Ornitz, D.M., Quaife, C.J., Palmiter, R.D., Brinster, R.L., MacDonald, R.J., 1987. The rat elastase I regulatory element is an enhancer that directs correct cell specificity and developmental onset of expression in transgenic mice. *Molecular and Cellular Biology* 7(8), 2956–2967. https://doi.org/10.1128/mcb.7.8.2956-2967.1987

He, Q., Bardet, A.F., Patton, B., Purvis, J., Johnston, J., Paulson, A., Gogol, M., Stark, A., Zeitlinger, J., 2011. High conservation of transcription factor binding and evidence for combinatorial regulation across six *Drosophila* species. *Nature Genetics* 43(5), 414–420. https://doi.org/10.1038/ng.808

Hiscock, T.W., Megason, S.G., 2015. Mathematically guided approaches to distinguish models of periodic patterning. *Development* 142(3), 409–419. https://doi.org/10.1242/dev.107441

Ho, W.K.W., Freem, L., Zhao, D., Painter, K.J., Woolley, T.E., Gaffney, E.A., McGrew, M.J., Tzika, A., Milinkovitch, M.C., Schneider, P., Drusko, A., Matthäus, F., Glover, J.D., Wells, K.L., Johansson, J.A., Davey, M.G., Sang, H.M., Clinton, M., Headon, D.J., 2019. Feather arrays are patterned by interacting signalling and cell density waves. *PLOS Biology* 17(2), e3000132. https://doi.org/10.1371/journal.pbio.3000132

Hong, J.-W., Hendrix, D.A., Levine, M.S., 2008. Shadow enhancers as a source of evolutionary novelty. *Science* 321(5894), 1314. https://doi.org/10.1126/science.1160631

Hui, C.-C., Angers, S., 2011. Gli proteins in development and disease. *Annual Review of Cell and Developmental Biology* 27, 513–537. https://doi.org/10.1146/annurev-cellbio-092910-154048

Huynh-Thu, V.A., Irrthum, A., Wehenkel, L., Geurts, P., 2010. Inferring regulatory networks from expression data using tree-based methods. *PloS One* 5(9), e12776. https://doi.org/10.1371/journal.pone.0012776

Indjeian, V.B., Kingman, G.A., Jones, F.C., Guenther, C.A., Grimwood, J., Schmutz, J., Myers, R.M., Kingsley, D.M., 2016. Evolving New Skeletal Traits by cis-Regulatory Changes in Bone Morphogenetic Proteins. *Cell* 164(1), 45–56. https://doi.org/10.1016/j.cell.2015.12.007

Inoue, F., Kircher, M., Martin, B., Cooper, G.M., Witten, D.M., McManus, M.T., Ahituv, N., Shendure, J., 2017. A systematic comparison reveals substantial differences in chromosomal versus episomal encoding of enhancer activity. *Genome Research* 27(1), 38–52. https://doi.org/10.1101/gr.212092.116

Inoue, F., Kreimer, A., Ashuach, T., Ahituv, N., Yosef, N., 2019. Identification and Massively Parallel Characterization of Regulatory Elements Driving Neural Induction. *Cell Stem Cell* 25(5), 713–727.e10. https://doi.org/10.1016/j.stem.2019.09.010

Jacob, F., Monod, J., (1961). Genetic regulatory mechanisms in the synthesis of proteins. *Journal of Molecular Biology*, *3*, 318–356.

Jacques, P.-É., Jeyakani, J., Bourque, G., 2013. The majority of primate-specific regulatory sequences are derived from transposable elements. *PLoS Genetics* 9(5), e1003504. https://doi.org/10.1371/journal.pgen.1003504

Jaeger, J., Surkova, S., Blagov, M., Janssens, H., Kosman, D., Kozlov, K.N., Manu, N., Myasnikova, E., Vanario-Alonso, C.E., Samsonova, M., Sharp, D.H., Reinitz, J., 2004. Dynamic control of positional information in the early *Drosophila* embryo. *Nature* 430(6997), 368–371. https://doi.org/10.1038/nature02678

Janssens, H., Hou, S., Jaeger, J., Kim, A.-R., Myasnikova, E., Sharp, D., Reinitz, J., 2006. Quantitative and predictive model of transcriptional control of the *Drosophila* melanogaster even skipped gene. *Nature Genetics* 38(10), 1159–1165. https://doi.org/10.1038/ng1886

Janssens, J., Aibar, S., Taskiran, I.I., Ismail, J.N., Gomez, A.E., Aughey, G., Spanier, K.I., De Rop, F.V., González-Blas, C.B., Dionne, M., Grimes, K., Quan, X.J., Papasokrati, D., Hulselmans, G., Makhzami, S., De Waegeneer, M., Christiaens, V., Southall, T., Aerts, S., 2022. Decoding gene regulation in the fly brain. *Nature* 601(7894), 630–636. https://doi.org/10.1038/s41586-021-04262-z

Johnson, J.E., Wold, B.J., Hauschka, S.D., 1989. Muscle creatine kinase sequence elements regulating skeletal and cardiac muscle expression in transgenic mice. *Molecular and Cellular Biology* 9(8) 3393–3399. https://doi.org/10.1128/mcb.9.8.3393-3399.1989

Kamimoto, K., Stringa, B., Hoffmann, C.M., Jindal, K., Solnica-Krezel, L., Morris, S.A., 2023. Dissecting cell identity via network inference and in silico gene perturbation. *Nature* 614(7949) 742–751. https://doi.org/10.1038/s41586-022-05688-9

Kane, L., Williamson, I., Flyamer, I.M., Kumar, Y., Hill, R.E., Lettice, L.A., Bickmore, W. A., 2022. Cohesin is required for long-range enhancer action at the Shh locus. *Nature Structural and Molecular Biology* 29(9), 891–897. https://doi.org/10.1038/s41594-022-00821-8

Karlebach, G., Shamir, R., 2008. Modelling and analysis of gene regulatory networks. *Nature Reviews. Molecular Cell Biology* 9(10), 770–780. https://doi.org/10.1038/nrm2503

Karr, J.P., Ferrie, J.J., Tjian, R., Darzacq, X., 2022. The transcription factor activity gradient, (TAG) model: Contemplating a contact-independent mechanism for enhancer-promoter communication. *Genes and Development* 36(1–2), 7–16. https://doi.org/10.1101/gad.349160.121

Kaya-Okur, H.S., Wu, S.J., Codomo, C.A., Pledger, E.S., Bryson, T.D., Henikoff, J.G., Ahmad, K., Henikoff, S., 2019. CUTandTag for efficient epigenomic profiling of small samples and single cells. *Nature Communications* 10(1), 1930. https://doi.org/10.1038/s41467-019-09982-5

Kehayova, P., Monahan, K., Chen, W., Maniatis, T., 2011. Regulatory elements required for the activation and repression of the protocadherin-alpha gene cluster. *Proceedings of the National Academy of Sciences of the United States of America* 108(41), 17195–17200. https://doi.org/10.1073/pnas.1114357108

Kelley, D.R., Reshef, Y.A., Bileschi, M., Belanger, D., McLean, C.Y., Snoek, J., 2018. Sequential regulatory activity prediction across chromosomes with convolutional neural networks. *Genome Research* 28(5) 739–750. https://doi.org/10.1101/gr.227819.117

Kelley, D.R., Snoek, J., Rinn, J.L., 2016. Basset: Learning the regulatory code of the accessible genome with deep convolutional neural networks. *Genome Research* 26(7), 990–999. https://doi.org/10.1101/gr.200535.115

Khalil, A.S., Lu, T. K., Bashor, C.J., Ramirez, C.L., Pyenson, N.C., Joung, J.K., Collins, J.J., 2012. A synthetic biology framework for programming eukaryotic transcription functions. *Cell* 150(3), 647–658. https://doi.org/10.1016/j.cell.2012.05.045

Kicheva, A., Bollenbach, T., Ribeiro, A., Valle, H.P., Lovell-Badge, R., Episkopou, V., Briscoe, J., (2014). Coordination of progenitor specification and growth in mouse and chick spinal cord. *Science (New York, N.Y.), 345*(6204), 1254927. https://doi.org/10.1126/science.1254927

Kicheva, A., Briscoe, J., 2023. Control of Tissue Development by Morphogens. *Annual Review of Cell and Developmental Biology* 39, 91–121. https://doi.org/10.1146/annurev-cellbio-020823-011522

Kim, D.S., Risca, V.I., Reynolds, D.L., Chappell, J., Rubin, A.J., Jung, N., Donohue, L.K.H., Lopez-Pajares, V., Kathiria, A., Shi, M., Zhao, Z., Deep, H., Sharmin, M., Rao, D., Lin, S., Chang, H. Y., Snyder, M.P., Greenleaf, W.J., Kundaje, A., Khavari, P.A., 2021. The dynamic, combinatorial cis-regulatory lexicon of epidermal differentiation. *Nature Genetics* 53(11), 1564–1576. https://doi.org/10.1038/s41588-021-00947-3

Kim, J., Koo, B.-K., Knoblich, J.A., 2020. Human organoids: Model systems for human biology and medicine. *Nature Reviews Molecular Cell Biology* 21(10), 571–584. https://doi.org/10.1038/s41580-020-0259-3

Kim, S., Wysocka, J., 2023. Deciphering the multi-scale, quantitative cis-regulatory code. *Molecular Cell* 83(3), 373–392. https://doi.org/10.1016/j.molcel.2022.12.032

Kinney, J.B., Murugan, A., Callan, C.G., Cox, E.C., 2010. Using deep sequencing to characterize the biophysical mechanism of a transcriptional regulatory sequence. *Proceedings of the National Academy of Sciences*, 107(20) 9158–9163. https://doi.org/10.1073/pnas.1004290107

Kitadate, Y., Jörg, D.J., Tokue, M., Maruyama, A., Ichikawa, R., Tsuchiya, S., Segi-Nishida, E., Nakagawa, T., Uchida, A., Kimura-Yoshida, C., Mizuno, S., Sugiyama, F., Azami, T., Ema, M., Noda, C., Kobayashi, S., Matsuo, I., Kanai, Y., Nagasawa, T., … Yoshida, S., 2019. Competition for Mitogens Regulates Spermatogenic Stem Cell Homeostasis in an Open Niche. *Cell Stem Cell* 24(1), 79–92.e6. https://doi.org/10.1016/j.stem.2018.11.013

Klein, J.C., Agarwal, V., Inoue, F., Keith, A., Martin, B., Kircher, M., Ahituv, N., Shendure, J., 2020. A systematic evaluation of the design and context dependencies of massively parallel reporter assays. *Nature Methods* 17(11), 1083–1091. https://doi.org/10.1038/s41592-020-0965-y

Koch, A.J., Meinhardt, H., 1994. Biological pattern formation: From basic mechanisms to complex structures. *Reviews of Modern Physics* 66(4), 1481–1507. https://doi.org/10.1103/RevModPhys.66.1481

Krauss, S., Concordet, J.P., Ingham, P.W., 1993. A functionally conserved homolog of the *Drosophila* segment polarity gene hh is expressed in tissues with polarizing activity in zebrafish embryos. *Cell* 75(7), 1431–1444.

Kraut, R., Levine, M., 1991. Mutually repressive interactions between the gap genes giant and Krüppel define middle body regions of the *Drosophila* embryo. *Development* 111(2), 611–621. https://doi.org/10.1242/dev.111.2.611

Kutejova, E., Sasai, N., Shah, A., Gouti, M., Briscoe, J., 2016. Neural Progenitors Adopt Specific Identities by Directly Repressing All Alternative Progenitor Transcriptional Programs. *Developmental Cell* 36(6), 639–653. https://doi.org/10.1016/j.devcel.2016.02.013

Kvon, E.Z., Kazmar, T., Stampfel, G., Yáñez-Cuna, J.O., Pagani, M., Schernhuber, K., Dickson, B.J., Stark, A., 2014. Genome-scale functional characterization of *Drosophila* developmental enhancers in vivo. *Nature* 512(7512), 91–95. https://doi.org/10.1038/nature13395

Kvon, E.Z., Zhu, Y., Kelman, G., Novak, C.S., Plajzer-Frick, I., Kato, M., Garvin, T.H., Pham, Q., Harrington, A.N., Hunter, R.D., Godoy, J., Meky, E.M., Akiyama, J.A., Afzal, V., Tran, S., Escande, F., Gilbert-Dussardier, B., Jean-Marçais, N., Hudaiberdiev, S., … Pennacchio, L.A., 2020. Comprehensive In Vivo Interrogation Reveals Phenotypic Impact of Human Enhancer Variants. *Cell* 180(6), 1262–1271.e15. https://doi.org/10.1016/j.cell.2020.02.031

La Fleur, A., Shi, Y., Seelig, G., 2024. Decoding biology with massively parallel reporter assays and machine learning. *Genes and Development* 38(17–20), 843–865. https://doi.org/10.1101/gad.351800.124

Lähdesmäki, H., Shmulevich, I., Yli-Harja, O., 2003. On Learning Gene Regulatory Networks Under the Boolean Network Model. *Machine Learning* 52(1), 147–167. https://doi.org/10.1023/A:1023905711304

Lalanne, J.-B., Regalado, S.G., Domcke, S., Calderon, D., Martin, B.K., Li, X., Li, T., Suiter, C.C., Lee, C., Trapnell, C., Shendure, J., 2024. Multiplex profiling of developmental cis-regulatory elements with quantitative single-cell expression reporters. *Nature Methods* 21(6), 983–993. https://doi.org/10.1038/s41592-024-02260-3

Lambert, J.T., Su-Feher, L., Cichewicz, K., Warren, T. L., Zdilar, I., Wang, Y., Lim, K.J., Haigh, J.L., Morse, S.J., Canales, C.P., Stradleigh, T.W., Castillo Palacios, E., Haghani, V., Moss, S.D., Parolini, H., Quintero, D., Shrestha, D., Vogt, D., Byrne, L.C., Nord, A.S., 2021. Parallel functional testing identifies enhancers active in early postnatal mouse brain. *eLife* 10, e69479.

Lambert, S.A., Jolma, A., Campitelli, L.F., Das, P.K., Yin, Y., Albu, M., Chen, X., Taipale, J., Hughes, T. R., Weirauch, M.T., 2018. The Human Transcription Factors. *Cell* 172(4), 650–665. https://doi.org/10.1016/j.cell.2018.01.029

Larson, M.H., Gilbert, L.A., Wang, X., Lim, W.A., Weissman, J.S., Qi, L.S., 2013. CRISPR interference, (CRISPRi) for sequence-specific control of gene expression. *Nature Protocols* 8(11), 2180–2196. https://doi.org/10.1038/nprot.2013.132

Lee, J., Platt, K.A., Censullo, P., Altaba, A.R., 1997. Gli1 is a target of Sonic hedgehog that induces ventral neural tube development. *Development* 124(13), 2537–2552. https://doi.org/10.1242/dev.124.13.2537

Lettice, L.A., Heaney, S.J.H., Purdie, L.A., Li, L., de Beer, P., Oostra, B.A., Goode, D., Elgar, G., Hill, R.E., de Graaff, E., 2003. A long-range Shh enhancer regulates expression in the developing limb and fin and is associated with preaxial polydactyly. *Human Molecular Genetics* 12(14), 1725–1735. https://doi.org/10.1093/hmg/ddg180

Levine, M., Davidson, E.H., 2005. Gene regulatory networks for development. *Proceedings of the National Academy of Sciences of the United States of America* 102(14), 4936–4942. https://doi.org/10.1073/pnas.0408031102

Levo, M., Raimundo, J., Bing, X.Y., Sisco, Z., Batut, P.J., Ryabichko, S., Gregor, T., Levine, M.S., 2022. Transcriptional coupling of distant regulatory genes in living embryos. *Nature* 605(7911), 754–760. https://doi.org/10.1038/s41586-022-04680-7

Li, Y.E., Preissl, S., Miller, M., Johnson, N.D., Wang, Z., Jiao, H., Zhu, C., Wang, Z., Xie, Y., Poirion, O., Kern, C., Pinto-Duarte, A., Tian, W., Siletti, K., Emerson, N., Osteen, J., Lucero, J., Lin, L., Yang, Q., … Ren, B., 2023. A comparative atlas of single-cell chromatin accessibility in the human brain. *Science* 382(6667), eadf7044. https://doi.org/10.1126/science.adf7044

Liem, K.F., Tremml, G., Jessell, T.M., 1997. A role for the roof plate and its resident TGFbeta-related proteins in neuronal patterning in the dorsal spinal cord. *Cell* 91(1), 127–138.

Lienert, F., Lohmueller, J.J., Garg, A., Silver, P.A., 2014. Synthetic biology in mammalian cells: Next generation research tools and therapeutics. *Nature Reviews Molecular Cell Biology* 15(2), 95–107. https://doi.org/10.1038/nrm3738

Lynch, V.J., Nnamani, M.C., Kapusta, A., Brayer, K., Plaza, S. L., Mazur, E.C., Emera, D., Sheikh, S.Z., Grützner, F., Bauersachs, S., Graf, A., Young, S.L., Lieb, J.D., DeMayo, F.J., Feschotte, C., Wagner, G.P., 2015. Ancient transposable elements transformed the uterine regulatory landscape and transcriptome during the evolution of mammalian pregnancy. *Cell Reports* 10(4), 551–561. https://doi.org/10.1016/j.celrep.2014.12.052

Ma, W., Trusina, A., El-Samad, H., Lim, W.A., Tang, C., 2009. Defining network topologies that can achieve biochemical adaptation. *Cell* 138(4), 760–773. https://doi.org/10.1016/j.cell.2009.06.013

Maeda, Y.T., Sano, M., (2006). Regulatory dynamics of synthetic gene networks with positive feedback. *Journal of Molecular Biology*, *359*(4), 1107–1124. https://doi.org/10.1016/j.jmb.2006.03.064

Marcon, L., Diego, X., Sharpe, J., Müller, P., 2016. High-throughput mathematical analysis identifies Turing networks for patterning with equally diffusing signals. *eLife* 5, e14022.

Martínez-Ara, G., Stapornwongkul, K.S., Ebisuya, M., 2022. Scaling up complexity in synthetic developmental biology. *Science* 378(6622), 864–868. https://doi.org/10.1126/science.add9666

Mateo, L.J., Murphy, S.E., Hafner, A., Cinquini, I.S., Walker, C.A., Boettiger, A.N., 2019. Visualizing DNA folding and RNA in embryos at single-cell resolution. *Nature* 568(7750), 49–54. https://doi.org/10.1038/s41586-019-1035-4

McGhee, G.R., 2011. *Convergent Evolution: Limited Forms Most Beautiful*. The MIT Press. https://www.jstor.org/stable/j.ctt5hhhwt

Milo, R., Shen-Orr, S., Itzkovitz, S., Kashtan, N., Chklovskii, D., Alon, U., 2002. Network motifs: Simple building blocks of complex networks. *Science* 298(5594), 824–827. https://doi.org/10.1126/science.298.5594.824

Minnoye, L., Taskiran, I.I., Mauduit, D., Fazio, M., Van Aerschot, L., Hulselmans, G., Christiaens, V., Makhzami, S., Seltenhammer, M., Karras, P., Primot, A., Cadieu, E., van Rooijen, E., Marine, J.-C., Egidy, G., Ghanem, G.-E., Zon, L., Wouters, J., Aerts, S., 2020. Cross-species analysis of enhancer logic using deep learning. *Genome Research* 30(12), 1815–1834. https://doi.org/10.1101/gr.260844.120

Mjolsness, E., Sharp, D.H., Reinitz, J., 1991. A connectionist model of development. *Journal of Theoretical Biology* 152(4), 429–453.

Moerman, T., Aibar Santos, S., Bravo González-Blas, C., Simm, J., Moreau, Y., Aerts, J., Aerts, S., 2019. GRNBoost2 and Arboreto: Efficient and scalable inference of gene regulatory networks. *Bioinformatics* 35(12), 2159–2161. https://doi.org/10.1093/bioinformatics/bty916

Montavon, T., Soshnikova, N., Mascrez, B., Joye, E., Thevenet, L., Splinter, E., de Laat, W., Spitz, F., Duboule, D., 2011. A regulatory archipelago controls Hox genes transcription in digits. *Cell* 147(5), 1132–1145. https://doi.org/10.1016/j.cell.2011.10.023

Moreau, P., Hen, R., Wasylyk, B., Everett, R., Gaub, M.P., Chambon, P., 1981. The SV40 72 base repair repeat has a striking effect on gene expression both in SV40 and other chimeric recombinants. *Nucleic Acids Research* 9(22), 6047–6068. https://doi.org/10.1093/nar/9.22.6047

Muroyama, Y., Fujihara, M., Ikeya, M., Kondoh, H., Takada, S., 2002. Wnt signaling plays an essential role in neuronal specification of the dorsal spinal cord. *Genes and Development* 16(5), 548–553. https://doi.org/10.1101/gad.937102

Nishi, Y., Zhang, X., Jeong, J., Peterson, K.A., Vedenko, A., Bulyk, M.L., Hide, W.A., McMahon, A.P., 2015. A direct fate exclusion mechanism by Sonic hedgehog-regulated transcriptional repressors. *Development* 142(19), 3286–3293. https://doi.org/10.1242/dev.124636

Novitch, B., Chen, A.I., Jessell, T.M., 2001. Coordinate Regulation of Motor Neuron Subtype Identity and Pan-Neuronal Properties by the bHLH Repressor Olig2. *Neuron* 31(5), 773–789.

Ochoa-Espinosa, A., Yucel, G., Kaplan, L., Pare, A., Pura, N., Oberstein, A., Papatsenko, D., Small, S., 2005. The role of binding site cluster strength in Bicoid-dependent patterning in *Drosophila. Proceedings of the National Academy of Sciences of the United States of America* 102(14), 4960–4965. https://doi.org/10.1073/pnas.0500373102

Oh, S., Shao, J., Mitra, J., Xiong, F., D'Antonio, M., Wang, R., Garcia-Bassets, I., Ma, Q., Zhu, X., Lee, J.-H., Nair, S.J., Yang, F., Ohgi, K., Frazer, K.A., Zhang, Z.D., Li, W., Rosenfeld, M.G., 2021. Enhancer release and retargeting activates disease-susceptibility genes. *Nature* 595(7869), 735–740. https://doi.org/10.1038/s41586-021-03577-1

Oosterveen, T., Kurdija, S., Alekseenko, Z., Uhde, C. W., Bergsland, M., Sandberg, M., Andersson, E., Dias, J.M., Muhr, J., Ericson, J., 2012. Mechanistic Differences in the Transcriptional Interpretation of Local and Long-Range Shh Morphogen Signaling. *Developmental Cell* 23(5), 1006–1019. https://doi.org/10.1016/j.devcel.2012.09.015

Osterwalder, M., Barozzi, I., Tissières, V., Fukuda-Yuzawa, Y., Mannion, B.J., Afzal, S.Y., Lee, E.A., Zhu, Y., Plajzer-Frick, I., Pickle, C.S., Kato, M., Garvin, T.H., Pham, Q.T., Harrington, A.N., Akiyama, J.A., Afzal, V., Lopez-Rios, J., Dickel, D.E., Visel, A., Pennacchio, L.A., 2018. Enhancer redundancy provides phenotypic robustness in mammalian development. *Nature* 554(7691), 239–243.

Oudelaar, A.M., Davies, J.O.J., Hanssen, L.L.P., Telenius, J.M., Schwessinger, R., Liu, Y., Brown, J. M., Downes, D.J., Chiariello, A.M., Bianco, S., Nicodemi, M., Buckle, V.J., Dekker, J., Higgs, D.R., Hughes, J.R., 2018. Single-allele chromatin interactions identify regulatory hubs in dynamic compartmentalized domains. *Nature Genetics* 50(12), 1744–1751. https://doi.org/10.1038/s41588-018-0253-2

Paliou, C., Guckelberger, P., Schöpflin, R., Heinrich, V., Esposito, A., Chiariello, A.M., Bianco, S., Annunziatella, C., Helmuth, J., Haas, S., Jerković, I., Brieske, N., Wittler, L., Timmermann, B., Nicodemi, M., Vingron, M., Mundlos, S., Andrey, G., 2019. Preformed chromatin topology assists transcriptional robustness of Shh during limb development. *Proceedings of the National Academy of Sciences* 116(25), 12390–12399. https://doi.org/10.1073/pnas.1900672116

Pampari, A., Shcherbina, A., Kvon, E.Z., Kosicki, M., Nair, S., Kundu, S., Kathiria, A.S., Risca, V.I., Kuningas, K., Alasoo, K., Greenleaf, W.J., Pennacchio, L.A., Kundaje, A.,

2025. ChromBPNet: Bias factorized, base-resolution deep learning models of chromatin accessibility reveal cis-regulatory sequence syntax, transcription factor footprints and regulatory variants, p. 2024.12.25.630221. bioRxiv. https://doi.org/10.1101/2024.12.25.630221

Pan, Y., Bai, C.B., Joyner, A.L., Wang, B., 2006. Sonic hedgehog signaling regulates Gli2 transcriptional activity by suppressing its processing and degradation. *Molecular and Cellular Biology* 26(9), 3365–3377. https://doi.org/10.1128/MCB.26.9.3365-3377.2006

Pang, B., Snyder, M.P., 2020. Systematic identification of silencers in human cells. *Nature Genetics* 52(3), 254–263. https://doi.org/10.1038/s41588-020-0578-5

Papili Gao, N., Ud-Dean, S.M.M., Gandrillon, O., Gunawan, R., 2018. SINCERITIES: Inferring gene regulatory networks from time-stamped single cell transcriptional expression profiles. *Bioinformatics* 34(2), 258–266. https://doi.org/10.1093/bioinformatics/btx575

Pennacchio, L.A., Bickmore, W., Dean, A., Nobrega, M.A., Bejerano, G., 2013. Enhancers: Five essential questions. *Nature Reviews. Genetics* 14(4), 288–295. https://doi.org/10.1038/nrg3458

Perez-Carrasco, R., Barnes, C.P., Schaerli, Y., Isalan, M., Briscoe, J., Page, K.M., 2018. Combining a Toggle Switch and a Repressilator within the AC-DC Circuit Generates Distinct Dynamical Behaviors. *Cell Systems* 6(4), 521–530.e3. https://doi.org/10.1016/j.cels.2018.02.008

Perez-Carrasco, R., Guerrero, P., Briscoe, J., Page, K.M., 2016. Intrinsic Noise Profoundly Alters the Dynamics and Steady State of Morphogen-Controlled Bistable Genetic Switches. *PLOS Computational Biology* 12(10), e1005154. https://doi.org/10.1371/journal.pcbi.1005154

Perez-Pinera, P., Kocak, D.D., Vockley, C.M., Adler, A.F., Kabadi, A.M., Polstein, L.R., Thakore, P.I., Glass, K.A., Ousterout, D.G., Leong, K.W., Guilak, F., Crawford, G.E., Reddy, T.E., Gersbach, C.A., 2013. RNA-guided gene activation by CRISPR-Cas9–based transcription factors. *Nature Methods* 10(10), 973–976. https://doi.org/10.1038/nmeth.2600

Peterson, K.A., Nishi, Y., Ma, W., Vedenko, A., Shokri, L., Zhang, X., McFarlane, M., Baizabal, J.-M., Junker, J.P., Van Oudenaarden, A., Mikkelsen, T., Bernstein, B.E., Bailey, T.L., Bulyk, M.L., Wong, W.H., McMahon, A.P., 2012. Neural-specific Sox2 input and differential Gli-binding affinity provide context and positional information in Shh-directed neural patterning. *Genes and Development* 26(24), 2802–2816. https://doi.org/10.1101/gad.207142.112

Pezzotta, A., Briscoe, J., 2023. Optimal control of gene regulatory networks for morphogen-driven tissue patterning. *Cell Systems* 14(11), 940–952.e11. https://doi.org/10.1016/j.cels.2023.10.004

Pinkert, C.A., Ornitz, D.M., Brinster, R.L., Palmiter, R.D., 1987. An albumin enhancer located 10 kb upstream functions along with its promoter to direct efficient, liver-specific expression in transgenic mice. *Genes and Development* 1(3), 268–276. https://doi.org/10.1101/gad.1.3.268

Plum, A.M., Serra, M., 2025. *Morphogen Patterning in Dynamic Tissues*, p. 2025.01.04.631293. bioRxiv. https://doi.org/10.1101/2025.01.04.631293

Pownall, M.E., Miao, L., Vejnar, C.E., M'Saad, O., Sherrard, A., Frederick, M.A., Benitez, M.D.J., Boswell, C.W., Zaret, K.S., Bewersdorf, J., Giraldez, A.J., 2023. Chromatin expansion microscopy reveals nanoscale organization of transcription and chromatin. *Science* 381(6653), 92–100.

Rahimi, N., Carmon, S., Averbukh, I., Khajouei, F., Sinha, S., Schejter, E. D., Barkai, N., Shilo, B.-Z., 2020. Global shape of Toll activation is determined by wntD enhancer properties. *Proceedings of the National Academy of Sciences* 117(3), 1552–1558.

Raspopovic, J., Marcon, L., Russo, L., Sharpe, J. 2014. Modeling digits. Digit patterning is controlled by a Bmp-Sox9-Wnt Turing network modulated by morphogen gradients. *Science* 345(6196), 566–570.

Rayon, T., Stamataki, D., Perez-Carrasco, R., Garcia-Perez, L., Barrington, C., Melchionda, M., Exelby, K., Lazaro, J., Tybulewicz, V.L.., Fisher, E.M.C., Briscoe, J., 2020. Species-specific pace of development is associated with differences in protein stability. *Science* 369(6510), eaba7667.

Rebeiz, M., Jikomes, N., Kassner, V.A., Carroll, S.B. 2011. Evolutionary origin of a novel gene expression pattern through co-option of the latent activities of existing regulatory sequences. *Proceedings of the National Academy of Sciences of the United States of America* 108(25), 10036–10043.

Rebeiz, M., Tsiantis, M., 2017. Enhancer evolution and the origins of morphological novelty. *Current Opinion in Genetics and Development* 45, 115–123.

Reilly, S.K., Gosai, S.J., Gutierrez, A., Mackay-Smith, A., Ulirsch, J.C., Kanai, M., Mouri, K., Berenzy, D., Kales, S., Butler, G.M., Gladden-Young, A., Bhuiyan, R.M., Stitzel, M.L., Finucane, H.K., Sabeti, P.C., Tewhey, R., 2021. Direct characterization of cis-regulatory elements and functional dissection of complex genetic associations using HCR-FlowFISH. *Nature Genetics* 53(8), 1166–1176.

Rekaik, H., Lopez-Delisle, L., Hintermann, A., Mascrez, B., Bochaton, C., Mayran, A., Duboule, D., 2023. Sequential and directional insulation by conserved CTCF sites underlies the Hox timer in stembryos. *Nature Genetics* 55(7), 1164–1175.

Rengachari, S., Schilbach, S., Aibara, S., Dienemann, C., Cramer, P., 2021. Structure of the human Mediator–RNA polymerase II pre-initiation complex. *Nature* 594(7861), 129–133.

Richards, D.M., Saunders, T.E., 2015. Spatiotemporal analysis of different mechanisms for interpreting morphogen gradients. *Biophysical Journal* 108(8), 2061–2073. https://doi.org/10.1016/j.bpj.2015.03.015

Rogers, K.W., Schier, A. F., 2011. Morphogen gradients: From generation to interpretation. *Annual Review of Cell and Developmental Biology* 27, 377–407.

Rosenfeld, N., Young, J.W., Alon, U., Swain, P.S., Elowitz, M.B., 2005. Gene regulation at the single-cell level. *Science* 307(5717), 1962–1965. https://doi.org/10.1126/science.1106914

Sabari, B.R., Dall'Agnese, A., Boija, A., Klein, I.A., Coffey, E.L., Shrinivas, K., Abraham, B.J., Hannett, N.M., Zamudio, A.V., Manteiga, J.C., Li, C.H., Guo, Y.E., Day, D.S., Schuijers, J., Vasile, E., Malik, S., Hnisz, D., Lee, T.I., Cisse, I.I., … Young, R.A., 2018. Coactivator condensation at super-enhancers links phase separation and gene control. *Science* 361(6400), eaar3958. https://doi.org/10.1126/science.aar3958

Sagner, A., Briscoe, J., 2019. Establishing neuronal diversity in the spinal cord: A time and a place. *Development* 146(22), dev182154. https://doi.org/10.1242/dev.182154

Sahu, B., Hartonen, T., Pihlajamaa, P., Wei, B., Dave, K., Zhu, F., Kaasinen, E., Lidschreiber, K., Lidschreiber, M., Daub, C.O., Cramer, P., Kivioja, T., Taipale, J., 2022. Sequence determinants of human gene regulatory elements. *Nature Genetics* 54(3), 283–294. https://doi.org/10.1038/s41588-021-01009-4

Santos Moreno, J., Tasiudi, E., Stelling, J., Schaerli, Y., 2020. Multistable and dynamic CRISPRi-based synthetic circuits. *Nature Communications* 11(1), 2746. https://doi.org/10.1038/s41467-020-16574-1

Schaerli, Y., Munteanu, A., Gili, M., Cotterell, J., Sharpe, J., Isalan, M., 2014. A unified design space of synthetic stripe-forming networks. *Nature Communications* 5(1), 4905. https://doi.org/10.1038/ncomms5905

Schaffner, W., 1988. Gene regulation. A hit-and-run mechanism for transcriptional activation? *Nature* 336(6198), 427–428. https://doi.org/10.1038/336427a0

Segal, E., Raveh-Sadka, T., Schroeder, M., Unnerstall, U., Gaul, U., 2008. Predicting expression patterns from regulatory sequence in *Drosophila* segmentation. *Nature* 451(7178), 535–540. https://doi.org/10.1038/nature06496

Segal, E., Shapira, M., Regev, A., Pe'er, D., Botstein, D., Koller, D., Friedman, N., 2003. Module networks: Identifying regulatory modules and their condition-specific regulators from gene expression data. *Nature Genetics* 34(2), 166–176. https://doi.org/10.1038/ng1165

Segert, J.A., Gisselbrecht, S.S., Bulyk, M.L., 2021. Transcriptional silencers: Driving gene expression with the brakes on. *Trends in Genetics* 37(6), 514–527. https://doi.org/10.1016/j.tig.2021.02.002

Sekine, R., Shibata, T., Ebisuya, M., 2018. Synthetic mammalian pattern formation driven by differential diffusivity of Nodal and Lefty. *Nature Communications* 9(1), 5456. https://doi.org/10.1038/s41467-018-07847-x

Setty, Y., Mayo, A.E., Surette, M.G., Alon, U., 2003. Detailed map of a cis-regulatory input function. *Proceedings of the National Academy of Sciences of the United States of America* 100(13), 7702–7707. https://doi.org/10.1073/pnas.1230759100

Sharon, E., Kalma, Y., Sharp, A., Raveh-Sadka, T., Levo, M., Zeevi, D., Keren, L., Yakhini, Z., Weinberger, A., Segal, E., 2012. Inferring gene regulatory logic from high-throughput measurements of thousands of systematically designed promoters. *Nature Biotechnology* 30(6), 521–530. https://doi.org/10.1038/nbt.2205

Shea, M.A., Ackers, G.K., 1985. The OR control system of bacteriophage lambda. A physical-chemical model for gene regulation. *Journal of Molecular Biology* 181(2), 211–230.

Sick, S., Reinker, S., Timmer, J., Schlake, T., 2006. WNT and DKK determine hair follicle spacing through a reaction-diffusion mechanism. *Science* 314(5804), 1447–1450. https://doi.org/10.1126/science.1130088

Skene, P.J., Henikoff, S., 2017. An efficient targeted nuclease strategy for high-resolution mapping of DNA binding sites. *eLife* 6, e21856. https://doi.org/10.7554/eLife.21856

Smith, J., Theodoris, C., Davidson, E.H., 2007. A gene regulatory network subcircuit drives a dynamic pattern of gene expression. *Science* 318(5851), 794–797. https://doi.org/10.1126/science.1146524

Smith, R.P., Riesenfeld, S.J., Holloway, A.K., Li, Q., Murphy, K.K., Feliciano, N.M., Orecchia, L., Oksenberg, N., Pollard, K.S., Ahituv, N., 2013. A compact, in vivo screen of all 6-mers reveals drivers of tissue-specific expression and guides synthetic regulatory element design. *Genome Biology* 14(7), R72. https://doi.org/10.1186/gb-2013-14-7-r72

Smith, R.P., Taher, L., Patwardhan, R.P., Kim, M.J., Inoue, F., Shendure, J., Ovcharenko, I., Ahituv, N., 2013. Massively parallel decoding of mammalian regulatory sequences supports a flexible organizational model. *Nature Genetics* 45(9), 1021–1028. https://doi.org/10.1038/ng.2713

Soutourina, J., 2018. Transcription regulation by the Mediator complex. *Nature Reviews. Molecular Cell Biology* 19(4), 262–274. https://doi.org/10.1038/nrm.2017.115

Specht, A.T., Li, J., 2017. LEAP: Constructing gene co-expression networks for single-cell RNA-sequencing data using pseudotime ordering. *Bioinformatics* 33(5), 764–766. https://doi.org/10.1093/bioinformatics/btw729

Strogatz, S., 2007. *Nonlinear dynamics and chaos: With applications to physics, biology, chemistry, and engineering*, Repr. Westview Press.

Struhl, G., Struhl, K., Macdonald, P.M., 1989. The gradient morphogen bicoid is a concentration-dependent transcriptional activator. *Cell* 57(7), 1259–1273.

Su, M., Han, D., Boyd-Kirkup, J., Yu, X., Han, J.-D.J., 2014. Evolution of Alu elements toward enhancers. *Cell Reports* 7(2), 376–385. https://doi.org/10.1016/j.celrep.2014.03.011

Swift, G.H., Hammer, R.E., MacDonald, R.J., Brinster, R.L., 1984. Tissue-specific expression of the rat pancreatic elastase I gene in transgenic mice. *Cell* 38(3), 639–646.

Taskiran, I.I., Spanier, K.I., Dickmänken, H., Kempynck, N., Pančíková, A., Ekşi, E.C., Hulselmans, G., Ismail, J.N., Theunis, K., Vandepoel, R., Christiaens, V., Mauduit, D., Aerts, S., 2024. Cell-type-directed design of synthetic enhancers. *Nature* 626(7997), 212–220. https://doi.org/10.1038/s41586-023-06936-2

Thomas, R., 1973. Boolean formalization of genetic control circuits. *Journal of Theoretical Biology* 42(3), 563–585.

Thurman, R.E., Rynes, E., Humbert, R., Vierstra, J., Maurano, M.T., Haugen, E., Sheffield, N.C., Stergachis, A.B., Wang, H., Vernot, B., Garg, K., John, S., Sandstrom, R., Bates, D., Boatman, L., Canfield, T.K., Diegel, M., Dunn, D., Ebersol, A.K., … Stamatoyannopoulos, J.A., 2012. The accessible chromatin landscape of the human genome. *Nature* 489(7414), 75–82. https://doi.org/10.1038/nature11232

Tica, J., Oliver Huidobro, M., Zhu, T., Wachter, G.K.A., Pazuki, R.H., Bazzoli, D.G., Scholes, N.S., Tonello, E., Siebert, H., Stumpf, M.P.H., Endres, R.G., Isalan, M., 2024. A three-node Turing gene circuit forms periodic spatial patterns in bacteria. *Cell Systems* 15(12), 1123–1132.e3. https://doi.org/10.1016/j.cels.2024.11.002

Toda, S., Brunger, J.M., Lim, W.A., 2019. Synthetic development: Learning to program multicellular self-organization. *Current Opinion in Systems Biology* 14, 41–49. https://doi.org/10.1016/j.coisb.2019.02.008

Tozer, S., Le Dréau, G., Marti, E., Briscoe, J., 2013. Temporal control of BMP signalling determines neuronal subtype identity in the dorsal neural tube. *Development* 140(7), 1467–1474. https://doi.org/10.1242/dev.090118

Trizzino, M., Park, Y., Holsbach-Beltrame, M., Aracena, K., Mika, K., Caliskan, M., Perry, G.H., Lynch, V.J., Brown, C.D., 2017. Transposable elements are the primary source of novelty in primate gene regulation. *Genome Research* 27(10), 1623–1633. https://doi.org/10.1101/gr.218149.116

Tuan, D., Kong, S., Hu, K., 1992. Transcription of the hypersensitive site HS2 enhancer in erythroid cells. *Proceedings of the National Academy of Sciences of the United States of America* 89(23), 11219–11223. https://doi.org/10.1073/pnas.89.23.11219

Turing, A.M., 1952. The chemical basis of morphogenesis. *Philosophical Transactions of the Royal Society of London. Series B, Biological Sciences* 237(641), 37–72. https://doi.org/10.1098/rstb.1952.0012

Vaishnav, E.D., de Boer, C.G., Molinet, J., Yassour, M., Fan, L., Adiconis, X., Thompson, D.A., Levin, J.Z., Cubillos, F.A., Regev, A., 2022. The evolution, evolvability and engineering of gene regulatory DNA. *Nature* 603(7901), 455–463. https://doi.org/10.1038/s41586-022-04506-6

Verd, B., Monk, N.A., Jaeger, J., 2019. Modularity, criticality, and evolvability of a developmental gene regulatory network. *eLife* 8, e42832.

Villar, D., Berthelot, C., Aldridge, S., Rayner, T.F., Lukk, M., Pignatelli, M., Park, T.J., Deaville, R., Erichsen, J.T., Jasinska, A.J., Turner, J.M.A., Bertelsen, M.F., Murchison, E.P., Flicek, P., Odom, D.T., 2015. Enhancer evolution across 20 mammalian species. *Cell* 160(3), 554–566. https://doi.org/10.1016/j.cell.2015.01.006

Visel, A., Akiyama, J.A., Shoukry, M., Afzal, V., Rubin, E.M., Pennacchio, L.A., 2009. Functional autonomy of distant-acting human enhancers. *Genomics* 93(6), 509–513. https://doi.org/10.1016/j.ygeno.2009.02.002

Wang, B., Fallon, J.F., Beachy, P.A., 2000. Hedgehog-regulated processing of Gli3 produces an anterior/posterior repressor gradient in the developing vertebrate limb. *Cell* 100(4), 423–434.

Wolpert, L., 1969. Positional information and the spatial pattern of cellular differentiation. *Journal of Theoretical Biology* 25(1), 1–47.

Wong, F., Gunawardena, J., 2020. Gene Regulation in and out of Equilibrium. *Annual Review of Biophysics* 49, 199–226. https://doi.org/10.1146/annurev-biophys-121219-081542

Xiao, J.Y., Hafner, A., Boettiger, A.N., 2021. How subtle changes in 3D structure can create large changes in transcription. *eLife* 10, e64320.

Yamada, T., Pfaff, S.L., Edlund, T., Jessell, T.M., 1993. Control of cell pattern in the neural tube: Motor neuron induction by diffusible factors from notochord and floor plate. *Cell* 73(4), 673–686.

Yang, J. H., Hansen, A.S., 2024. Enhancer selectivity in space and time: From enhancer–promoter interactions to promoter activation. *Nature Reviews Molecular Cell Biology* 25(7), 574–591. https://doi.org/10.1038/s41580-024-00710-6

Zabidi, M.A., Arnold, C.D., Schernhuber, K., Pagani, M., Rath, M., Frank, O., Stark, A., 2015. Enhancer–core-promoter specificity separates developmental and housekeeping gene regulation. *Nature* 518(7540), 556–559. https://doi.org/10.1038/nature13994

Zagorski, M., Tabata, Y., Brandenberg, N., Lutolf, M.P., Tkačik, G., Bollenbach, T., Briscoe, J., Kicheva, A., 2017. Decoding of position in the developing neural tube from antiparallel morphogen gradients. *Science* 356(6345), 1379–1383. https://doi.org/10.1126/science.aam5887

Zaret, K.S., Carroll, J.S., 2011. Pioneer transcription factors: Establishing competence for gene expression. *Genes and Development* 25(21), 2227–2241. https://doi.org/10.1101/gad.176826.111

Zaret, K.S., Mango, S.E., 2016. Pioneer transcription factors, chromatin dynamics, and cell fate control. *Current Opinion in Genetics and Development* 37, 76–81. https://doi.org/10.1016/j.gde.2015.12.003

Zhang, F., Cong, L., Lodato, S., Kosuri, S., Church, G.M., Arlotta, P., 2011. Efficient construction of sequence-specific TAL effectors for modulating mammalian transcription. *Nature Biotechnology* 29(2), 149–153. https://doi.org/10.1038/nbt.1775

Zhang, I., Boezio, G. L., Cornwall-Scoones, J., Frith, T., Jiang, M., Howell, M., Lovell-Badge, R., Sagner, A., Briscoe, J., Delás, M.J., 2024. The cis regulatory logic integrating spatial and temporal patterning in the vertebrate neural tube, p. 2024.04.17.589864. bioRxiv. https://doi.org/10.1101/2024.04.17.589864

Zhang, K., Hocker, J.D., Miller, M., Hou, X., Chiou, J., Poirion, O.B., Qiu, Y., Li, Y.E., Gaulton, K.J., Wang, A., Preissl, S., Ren, B., 2021. A single-cell atlas of chromatin accessibility in the human genome. *Cell* 184(24), 5985–6001.e19. https://doi.org/10.1016/j.cell.2021.10.024

Zhao, S., Hong, C.K.Y., Myers, C.A., Granas, D.M., White, M.A., Corbo, J.C., Cohen, B.A., 2023. A single-cell massively parallel reporter assay detects cell-type-specific gene regulation. *Nature Genetics* 55(2), 346–354. https://doi.org/10.1038/s41588-022-01278-7

Zheng, W., Zhao, H., Mancera, E., Steinmetz, L., Snyder, M., 2010. Genetic analysis of variation in transcription factor binding in yeast. *Nature* 464(7292), 1187–1191. https://doi.org/10.1038/nature08934

Zhou, J., Troyanskaya, O.G., 2015. Predicting effects of noncoding variants with deep learning-based sequence model. *Nature Methods* 12(10), 931–934. https://doi.org/10.1038/nmeth.3547

Zhou, Q., Anderson, D.J., 2002. The bHLH transcription factors OLIG2 and OLIG1 couple neuronal and glial subtype specification. *Cell* 109(1), 61–73.

Zhu, R., Del Rio-Salgado, J.M., Garcia-Ojalvo, J., Elowitz, M.B., 2022. Synthetic multistability in mammalian cells. *Science* 375(6578), eabg9765. https://doi.org/10.1126/science.abg9765

Zinzen, R.P., Girardot, C., Gagneur, J., Braun, M., Furlong, E.E.M., 2009. Combinatorial binding predicts spatio-temporal cis-regulatory activity. *Nature* 462(7269), 65–70. https://doi.org/10.1038/nature08531

Section III

The Cell as the Unit of Development

5 Nothing in EvoDevo Makes Sense Except in the Light of Cell Biology

Günter P. Wagner

5.1 INTRODUCTION

According to lore, the phrase format "nothing in X makes sense, except in the light of Y" goes back to Theodosius Dobzhansky's presidential address to the American Zoologist's Society in 1964 and later to an article in 1973, where he famously declared "nothing in biology makes sense except in the light of evolution" (Dobzhansky, 1973). This phrase has been widely adopted by evolutionary biologists to defend their discipline against the marginalization it suffers from the biomedical disciplines. This phrase format is also used to argue for particular perspectives within evolutionary biology, such as Michael Lynch's claim that "nothing in evolutionary biology makes sense except in the light of population genetics" (Lynch, 2007), as well as other applications of this type of phrase (Reiskind et al., 2021). Of course, on face value, each of these phrases is an overstatement and should better be phrased as "not much makes sense …" since there is, of course, sense making within molecular and cellular biology, for example, how the structure of the nervous system generates behavior, and how biophysical principles of ion diffusion explain action potentials. To use this phrase format again, as I am doing in this short article, should be preceded by a few thoughts about "sense making" in biology in particular.

5.2 SENSE MAKING

Each branch of science aims at describing, understanding, and explaining a certain domain of reality. Sense making consists of a scientist's ability to explain the relevant facts of his/her domain of interest with the conceptual resources available to a branch of science. The facts of mechanics are explained with Newton's laws of motion; the facts of chemistry are explained by the atomic theory and the theory of valency, etc. Biology, however, is different in important ways. While different branches of physics and chemistry carve up the world in a way that guarantees a match between the domain of explanation and its conceptual resources, biology does not have this privilege, or to a much lesser extent. Biology, or rather, what eventually would be called biology, started as a collection of largely disconnected disciplines, such as anatomy, embryology, histology, physiology, genetics, and biochemistry, as well as

DOI: 10.1201/9781003560784-8

taxonomy and ecology. The fundamental truth, however, is that all of life is a unitary process, and ignoring that fact necessarily leads to deficits in explanatory power. Life is an emergent phenomenon and one that is also the product of a historical process, evolution. For that reason, each and every biological sub-discipline is vulnerable to explanatory incompleteness.

While the resources of biochemistry are excellent for explaining how energy is gained from metabolism, for instance, some facts of biochemistry remain unexplained if viewed from a purely chemical point of view. An example is the chirality of proteinogenic amino acids, which is likely a frozen evolutionary coincidence. Another example is the existence of iso-enzymes, which are the product of genome evolution rather than a biochemical necessity. And the same is, of course, true for all of the mechanistic branches of biology, such as genetics, physiology, and developmental biology, and so on. Hence, Dobzhansky's phrase is apt. But here I am going to argue that the inverse is also true. Nothing in evolutionary biology makes sense except in the light of the rest of biology (Reiskind et al., 2021), including ecology, genetics, as well as cell biology and biochemistry. I am not prepared to defend this most inclusive claim in this essay but want to focus on evolutionary developmental biology, or EvoDevo, and argue that the next step in a more comprehensive understanding of evolution will have to focus on the integration of cell biological principles into evolutionary developmental biology, or even better "developmental evolutionary biology," where the latter is the application of developmental biological knowledge to explain facts about evolution.

5.3 A CRITIQUE OF EVODEVO

EvoDevo arose through the re-integration of developmental biology into evolutionary biology (Carroll, 2008; Goodman and Coughlin, 2000; Hall, 2012; Love and Raff, 2003), or so the usual narrative goes. As such, it is the child of the developmental biology of the 1980s and 1990s, which in its core is developmental genetics, meaning the biology of gene regulation that controls morphogenesis and cell differentiation. During the late 1970s, a synthesis of developmental and evolutionary biology was already underway in evolutionary biology, as witnessed by the 1981 Dahlem Conference on Development and Evolution (Bonner, 1982). In contrast, the discovery of the "homeobox" occurred a few years later (Garber et al., 1983; Gehring, 1994), which is credited with precipitating the molecular-based research program of EvoDevo. This, of course, was a great breakthrough in understanding how evolutionary change at the phenotype level is underwritten by changes in gene regulation. EvoDevo, being derived from a branch of genetics, that is, developmental genetics, naturally fits into an understanding of evolution as change in the genetic composition of populations. The synthesis between developmental genetics and evolutionary genetics is perfect, perhaps too perfect.

When we step back and look at EvoDevo from the perspective of "sense making," as discussed in the previous section, we see that the primary theoretical resource of EvoDevo for explaining evolutionary outcomes (biological diversity as we know it from current fauna and flora) is a model of gene regulation, in particular the regulation of transcription (Carroll et al., 2001; Davidson, 2006; Wilkins, 2002). The loss of a character or the change of a pigment pattern is explainable by changes in the molecular machinery of gene regulation, mostly transcriptional regulation. There is consensus that most evolutionary changes in gene regulation are due to changes in cis-regulatory elements (Carroll, 2008). This is a powerful paradigm, and a great amount of biological understanding has come from this research program (Wilkins, 2005). So where is the problem, if there is one?

As mentioned above, the primary explanatory model of EvoDevo is the cis-regulatory paradigm, where evolution is conceptualized as changes in the structure of so-called gene regulatory networks, in particular changes in cis-regulatory elements. As a consequence, the explanans offered by EvoDevo research is often presented as a disembodied gene regulatory network hovering above the explanandum, that is, the morphology of the organism. Of course, in reality, gene regulatory networks exert their effects on the phenotype through the activity of cells rather than directly through their gene products. The role of the gene products is to influence the activities of cells, which in turn perform the developmental process (Prum, 2023), eventually leading to the phenotype of an organism that engages with its biotic and abiotic environment. Of course, every biologist knows that, and in fact, the role of cells is a background condition of EvoDevo theorizing that is acknowledged to be there but that does not itself contribute to the explanatory narrative. Hence, it is not that cells are ignored as such but that it is assumed that cell biology does not contribute anything specific to the explanation of evolutionary change. The situation is reminiscent of Ernst Mayr's claim that molecular biology does not contribute anything to evolutionary biology (Mayr, 1982). Now we know better when it comes to molecular biology, and I argue that we should learn to appreciate the importance of cell biology as well.

How then can the facts of cell biology move from a background condition to be a part of the explanatory model of evolutionary developmental biology? In what way could they be relevant beyond conveying the effects of gene regulatory changes? Below, I will discuss two examples where a deeper understanding of cell biology is necessary to understand evolutionary outcomes.

5.4 TWO PARADIGMS OF HOW CELL BIOLOGY MATTERS

The most straightforward way in which cell biology matters in evolutionary biology is the fact that the bodies of animals and plants are composed of cells themselves, and the complexity of body plans is, in part, derived from the diversity of cell types (Arendt et al., 2016; Bonner, 1982). Naturally, the evolution of body plan complexity then depends on the evolution of novel cell types, and thus understanding the nature of cell type identity and its evolution is a central aim of evolutionary cell biology. This is straightforward but still treats cells and cell types as an explanandum, that is, as a fact that is in need of explanation, rather than a part of the explanans.

Beyond the need to understand cell type evolution, an even more fundamental contribution of cell biology is the realization that cells are homeostatic "machines[1]" that are organized in a way such as to resist change, be it externally triggered or genetically caused. In addition, cells in a multicellular organism interact to maintain the homeostasis of the tissues, organs, and organismal structure and function, and thus also act to resist change even at supra-cellular levels. These facts determine what kind of genetic change qualifies as an explanation of evolutionary change at the cellular and organismal level. In short, mutations that effect evolutionary change at the cellular and organismal level must be able to either overcome the homeostatic tendencies of the cells or interact with these mechanisms to "manipulate" them into allowing change rather than compensating for it. Below I will sketch two examples from my and my collaborator's work that inspired the cell biological perspective advocated here.

5.4.1 Mammalian Embryo Implantation

Viviparity, in its various manifestations, evolved many times among animals, not only in mammals and squamate reptiles, where most of the research has focused (Blackburn, 2015; Ostrovsky et al., 2016). Also, the majority of invertebrate phyla have some species displaying one form of viviparity or another (Ostrovsky et al., 2016). However, viviparity with invasive, hemochorial placentation evolved only once in the stem lineage of eutherian mammals (Elliot and Crespi, 2009; Wildman et al., 2006). In hemochorial placentation, the fetal membranes, forming the placenta, are in direct contact with maternal blood, and thus require partial destruction of the maternal tissue due to the implantation of the embryo (Ramsey, 1982). In contrast, the large number of independently originating cases of squamate viviparity are almost exclusively based on various degrees of egg retention with no or only rudimentary placentation (Blackburn, 2006; Blackburn, 2015). Only a small number of squamate species have a definite placenta, and even fewer have at least some degree of erosion of the uterine epithelium (Blackburn and Flemming, 2009, 2012). But not a single case of hemochorial placentation has been reported for squamates or anywhere outside mammals. Why is the evolution of hemochorial placentation so rare?

A key factor in the difficulty of evolving hemochorial placentation is the fact that this form of placentation requires the embryo to impart an injury to the maternal tissue (Chavan et al., 2017). Under normal circumstances, any violation of tissue integrity elicits an inflammatory reaction. Inflammation is often viewed as a reaction to infection, but in fact has a much broader significance (Medzhitov, 2008). Inflammation is also caused by injuries without infection, as is the case with damaged connective tissue as a consequence of a sprained ankle, or when adipocytes come under physiological stress as a consequence of obesity (Meizlish et al., 2021). We can thus expect, with some certainty, that, during the very first steps of the evolution of invasive placentation, the maternal organism reacted with inflammation to the "nidatory injury." The main initial step in inflammation is the recruitment of neutrophil cells, which have the task of indiscriminately destroying invading cells, including collateral damage to the injured tissue. Hence, embryos that cause endometrial tissue injury, that is, embryos able to establish invasive placentation, should be attacked by neutrophils. An echo of this situation can still be observed during the brief fetal-maternal contact during the pregnancy of the opossum, a marsupial without sustained gestation beyond two days of contact. In the opossum, fetal-maternal contact coincides with the expression of many pro-inflammatory mediators and the recruitment of neutrophils (Griffith et al., 2017; Hansen et al., 2017). In the case of the opossum, the initial fetal-maternal contact continues with progressively increasing expression of inflammatory mediators and directly proceeds to parturition. In contrast, in humans and other eutherian mammals, there is also an inflammation-like process coincidentally with embryo implantation, but inflammation is then suppressed for the rest of gestation (Chavan et al., 2017; Mor et al., 2011). Toward the end of pregnancy, inflammatory processes are re-activated leading to birth (Romero et al., 2006).

From the outline in the previous paragraph, it can be inferred that a key innovation for eutherian pregnancy is the evolution of various mechanisms that suppress the inflammation caused by the nidatory injury from the embryo (Chavan et al., 2017; Griffith et al., 2017). One of the innovations that made this possible is the origin of a novel cell type, the so-called decidual cell (Chavan et al., 2021; Stadtmauer et al., 2025). In most eutherian mammals with hemochorial placentation, the differentiation of decidual cells, called decidualization, is induced by the embryo. In fact, in a mouse model, decidualization can be artificially induced during the luteal phase of the ovarian cycle by any injury to the endometrium or other forms of irritation, such as those caused by a bead or oil droplet (Finn, 1986). Decidual cells have various functions, including regulating embryo invasion, suppressing inflammation, activating of regulatory T-cells, and others.

At this point, it is worth noting that inflammation is one of the homeostatic mechanisms through which cells maintain the integrity of the body, and that the evolution of eutherian implantation can only be understood in the context of regulating inflammation. Mammalian embryo implantation evolved by overcoming tissue homeostasis mediated through inflammatory mechanisms.

From the above considerations, it can be understood that explaining the evolution of invasive placentation is not complete by arguing that viviparity is advantageous, for one reason or another, since other forms of viviparity do exist. It is also not explained by assuming that invasive placentation is advantageous over non-invasive placentation because it is more efficient in nutrient transfer. In fact, it is not (Martin, 2008). Explaining the evolution of invasive placentation requires to understand that this form of placentation is only possible after evolution has found a way to prevent or regulate the nidatory inflammation and its deleterious consequences (Chavan et al., 2016).[2]

The point here is this: the question of how the constraints from the nidatory inflammation have been overcome *would not even be asked without a cell biological perspective*. The inclusion of a cell biological perspective changes the nature of the explanandum (that which is to be explained) and what qualifies as an explanation, that is, as the explanans. In the cell biological perspective, the life-history advantages of viviparity do not qualify as an explanation for the evolution of hemochorial placentation, because viviparity can take different forms, and it does not address how the inflammation constraint has been overcome. Also, gene regulatory innovations only qualify as explanations if they contribute to explaining how the inflammation constraint was overcome. One also should note that eutherian pregnancy is special in terms of the form of the fetal-maternal relationship, and explaining its origin is not the same as explaining viviparity itself.

5.4.2 Cellular Self-Maintenance and the Origin of Tissue Level Organization

Another cell biological insight with interesting consequences for evolution is the fact that, whatever a cell is doing for the organism [what we call the "service function" of a cell, see (Pavlicev and Wagner, 2024)], it also has to take care of its own self-maintenance. Cells with dedicated service functions are usually called "parenchymatic" and are often, but not always, specialized epithelial cells. Self-maintenance includes replacing molecules that have been damaged and providing the resources to combat the reactive oxygen species (ROS) that are produced during metabolism, for instance, during oxidative phosphorylation or while breaking down fatty acids. The more a cell does for the organism, the higher the demand for self-repair and self-maintenance, which both draw upon the same supply of metabolic resources. One way out of this "catch 22" is to recruit supporting cells, which supplement the metabolism of the parenchymatic cells by donating metabolites that the parenchymatic cells can use (Pavlicev et al., 2024). A frequently donated metabolite is lactate, which is a three-carbon compound that is easily oxidized into pyruvate, which in turn can enter the Krebs cycle or other core metabolic pathways. Metabolic complementation is a well-known cell physiological phenomenon (Hermann, 1960) and is very well documented. Examples are the supplementation of central nervous system neurons by astroglia (Turner and Adamson, 2011), or that of hepatocytes by stellate cells, a type of fibroblast (Gasbarrini et al., 1992).

The ideas of the need for metabolic supplementation help explain an important evolutionary conundrum: why do more derived vertebrates evolve functionally specialized tissues when cells were already performing the same functional tasks before the tissue evolved? This is a well-established pattern, for instance for hepatocytes, which exist in lower chordates but are not organized into a specialized "liver tissue" (Wu et al., 2024). The idea is that tissue-level organization arises when the functional demands on the cells exceed the level compatible with self-maintenance of the cell itself (Pavlicev et al., 2024). Increasing the intensity of a service function beyond that limit can be achieved through the recruitment of supportive cells, such as stellate cells or astroglia, in the liver and the brain, respectively. Metabolic complementation leads to the assembly of parenchymatic cells with their support cells into specialized tissues.

In this example, we are confronted with a situation where the homeostatic nature of cells is not so much resisting evolutionary change but imposing physiological constraints on what a cell can contribute. Overcoming these constraints explains the origin of tissue-level organization (Pavlicev et al., 2024).

5.5 CONCLUSIONS

Genes are not the agents of evolutionary change. They are informational resources that cells can use to perform cellular functions, and the evolution of genes leads to changes in cellular behavior, which, in turn, effect evolutionary change at the level of the organism. Yet, cells are not passive conduits of genetic information to the phenotype but are living homeostatic and self-maintaining systems in their own account.

The latter means that mutations that are successfully established in a population are filtered through cell biological constraints, and the kind of genetic changes that lead to evolutionary novelty have to "make sense" with respect to the intrinsic behavioral tendencies and constraints that cells generically display. From that perspective, "nothing in EvoDevo makes sense except in the light of cell biology."

Writing the previous paragraph with its appeal to the importance of cell biology in explaining evolutionary change, I am aware that it is easy to misunderstand. In particular, the appeal to the quasi-agential properties of cells (e.g., homeostasis) brings this essay in close conceptual proximity to recent proposals about an agential view of living systems and evolution in particular (Levin, 2022; Moczek and Sultan, 2023; Newman, 2023; Potter and Mitchell, 2022; Sultan et al., 2022). An agential view has its justification in the fact of the quasi-agential properties of cells and organisms (Okasha, 2018, 2024), but becomes problematic when agency, or the agents' "goals," are treated as if they provide an explanans of evolution. For a deeper treatment of this problem, see the incisive critique by James DiFrisco and Richard Gawne (DiFrisco and Gawne, 2025). Instead, what is advocated here is a model where cell biology and with it the quasi-agential properties of cells are part of a complex explanans, i.e., a system of models that together explain evolutionary change. Nothing said in this essay denies that evolutionary change in most populations and situations is realized by mutations and recombination, the fixation of genetic factors by population genetic processes (natural selection, genetic drift, etc.), and that competition among organisms plays a causal role in evolution. What is added to this narrative is a view about what kind of mutational changes are necessary to effect the evolutionary outcome we seek to explain, be it the origin of tissues, mammalian embryo implantation, or anything else. It is not a metaphysical statement about a new kind of process evolution is as such, but one aiming at what kinds of questions biologists have to ask in order to gain a deeper understanding of genetic change.

5.6 PERSONAL OUTLOOK

As requested by the editors, I am adding here a few words about how I reached the position advocated here. My personal journey started as a biochemistry student who was fascinated by the creativity of biological evolution in coming up with ever new forms of life, from bacterial biofilms to human culture and everything in between. With that outlook, I was attracted to the views of Rupert Riedl, at that time professor of zoology who had just returned from a professorship in the United States, whose anti-reductionist but still mechanistic views felt true to both my anti-reductionist sensibilities and those attitudes acquired through the study of chemistry. For the next step, I have to thank my postdoc advisor Professor Joachim Wolf (Max-Planck-Institute for Biophysical Chemistry, and later the University of Göttingen, Department of Anatomy), who taught me how to think in terms of cell behavior. These points of view converged when my lab started to study reproduction in the marsupial, *Monodelphis domestica*, and we discovered intra-uterine inflammatory processes after the fetus hatches from the egg cover. The question of how we ended up studying opossum reproduction is a longer story that does not belong here, except to say that the opossum is an outgroup for studying the origin of eutherian placentation and pregnancy. More important is that learning about inflammation as a normal part of physiology raises any number of questions about the relationship between so-called "pathological processes" and evolution. The most salient of these was the realization that I would not have asked questions about the reason for the origin of decidual cells, according to our theory, the key innovation for the origin of eutherian pregnancy, had I and my collaborators not encountered and appreciated the importance of inflammation in female reproductive biology. Hence, the opossum forced me to look at evolution, not from the vantage point of adaptive utility, but from the point of view of mechanistic, cell-biological obstacles to evolutionary change. In a way that is just a special example of what we called in the 1990s "developmental constraints," but this notion remained somewhat abstract until I understood more about how an organism is made and kept functioning. If you want, this can be seen as an agential view of the organisms, but one where the "goals" of the organism are *not* sold as *a causal factor*, as discussed in the conclusions section, but as a factor that shapes the trajectory of evolution in the same way as the environment does in ordinary adaptive scenarios. It points toward a kind of internal selection, meaning selective pressures that emanate from the structure, development, and physiology of the organism and not directly from specific environmental factors. The view can thus be called structuralist, but one that embraces the notion of molecular and population genetic mechanisms as well.

NOTES

1 If one does not like the machine metaphor, the term "machine" can be substituted with any other metaphor, like agent or others, without changing the direction of the argument.

2 This narrative answers a slightly different question than is usually asked in evolutionary biology. Usually, evolutionary biologists ask "why" questions, while this narrative is one about "how did implantation evolve?" rather than why. Interestingly, the why question is harder to answer. One could imagine that the initial nidatory inflammation was advantageous to the fetus because it also included processes that were beneficial for fetal growth, like vascularization and edema, making nutrients more readily available to the fetus (Chavan et al., 2016). However, this idea turned out to be hard to test experimentally, at least in our hands. We did not succeed in showing that inhibiting inflammatory processes leads to lower fetal biomass. Another perspective is that nidatory inflammation was originally, and is still in the opossum, a way for the fetus to initiate parturition (Hansen et al., 2017).

REFERENCES

Arendt, D., Musser, J.M., Baker, C.V.H., Bergman, A., Cepko, C., Erwin, D.H., Pavlicev, M., Schlosser, G., Widder, S., Laubichler, M.D., Wagner, G.P., 2016. The origin and evolution of cell types. *Nat Rev Genet* 17, 744–757.

Blackburn, D.G., 2006. Squamate reptiles as model organisms for the evolution of vivipary. *Herpetol Monogr* 20, 131–146.

Blackburn, D.G., 2015. Evolution of viviparity in squamate reptiles: reversibility reconsidered. *J Exp Zool B Mol Dev Evol* 324, 473–486.

Blackburn, D.G., Flemming, A.F., 2009. Morphology, development, and evolution of fetal membranes and placentation in squamate reptiles. *J Exp Zool B Mol Dev Evol* 312, 579–589.

Blackburn, D.G., Flemming, A.F., 2012. Invasive implantation and intimate placental associations in a placentotrophic African lizard, Trachylepis ivensi (scincidae). *J Morphol* 273, 137–159.

Bonner, J.T., 1982. Evolution and development; life sciences research reports, in: Bernhard, S. (Ed.), *Dahlem Workshop Reports*. Springer-Verlag, Berlin, Heidelberg, New York.

Carroll, S.B., 2008. Evo-Devo and an expanding evolutionary synthesis: a genetic theory of morphological evolution. *Cell* 134, 25–34.

Carroll, S.B., Grenier, J.K., Weatherbee, S.D., 2001. *From DNA to Diversity*. Blackwell Science, Malden, MA.

Chavan, A.R., Bhullar, B.-A.S., Wagner, G.P., 2016. What was the ancestral function of decidual stromal cells? A model for the evolution of eutherian pregnancy. *Placenta* 40, 40–51.

Chavan, A.R., Griffith, O.W., Stadtmauer, D.J., Maziarz, J., Pavlicev, M., Fishman, R., Koren, L., Romero, R., Wagner, G.P., 2021. Evolution of embryo implantation was enabled by the origin of Decidual Stromal cells in Eutherian Mammals. *Mol Biol Evol* 38, 1060–1074.

Chavan, A.R., Griffith, O.W., Wagner, G.P., 2017. The inflammation paradox in the evolution of mammalian pregnancy: turning a foe into a friend. *Curr Opin Genet Dev* 47, 24–32.

Davidson, E.H., 2006. *The Regulatory Genome: Gene Regulatory Networks in Development and Evolution*. Academic Press, Amsterdam.

DiFrisco, J., Gawne, R., 2025. Biological agency: a concept without a research program. *J Evol Biol* 38, 143–156.

Dobzhansky, T., 1973. Nothing in biology makes sense except in the light of evolution. *Am Biol Teach* 75, 87–91.

Elliot, M.G., Crespi, B.J., 2009. Phylogenetic evidence for early hemochorial placentation in eutheria. *Placenta* 30, 949–967.

Finn, C.A., 1986. Implantation, menstruation and inflammation. *Biol Rev Camb Philos Soc* 61, 313–328.

Garber, R.L., Kuroiwa, A., Gehring, W.J., 1983. Genomic and cDNA clones of the homeotic locus Antennapedia in Drosophila. *EMBO J* 2, 2027–2036.

Gasbarrini, A., Borle, A.B., Farghali, H., Bender, C., Francavilla, A., Van Thiel, D., 1992. Effect of anoxia on intracellular ATP, Na+i, Ca2+i, Mg2+i, and cytotoxicity in rat hepatocytes. *J Biol Chem* 267, 6654–6663.

Gehring, W.J., 1994. A history of the homeobox., in: Duboule, D. (Ed.), *Guedebook to the Homeobox Genes*. Oxford University Press, Oxford, pp. 1–10.

Goodman, C.S., Coughlin, B.C., 2000. Introduction. The evolution of evo-devo biology. *Proc Natl Acad Sci U S A* 97, 4424–4425.

Griffith, O.W., Chavan, A.R., Protopapas, S., Maziarz, J., Romero, R., Wagner, G.P., 2017. Embryo implantation evolved from an ancestral inflammatory attachment reaction. *Proc Natl Acad Sci U S A* 114, E6566–E6575.

Hall, B.K., 2012. Evolutionary developmental biology (Evo-Devo): past, present, and future. *Evo Edu Outreach* 5, 184–193.

Hansen, V.L., Faber, L.S., Salehpoor, A.A., Miller, R.D., 2017. A pronounced uterine pro-inflammatory response at parturition is an ancient feature in mammals. *Proc Biol Sci* 284, 20171694.

Hermann, H., 1960. Direct metabolic interactions between animal cells. their role in tissue function and development is reconsidered. *Science* 132, 529–532.

Levin, M., 2022. Technological approach to mind everywhere: an experimentally-grounded framework for understanding diverse bodies and minds. *Front Syst Neurosci* 16, 768201.

Love, A.C., Raff, R.A., 2003. Knowing your ancestors: themes in the history of evo-devo. *Evol Dev* 5, 327–330.

Lynch, M., 2007. The frailty of adaptive hypotheses for the origins of organismal complexity. *Proc Natl Acad Sci U S A* 104(Suppl 1), 8597–8604.

Martin, R.D., 2008. Evolution of placentation in primates: Implications of mammalian phylogeny. *Evol Biol* 35, 125–145.

Mayr, E., 1982. *The Growth of Biological Thought*. The Belknap Press, Cambridge, London.

Medzhitov, R., 2008. Origin and physiological roles of inflammation. *Nature* 454, 428–435.

Meizlish, M.L., Franklin, R.A., Zhou, X., Medzhitov, R., 2021. Tissue Homeostasis and Inflammation. *Annu Rev Immunol* 39, 557–581.

Moczek, A.P., Sultan, S.E., 2023. Agency in living systems. *Evol Dev* 25, 331–334.

Mor, G., Cardenas, I., Abrahams, V., Guller, S., 2011. Inflammation and pregnancy: the role of the immune system at the implantation site. *Ann N Y Acad Sci* 1221, 80–87.

Newman, S.A., 2023. Inherency and agency in the origin and evolution of biological functions. *Biol J Linn Soc* 139, 487–502.

Okasha, S., 2018. *Agents and Goals in Evolution*. Oxford University Press.

Okasha, S., 2024. The concept of agency in biology: motivations and meanings. *Biol Theory* 19, 6–10.

Ostrovsky, A.N., Lidgard, S., Gordon, D.P., Schwaha, T., Genikhovich, G., Ereskovsky, A.V., 2016. Matrotrophy and placentation in invertebrates: a new paradigm. *Biol Rev Camb Philos Soc* 91, 673–711.

Pavlicev, M., DiFrisco, J., Love, A.C., Wagner, G.P., 2024. Metabolic complementation between cells drives the evolution of tissues and organs. *Biol Lett* 20, 20240490.

Pavlicev, M., Wagner, G.P., 2024. Reading the palimpsest of cell interactions: What questions may we ask of the data? *iScience* 27, 109670.

Potter, H.D., Mitchell, K.J., 2022. Naturalising Agent Causation. *Entropy (Basel)* 24, 472.

Prum, R.O., 2023. *Performance all the way down: Genes, Development and Sexual Difference*. The University of Chicago Press, Chicago and London.

Ramsey, E.M., 1982. *The Placenta. Human and Animal*. Praeger Inc, Santa Barbara, CA.

Reiskind, M.O.B., Moody, M.L., Bolnick, D.I., Hanifin, C.T., Farrior, C.E., 2021. Nothing in evolution makes sense except in the light of biology. *Bioscience* 71, 370–382.

Romero, R., Espinoza, J., Kusanovic, J.P., Gotsch, F., Hassan, S., Erez, O., Chaiworapongsa, T., Mazor, M., 2006. The preterm parturition syndrome. *BJOG* 113 Suppl 3, 17–42.

Stadtmauer, D.J., Basanta, S., Maziarz, J.D., Cole, A.G., Dagdas, G., Smith, G.R., van Breukelen, F., Pavlicev, M., Wagner, G.P., 2025. Cell type and cell signalling innovations underlying mammalian pregnancy. *Nat Ecol Evol* 9, 1469–1486.

Sultan, S.E., Moczek, A.P., Walsh, D., 2022. Bridging the explanatory gaps: What can we learn from a biological agency perspective? *Bioessays* 44, e2100185.

Turner, D.A., Adamson, D.C., 2011. Neuronal-astrocyte metabolic interactions: understanding the transition into abnormal astrocytoma metabolism. *J Neuropathol Exp Neurol* 70, 167–176.

Wildman, D.E., Chen, C., Erez, O., Grossman, L.I., Goodman, M.R., 2006. Evolution of the mammalian placenta revealed by phylogenetic analysis. *Proc Natl Acad Sci USA* 103, 3203–3208.

Wilkins, A.S., 2002. *The Evolution of Developmental Pathways*. Sinauer Assoc., Sunderland, MA.

Wilkins, A.S., 2005. Recasting developmental evolution in terms of genetic pathway and network evolution ... and the implications for comparative biology. *Brain Res Bull* 66, 495–509.

Wu, B., Xu, W., Wu, K., Li, Y., Hu, M., Feng, C., Zhu, C., Zheng, J., Cui, X., Li, J., Fan, D., Zhang, F., Liu, Y., Chen, J., Liu, C., Li, G., Qiu, Q., Qu, K., Wang, W., Wang, K., 2024. Single-cell analysis of the amphioxus hepatic caecum and vertebrate liver reveals genetic mechanisms of vertebrate liver evolution. *Nat Ecol Evol* 8, 1972–1990.

6 Developmental Biology at the Crossroads

Classical Puzzles and Future Directions

Roberto Mayor

6.1 INTRODUCTION: THE FOREST AND THE TREES

In contemporary scientific inquiry, a strong tendency towards specialised analysis has fostered a certain prejudice against general or integrative questions. This is evident across disciplines, where the study of complex systems is often displaced by the meticulous dissection of their constituent elements. The metaphor of the forest and the trees is apt: the forest as a whole is neglected, while attention is directed towards individual trees, or even further reduced to the analysis of leaves and their molecular details.

Embryo development illustrates this paradigm in a particularly acute way. Research on embryogenesis has been dominated by increasingly fine-grained analyses of gene regulatory networks, signalling pathways, and molecular interactions. The field has produced an extraordinary catalogue of mechanisms, each described in excruciating detail. Yet the integrative understanding of how these components collectively generate a coherent organism remains underdeveloped. In many cases, this synthesis is deferred to others or tacitly assumed to be reducible to the sum of its molecular parts. Such a reductionist orientation has clear strengths: it enables mechanistic precision and predictive modelling at the molecular level. However, it risks obscuring the emergent properties of developmental systems, where spatial organisation, temporal coordination, and dynamic feedback produce phenomena that cannot be straightforwardly inferred from isolated components. The embryo is not simply a molecular aggregate, but a self-organising system in which multi-level interactions shape the emergence of form and function.

A critical task for developmental biology, therefore, is to rebalance the emphasis between detailed analysis and integrative synthesis. Without attention to the forest as well as the trees, the science of embryogenesis risks fragmenting into ever smaller scales of description, while leaving unanswered the central question of how embryos develop as unified, living wholes.

DOI: 10.1201/9781003560784-9

In this chapter, I propose to revisit a set of general questions that, in my view, remain unanswered, or at least not satisfactorily addressed, despite having been formulated centuries ago. In contemporary developmental biology, many of these problems are encountered only in historical accounts, if at all. Indeed, they are often treated as outdated curiosities, relegated to the domain of the history of science rather than considered as active scientific challenges. A striking example is the "Theory of the Germ Layers" (Collins and Billett, 1995). Today, this theory is commonly presented as a milestone in the history of embryology, with little more than historical significance. Yet a modern analysis reveals that it continues to raise unresolved and fundamental questions about the organisation of embryonic tissues, their interactions, and the principles by which complex forms arise from simple beginnings. What may appear to be a relic of nineteenth-century thought still has the potential to inform and challenge contemporary developmental biology.

I will argue that a re-examination of such concepts is not merely of antiquarian interest. Rather, it highlights the extent to which certain basic issues—issues of organisation, integration, and the emergence of form—have been displaced rather than resolved by molecular and reductionist approaches. By bringing these "historical" questions back into focus, we can re-engage with the conceptual foundations of the field and confront problems that remain central, even if obscured.

To frame these questions properly, however, it is necessary to look back at the history of developmental biology itself. This history is rich in both successes and failures, and in debates that continue to resonate today. For instance, the long-standing controversy between preformation and epigenesis (Callebaut, 2008) shaped the very possibility of thinking about how form arises during development. The emergence of the germ layer theory in the nineteenth century provided a unifying framework but also raised new questions about tissue organisation that were never fully settled. The experimental embryology of the early twentieth century, through the work of Roux, Driesch, Spemann, and others, demonstrated the remarkable regulative capacities of embryos, yet at the same time exposed the limits of purely experimental manipulation in explaining developmental processes. Later, the incorporation of genetics and molecular biology into embryology was an extraordinary achievement, but it also narrowed the focus of inquiry, privileging mechanistic dissection over conceptual synthesis.

In the next section, I will briefly survey this historical trajectory, highlighting key advances and setbacks, to situate the "historical" questions that remain pressing for developmental biology today.

6.2 A BRIEF HISTORY OF DEVELOPMENT

The general problem of developmental biology is, without doubt, among the most difficult in the entire field of knowledge. That from a minute germ, apparently simple, there arises through a sequence of processes, beautifully orchestrated in both space and time, the complex organisation of the adult organism, is a fact that has never ceased to excite human wonder. The challenge of explaining how ordered structure and function emerge from such humble beginnings has occupied thinkers for millennia.

6.2.1 Aristotle

One of the first to address the problem of development systematically was Aristotle (Horder, 2010). His *Historia Animalium* presents a wealth of descriptive observations, which he later drew upon in his more theoretical treatment of development in *De Generatione Animalium.* Aristotle provided the first recorded account of chick development, noting that visible indications of the embryo appeared after three days of incubation. At this stage, he observed the heart as a tiny red spot, pulsating, with vessels—what we now identify as the vitelline veins—carrying blood. Subsequently, the body became differentiated, and the head with its disproportionately large eyes could be distinguished. From these observations, Aristotle concluded that blood is the first structure to appear, formed in the heart before the rest of the body differentiates.

His inquiry was not limited to descriptive accounts but extended to comparative studies across different bird species. In doing so, he sought not only to collect data but to organise it in such a way as to answer specific questions and derive general principles of development. One of his central concerns was the question of inheritance: what exactly do parents transmit to their offspring? He accepted that it must be a substance, which he termed "semen," present in both fathers and mothers. A prevailing view at the time was that each part of the adult body contributed specific material to the semen, an idea that would later resurface in the seventeenth century as the Theory of Pangenesis. Aristotle, however, argued that what is transmitted cannot be the physical parts themselves, but rather something more fundamental. In modern terms, one can interpret his view as anticipating the notion that parents transmit not structures but "information" by which structures are generated during development.

Aristotle also took a clear position in one of the longest-running debates in embryology: preformation versus epigenesis. Some philosophers held that all parts of the embryo exist simultaneously from the beginning, requiring only growth. Aristotle rejected this hypothesis, basing his refutation on observation. He pointed out that the chick's heart appears before the lungs. If one were to claim that the lungs existed but were too small to see, the argument collapses, since the lungs are in fact larger than the heart and therefore should be visible earlier if preformed. On this basis, Aristotle concluded that new structures arise sequentially during development. He thus aligned himself with epigenesis, the view that the embryo is progressively formed rather than simply unfolded from a miniature pre-existing version.

It must be acknowledged that Aristotle was not always correct in his biological claims. Notably, he accepted spontaneous generation as a common reproductive mode for certain organisms, a view that would persist well into early modern science before being definitively disproven. Nevertheless, his contribution to the field of developmental biology remains foundational. By combining careful empirical observation with systematic theoretical reflection, he not only initiated the study of embryology but also laid down questions that continue to frame the field: how does form arise from formless beginnings; what is transmitted from parent to offspring; and by what principles are developmental processes ordered in time and space? In this respect, Aristotle's work can rightly be regarded as the starting point of developmental biology.

6.2.2 Galen

Several centuries after Aristotle, Galen (129–c 216 CE) contributed further to the study of development, though his focus was often intertwined with broader concerns of anatomy and physiology (Cosans, 2010). Galen accepted and extended many Aristotelian principles, including the notion that semen carried the formative capacity for development. He believed, however, that both parents contributed material principles to the embryo: the father supplied the "formative" semen, while the mother contributed not only matter but also a kind of "female semen." Galen described the formation of organs as a process of "pneuma" (vital spirit) shaping the material of the embryo, a concept consistent with the medical doctrines of his time. His accounts were less observational than Aristotle's, but his authority as a physician ensured that his ideas dominated Western and Islamic thought for over a millennium.

6.2.3 William Harvey

The Aristotelian–Galenic synthesis remained influential until the early modern period, when the rise of experimental science reshaped embryological thought (Silverman, 2007). William Harvey (1578–1657), renowned for his discovery of the circulation of the blood, also made significant contributions to developmental biology. In his *Exercitationes de Generatione Animalium* (1651), Harvey carefully studied the development of chick and deer embryos. He famously coined the phrase *ex ovo omnia* ("everything comes from the egg"), rejecting the idea of spontaneous generation for higher organisms and asserting that all animals originate from eggs. Harvey was deeply committed to epigenesis, describing development as a progressive unfolding of structures, orchestrated by what he termed the "divine architect." While his reliance on metaphysical explanations limited the mechanistic scope of his theory, his careful observations reinforced the idea that development is a sequential and dynamic process, not the mere enlargement of preformed structures.

6.2.4 Caspar Friedrich Wolff

A decisive step forward came in the eighteenth century with Caspar Friedrich Wolff (1734–1794), who is often credited as the true founder of modern embryology (Dye, 2002). Wolff directly challenged the prevailing doctrine of preformation, which by his time had been reformulated into two competing schools: the "ovists," who claimed the preformed organism resided in the egg, and the "spermists," who placed it in the sperm. Through meticulous microscopic studies of chick embryos, Wolff demonstrated that tissues and organs arise progressively from undifferentiated material. He described the folding of embryonic layers, showing how the intestines, for example, are formed through morphogenetic processes rather than pre-existing structures. Wolff thus provided compelling empirical support for epigenesis, reviving and strengthening Aristotle's position with modern observational tools.

Despite Wolff's work, preformation retained strong appeal throughout the eighteenth century, in part because it seemed to align with mechanistic philosophy and avoided the problem of explaining how new structures arise. Nevertheless, Wolff's

Theoria Generationis (1759) laid the foundation for a developmental framework in which form and complexity emerge progressively from simpler states. His insights foreshadowed later understandings of differentiation and morphogenesis, concepts that would dominate developmental biology in the centuries to come.

6.2.5 Karl Ernst von Baer and the Nineteenth-Century Foundations

The nineteenth century marked a decisive transformation in embryology, as careful observation and comparative analysis began to establish developmental biology as a rigorous scientific discipline. Among the most influential figures was Karl Ernst von Baer (1792–1876) (Collins and Billett, 1995). Through painstaking studies of vertebrate embryos, von Baer formulated a set of generalisations that became known as "von Baer's laws." He observed that embryos of different species within a class initially resemble one another closely, diverging only later in development, and that general features precede the appearance of specialised characteristics. In other words, development proceeds from the general to the particular.

Von Baer's most celebrated contribution was his formulation of the germ layer theory. Building upon earlier suggestions by Christian Pander, von Baer demonstrated that the embryo differentiates into three primary layers—the ectoderm, mesoderm, and endoderm—each giving rise to specific tissues and organs. This framework provided embryology with an organising principle of extraordinary explanatory power, linking diverse developmental phenomena across species and offering a systematic account of tissue origins. The germ layer concept quickly became central to embryology, and while often treated today as a historical relic, it continues to raise profound questions about developmental organisation, tissue interactions, and evolutionary homology.

6.2.6 Cell Theory and the Rise of Microscopy

Closely connected with these advances was the emergence of cell theory, formulated by Matthias Schleiden and Theodor Schwann in the 1830s. The recognition that all organisms are composed of cells, and that cells arise from pre-existing cells, had immediate implications for embryology. Development could now be studied as a process of cellular division, growth, and differentiation. Improved microscopic techniques allowed embryologists to describe cleavage, gastrulation, and organogenesis with increasing precision. The embryo was no longer conceived as a mystical unfolding of "vital forces" but as a material process accessible to direct observation and experimental intervention.

6.2.7 The Rise of Experimental Embryology

By the end of the nineteenth century, embryology had begun to move beyond description into an experimental science. Wilhelm Roux (1850–1924) pioneered this approach with his studies of frog embryos (Hamburger, 1997). By destroying one blastomere at the two-cell stage, he attempted to show that development was a mosaic process, with each part of the embryo already determined to give rise to a specific structure. This "mosaic theory" of development reflected a mechanistic view of embryogenesis, in which the fate of each cell was fixed early and the embryo assembled like a prearranged machine. Hans Driesch (1867–1941), working independently, arrived at very different conclusions through his experiments on sea urchins (Chanoine, 2022). When he separated blastomeres at the two- and four-cell stages, each was capable of giving rise to a complete larva. Driesch argued that this demonstrated the embryo's *regulative* capacity: its ability to reorganise and produce a whole organism even after disruption. For Driesch, development was not strictly a mosaic process but rather a dynamic and adaptive one. These contrasting results inaugurated one of the most enduring debates in developmental biology: is development best understood as a predetermined, mechanistic sequence, or as a self-regulating, flexible system? And it was followed by a spectacular sequence of discoveries led by Spemann, Holtfreter, Nieuwkoop, and many others.

6.2.8 Towards the Molecular Era

The early twentieth-century achievements of experimental embryology laid the foundation for the molecular approaches that would follow. The mid-twentieth century saw the rise of molecular embryology, which sought to identify the chemical nature of the inductive signals and the genetic programmes they activated. While this shift yielded extraordinary discoveries of signalling pathways and gene regulatory networks, it also gradually displaced the broader, integrative questions raised by experimental embryology.

6.2.9 Conrad Hal Waddington and the Epigenetic Landscape

Conrad Hal Waddington (1905–1975) sought to integrate embryology with emerging genetics (Alan, 1977). In his influential metaphor of the "epigenetic landscape," he represented development as a ball rolling down a landscape of branching valleys, with each bifurcation representing a developmental decision. While the precise path was shaped by genetic determinants, the topology of the landscape reflected the system's capacity for regulation and canalisation. Waddington's framework emphasised that genes act not in isolation but within networks that generate robustness in developmental outcomes. Though metaphorical, the epigenetic landscape anticipated modern systems biology and remains a powerful conceptual tool for thinking about developmental dynamics and plasticity.

6.2.10 The Gene-Centric Era

By the mid-twentieth century, developmental biology entered the molecular era, driven by advances in genetics, biochemistry, and molecular biology. The discovery of DNA as the hereditary material, the elucidation of the genetic code, and the development of techniques for gene manipulation shifted the field's centre of gravity. The long-standing question of what parents transmit to their offspring now seems to have a definitive answer: genes. The rise of molecular embryology in the latter half of the century brought spectacular successes. The identification of homeotic

(Hox) genes in *Drosophila*, and later their conservation across animals, revealed a deep genetic homology in the specification of body plans (Hubert and Wellik, 2023). The discovery of signalling pathways such as Wnt, Hedgehog, transforming growth factor β (TGF-β), and FGF provided molecular correlates of the inductive signals. Developmental biology thus became increasingly gene-centric, with embryos conceptualised as the executioners of a genetic programme.

6.2.11 From Developmental Genetics to Genomics

With the sequencing of genomes and the rise of high-throughput technologies at the turn of the twenty-first century, the scope of developmental biology expanded yet again. Gene regulatory networks were mapped with increasing precision, and the molecular choreography of embryogenesis began to be described in unprecedented detail. The embryo became a model for systems-level analysis, with computational approaches used to reconstruct networks of gene expression, signalling cascades, and feedback loops. Yet this molecular and genomic revolution also brought with it a narrowing of perspective. The role of dissection in molecular mechanisms often overshadows the integrative, organism-level molecular mechanisms, once central to embryology. The focus on cataloguing pathways and regulatory networks sometimes obscured the larger question of how the embryo as a whole achieves coherence, robustness, and form. In this respect, the field risks returning to the problem identified earlier: *the forest of development is neglected in favour of its molecular trees.*

6.3 OPEN QUESTIONS IN DEVELOPMENTAL BIOLOGY

I sustain here that although considerable progress has been made in understanding many mechanisms of developmental biology, many of the classical "historical" questions remain unanswered. However, before providing some of these unanswered questions, it is useful to reflect on an example of a question that has been, to a large extent, satisfactorily answered.

6.3.1 The Saga of the Organiser, a Successful Story

The Mangold and Spemann experiment (Figure 6.1), published a hundred years ago (1924), is probably one of the most influential experiments in developmental biology, as it revealed one of the core principles of this discipline and continues to inspire research on self-organisation to this day (Spemann and Mangold, 2024).

Hilde Mangold (1898–1924) was a PhD student working under Hans Spemann (1869–1941), initially using *Hydra* as the model for her research. As her experiments with *Hydra* did not progress as expected, she moved to amphibians on Spemann's advice. Mangold grafted tissue from the dorsal lip of the blastopore of one embryo onto the ventral side of another. Remarkably, this transplanted tissue induced the formation of a secondary body axis, complete with neural tube, notochord, and somites.

This discovery of the "organiser"—a region of the embryo with the ability to direct the development of surrounding tissues—was a turning point in embryology.

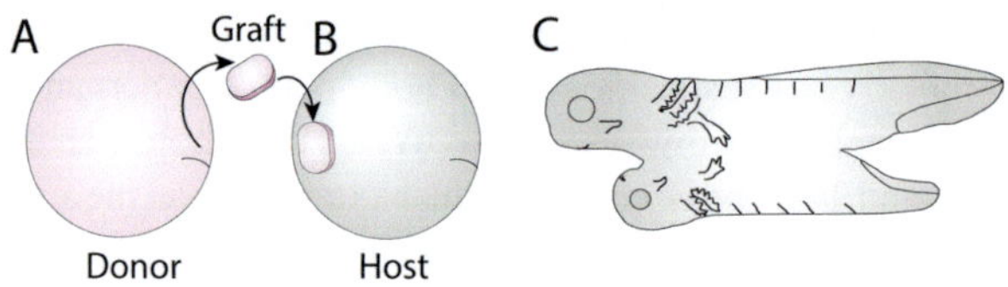

FIGURE 6.1 Mangold and Spemann's organiser experiment. (A) Diagram of the Mangold and Spemann experiment (1924). A piece of dorsal blastopore lip (indicated by the curved line at the edge of the embryo) is taken from a donor embryo (pink) and grafted onto the ventral side of a host embryo (grey). (B) The host embryo develops a secondary embryo attached to the primary embryo (the host). The secondary embryo is induced by signals produced by the dorsal blastopore lip (the Spemann organiser). Consequently, most of the tissue in the secondary embryo originates from the host embryo. To distinguish host from donor tissue, Mangold and Spemann used amphibian embryos with different pigmentation.

It demonstrated that embryonic development is coordinated not only by intrinsic cellular properties but also by inductive interactions between tissues. The concept of induction, formalised through these experiments, provided a framework for understanding how local signals and interactions guide the formation of complex structures. The organiser experiment had profound implications. It showed that development could not be reduced to the intrinsic fate of isolated parts, nor explained solely by mechanical arrangements. Instead, development emerged from dynamic interactions and hierarchies of control, in which certain regions of the embryo orchestrate the formation of the body plan. The conceptual legacy of Spemann and Mangold's work is evident in modern developmental biology, where organisers, signalling centres, and inductive fields remain central to explaining morphogenesis.

Johannes Holtfreter (1901–1992), working in the tradition of Spemann, extended the study of induction by demonstrating that not only living tissues but also killed or chemically treated tissues could elicit inductive effects in amphibian embryos (Holtfreter, 1925). He introduced the concept of "organiser gradients" and showed that embryonic cells were sensitive to graded signals that influenced their differentiation. Although the chemical identity of these signals remained elusive at the time, Holtfreter's experiments suggested that induction could be mediated by diffusible substances—an idea that foreshadowed the discovery of morphogens later in the century. The finding that killed cells or chemicals could reproduce the activities of the organiser fuelled the field with enthusiasm and generated a frenetic search for the chemical nature of the organiser. However, this initial enthusiasm faded as more and more unrelated factors were found to elicit organiser-like responses. Thus, from the initial excitement following the Mangold and Spemann discovery in the early 1920s, the field of organisers entered a dark period of disbelief, in which some even questioned whether the organiser was real (Williamson, 2024). We had to wait until the advent of the molecular biology era, 70 years later, to confirm the organiser's existence and identify its molecular nature.

As often happens in science, the question about the molecular basis of the organiser was asked at the wrong time: the necessary techniques simply did not yet exist. But when the field was both intellectually and technologically ready, not one, but three different organiser molecules were identified independently by three groups headed by Eddy De Robertis (Sasai et al., 1994), Richard Harland (Zimmerman et al., 1996), and Doug Melton (Hemmati-Brivanlou et al., 1994). Remarkably, these organiser molecules all employed the same molecular mechanism for their activity.

The current model of the organiser, with some variations across animal models, explains the patterned formation of the central nervous system (neural plate and neural tube) and the mesoderm. This model is known as the "default model" because it proposes that the default fate of the ectoderm is to become neural tissue (Hemmati-Brivanlou and Melton, 1997), and that the reason the entire ectoderm does not develop into neural tissue is that it secretes its own neural inhibitors: bone morphogenetic proteins (BMPs), members of the TGF-β family of signalling molecules. The role of the organiser is to inhibit BMP activity, which it does by secreting three proteins—noggin, chordin, and follistatin—that directly bind to BMP, preventing receptor binding and thus blocking the neural-inhibitory activity of BMPs. In this way, the organiser induces the formation of neural tissue, or a neural plate, which later develops into the neural tube.

An important aspect of the organiser is that it not only changes the fate of the ectoderm to neural, but also *organises* or patterns the responding tissue, so that different ectodermal derivatives are specified at precise locations within the embryo. Today we know that the subdivision of the ectoderm along the dorsal–ventral axis into neural plate, neural crest, placodes, and epidermis depends on graded levels of BMP activity, which are themselves controlled by a gradient of noggin, chordin, and follistatin secreted from the organiser (dorsal) and diffusing towards the ventral side of the embryo (Marchant et al., 1998; Nguyen et al., 1998). A similar mechanism operates to control the distinct types of mesoderm induced along the dorsal–ventral axis (Smith and Harland, 1992). Since the discovery of these inductive molecules, many molecular details have been added, showing that the expression and activity of these secreted proteins are under complex and tight regulation by a plethora of additional factors (De Robertis and Moriyama, 2016).

The story of the organiser illustrates how a classical problem in developmental biology can be resolved through the convergence of conceptual innovation and technological advance. It is an example of a historical question that has reached a satisfactory answer, and its centennial was fittingly celebrated in 2024 by the "*Self-Organisation in Biology: Freiburg Spemann-Mangold Centennial Symposium*" at the University of Freiburg (De Robertis et al., 2024), in the very hall where Spemann himself had attended academic events.

Now I will describe and discuss other important developmental biology questions that remain unanswered.

6.3.2 The Entelechy of Hans Driesch

Wilhelm Roux, the editor of the first important journal devoted to analytical embryology, *Wilhelm Roux's Archiv für Entwicklungsmechanik*, was a skilful experimentalist interested in the role of determinants located in specific regions of the embryo and required for the differentiation of its parts. Roux reasoned that if he destroyed some of these determinants, the resulting embryo should lack the corresponding structures.

To test this, he destroyed one of the blastomeres of a two-cell amphibian embryo with a hot needle (Sander, 1997). The results showed that only half an embryo developed, confirming Roux's prediction. This was an important achievement at the time, but not all of the observations fit the hypothesis. For example, in some embryos where Roux destroyed one cell, development still produced a normal—though smaller—embryo. Roux called this "postgeneration," but the idea was not easily accepted by the scientific community.

Years later, another German developmental biologist, Hans Driesch, repeated an equivalent experiment but used sea urchins and a different approach (Figure 6.2). Instead of killing one of the first two cells, Driesch separated them (Driesch, 1891). He placed sea urchin embryos in a tube of seawater and shook it vigorously until the cells separated. Driesch then observed their early cleavages and left them overnight

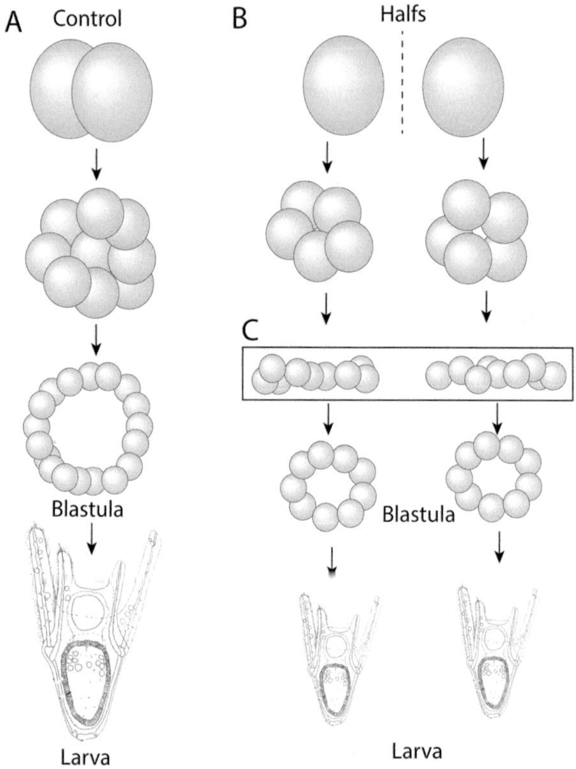

FIGURE 6.2 Hans Driesch's sea urchin cell-separation experiment. (A) Normal development of a sea urchin embryo from the two-cell stage to the larval stage, including the blastula. (B) Driesch separated the two-cell embryo into two individual blastomeres, which were cultured until the next day, when each developed into a normal—although half-sized—larva. (C) Recent analysis of this experiment (Suzuki et al., 2025) has shown that the half-embryo exhibits a transient structure composed of a flat layer of cells before rounding up into a blastula-like structure. This flat layer of cells is never observed during normal development.

for further examination. Based on Roux's results, he expected to see half a sea urchin embryo the next day. To his surprise, many of the separated blastomeres developed into completely normal pluteus larvae, albeit at half the usual size. From these experiments, Driesch concluded that embryonic cells are totipotent in their early stages of development, each with the capacity to form a complete organism.

This striking difference from Roux's results fuelled a rivalry between them, reviving the long-standing debate between epigenetic developmentalists (supported by Driesch) and preformationists (backed by Roux). Their work also led to the recognition that embryonic development can be classified into two general types: *regulative* and *mosaic*. In regulative development, embryos can adapt to perturbations and still form normal organisms, while in mosaic (or non-regulative) development, the loss of a part of the embryo results in the permanent absence of the structures derived from that part.

At the time of the Roux–Driesch experiments, development was conceived in mechanistic terms: embryos were thought of as small, complex machines that assembled their parts to generate a normal organism. Yet no machine, once cut in half, can regenerate its missing half. This paradox led Driesch to embrace vitalism. He proposed that a "vital force," which he called *entelechy*, guided embryonic cells towards normal development (Chanoine, 2022). He eventually left biology to become a professor of philosophy.

The Roux–Driesch experiments are generally analysed in the light of mosaic versus regulative development, a framework that has been the focus of intense research for over a century and has been revitalised in recent years by the remarkable self organising capacity observed in organoids. However, behind these classical experiments lies another, in my view, more pressing developmental biology question that remains unresolved. The mechanistic explanation of embryonic regulation has traditionally been framed in terms of inductive signals, cell adhesion mechanisms, and molecular gradients, all of which inform cells of their relative position and the absence of neighbouring tissues. From this perspective, isolated blastomeres are thought to compensate by adjusting proliferation, morphogenesis, and differentiation programmes to restore the missing half of the embryo. Yet this explanation remains unsatisfactory. When Driesch's experiment was repeated with modern live-imaging and molecular tools, researchers observed that the half-embryos initially formed an atypical flat epithelial sheet before rounding up into a blastula-like structure and proceeding towards normal development (Suzuki et al., 2025). Such a trajectory is never observed in intact embryos.

This raises profound questions. How do half-embryos "know" that they must proliferate more than normal to restore cell numbers, and equally importantly, how do they "know" when to stop? What kind of feedback mechanisms allow them to count cells or sense tissue proportions to achieve the correct size? Similarly, how can an embryo execute an entirely different morphogenetic route—via a flat sheet rather than a spherical blastula—and nevertheless converge on the same final architecture?

These questions suggest that embryonic regulation is not simply a matter of cells responding to local cues or missing neighbours. Instead, embryos appear to possess a global capacity for error correction, one that ensures robustness of form despite perturbations. Such capacity hints at principles of biological organisation that go beyond molecular pathways: principles of self-organisation, emergent order, and possibly higher-level constraints that guide development towards a viable whole.

In this sense, Driesch's intuition that something "extra-mechanical" governs development remains strikingly relevant, even if his invocation of Entelechy is no longer accepted in modern biology. Today, the challenge is to uncover how embryos integrate local interactions into global coherence—how molecular networks, mechanical forces, and geometry combine to produce a system that is not only self-assembling but also self-correcting.

The Roux–Driesch debate, therefore, foreshadowed one of the central questions of contemporary developmental biology: *what is the nature of developmental robustness, and how is it encoded in living systems?* The emergence of stem-cell–derived organoids (Zhao et al., 2022), which can spontaneously generate complex tissues and even rudimentary organs *in vitro*, shows that self-organisation is not an artefact of intact embryos but a fundamental property of developmental systems. Yet, just as with Driesch's half-embryos, the principles by which organoids "know" their scale, proportions, and final form remain largely mysterious.

6.3.3 Size Determination

A related but distinct problem to the one posed by Driesch more than a century ago is *what determines the size of organs during development*. Lewis Wolpert famously illustrated this by stretching out both arms and touching his fingertips together, asking: *What ensures that both arms grow to the same length?* At first glance, this may seem trivial, but it is far from simple. Each limb develops independently, without any direct communication with its counterpart, yet both achieve virtually identical lengths with astonishing precision (Wolpert, 2010).

The prevailing view is that each limb grows according to an intrinsic, genetically encoded programme, which is subsequently fine-tuned by systemic hormonal cues (Capdevila and Izpisua Belmonte, 2001). Limb length is not dictated by a central measuring device but rather emerges from the autonomous dynamics of *growth plate cartilage* in the long bones. A key set of experiments demonstrated this principle: when growth plates were transplanted between animals of different ages, their growth rates followed the *donor's intrinsic timeline* rather than that of the host, indicating that growth control resides within the cells themselves (Summerbell et al., 1973; Roselló-Díez and Torres, 2011). Growth plate chondrocytes divide rapidly over the years, yet bilateral limbs consistently attain matched lengths, despite inevitable stochastic variation in cellular behaviour. This extends to other paired structures, such as the eyes or ears, where symmetry arises despite each organ developing independently on opposite sides of the body.

The question of *size control* is not confined to limbs and organs but has fascinated embryologists since the earliest stages of development. In now-classical work, Patrick Tam and Janet Rossant investigated size regulation in mouse embryos (Figure 6.3) (Snow and Tam, 1979; Lewis and Rossant, 1982). They showed that reducing early embryo size with chemical inhibitors did not prevent later embryos from attaining normal size. Conversely, aggregating multiple embryos to produce an

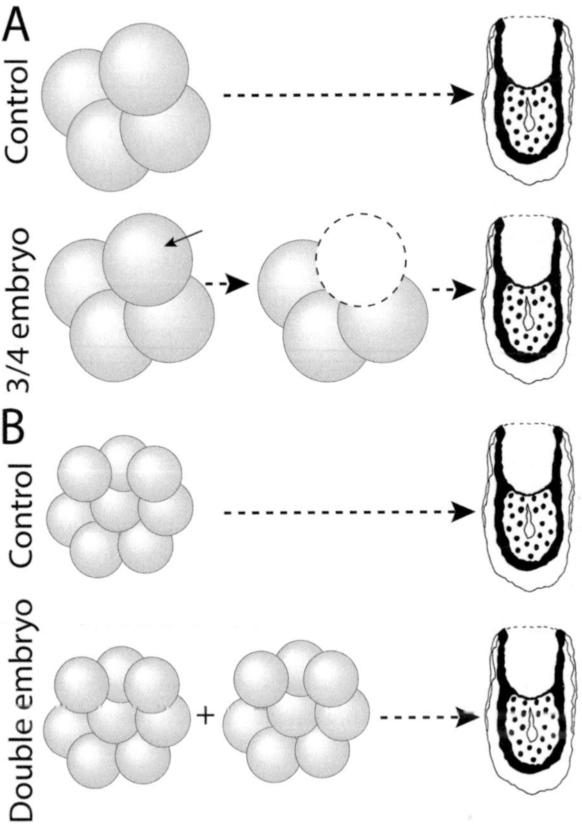

FIGURE 6.3 Size regulation in mammalian embryos. (A) Diagram of a 4-cell preimplantation mouse embryo developed until 5 dpc. When one of the four cells is removed, generating a smaller embryo, development proceeds normally, producing a 5 dpc embryo equal in size to the control, indicating that the ¾-size embryo is able to regulate its size. (B) Diagram of an eight-cell preimplantation mouse embryo developed until 5 dpc. When double embryos are created by fusing two 8-cell embryos, development proceeds normally, and the resulting 5 dpc embryo is the same size as control embryos, demonstrating size regulation.

oversized early embryo also resulted in a nearly normal-sized organism at later stages. In their striking summary, *a mouse embryo can be reduced to around 10% of its normal size at the onset of organogenesis, yet by completion of that stage, it is nearly indistinguishable from a control embryo*.

Subsequent studies have provided mechanistic insights into this phenomenon (Power and Tam, 1993; Hisaki et al., 2014; Orietti et al., 2021). Size regulation in mouse embryos appears to involve dynamic adjustments of the *cell cycle*: when cell numbers are increased, cell cycle lengthens, slowing proliferation; when cell numbers are reduced, cell cycles shorten, accelerating growth. This plasticity is complemented by modulation of cell size and apoptosis, with programmed cell death ensuring the removal of excess cells and contributing to morphogenesis.

Despite these advances, the fundamental question remains unresolved: *how does an embryo "know" its size?* What sensing mechanism informs cells that they are too few or too many, prompting compensatory proliferation or apoptosis? Unlike Roux and Driesch's experiments, which framed development as a problem of proportion and completeness, the problem of size regulation highlights a deeper principle of *scaling*. Somehow, embryos not only build structures with the correct shape but also with the correct *dimensions*, ensuring that an arm is neither too long nor too short, and that both arms match.

One possibility is that embryos use a form of *mechanical feedback*, where tissue tension or pressure provides size information (Parada et al., 2022). Another is that molecular gradients—classically invoked in pattern formation—may also scale dynamically with tissue size, providing proportional cues. Alternatively, information may be encoded in the *global dynamics of gene regulatory networks* or through emergent systems-level properties that integrate cell proliferation, death, and morphogenesis into a coherent outcome.

Thus, the problem of size determination, like that of regulative versus mosaic development, points beyond local cell behaviour to the question of how biological systems achieve *robust global coordination*. The embryo does not merely assemble parts; it measures, corrects, and scales them, ensuring symmetry and functionality. In this sense, size regulation may be one of the clearest examples of biological systems operating as *self-organising, self-correcting entities*, a theme that links Driesch's half-embryos, Wolpert's thought experiments, and modern organoid research into a continuous thread of inquiry about the nature of developmental robustness.

6.3.4 Anterior–Posterior Axis Specification in Vertebrates

How the anterior and posterior ends, and everything in between, are positioned at precisely the right place in the embryo has long fascinated developmental biologists. The openness of this question differs from those of embryonic regulation and size determination described above. In those cases, we struggle with a lack of explanatory frameworks; here, we face the opposite situation: a surplus of models and mechanisms, none of which is yet supported by a truly definitive experimental demonstration.

The prevailing view is that vertebrate anterior–posterior (AP) axis specification emerges from the integration of signalling pathways and transcriptional networks during gastrulation (Stern, 2001). Distinct anterior (head) and posterior (trunk and tail) domains are thought to be established by concentration gradients of morphogens, secreted from anteriorising and posteriorising centres. Yet how these signals are coordinated in space and time remains unsettled.

One of the earliest frameworks for thinking about AP patterning came in the wake of Hilde Mangold and Hans Spemann's organiser experiments. Otto Mangold extended their work by transplanting different portions of involuted mesoderm from late-gastrula embryos into the blastocoel of early gastrula hosts (Mangold, 1933). He found that the anterior archenteron roof consistently induced head structures, while more posterior regions gave rise to trunk or tail structures. From this, he inferred that the organiser was not uniform but released sequential positional cues: early-involuting cells specified anterior fates, and later-involuting cells posterior ones. Although conceptually elegant, Mangold's interpretation was later undermined by serious methodological problems. The timing and identity of signals were poorly controlled, and it remained unclear whether his grafts released multiple signals simultaneously. Moreover, the notion that each AP position corresponded to a unique inducer implied the existence of an implausibly large set of inducers.

A partial resolution came with Pieter Nieuwkoop's Activation–Transformation model, which still underpins much of modern thinking (Nieuwkoop, 1999). Nieuwkoop proposed that the organiser first emits a global anteriorising (or "activating") signal, inducing ectoderm to adopt a neural fate with default forebrain identity. A second wave of posteriorising signals then "transforms" parts of this anterior neural tissue into progressively more caudal structures. This gradient-like mechanism elegantly reduced the infinite-inducer problem to a simpler scheme of sequential activation and transformation. However, the original experiments, transplanting ectoderm grafts onto different AP levels of the host (Nieuwkoop and Nigtevecht, 1954), lacked the molecular markers and lineage-tracing tools we take for granted today. Alternative explanations, such as migration of mesoderm into the grafted ectoderm, with the consequent local induction instead of an inducive gradient, could not be rigorously excluded.

With the rise of genetic and imaging approaches, more recent work has shifted focus to mammals. Here, the anterior visceral endoderm (AVE) has been identified as a transient extra-embryonic signalling centre that migrates towards the future anterior pole during early gastrulation (Beddington, 1994; Thomas and Beddington, 1996; Varlet et al., 1997). The AVE secretes antagonists of posteriorising pathways (notably Wnt and Nodal), thereby promoting head development. Intriguingly, lineage-tracing studies show that AVE cells are not static. As the AVE migrates, the leading cells are displaced laterally and replaced by new visceral endoderm cells (Foley et al., 2000), suggesting that its signalling properties evolve dynamically.

Yet the AVE is not the sole anteriorising centre. Mouse experiments have shown that the anterior region of the organiser itself can induce anterior neural tissue even when the AVE is genetically ablated (Knoetgen et al., 2000). This points to redundancy and distributed control, where multiple signalling hubs ensure robust AP patterning. Comparative studies in zebrafish implicate the anterior neural fold (ANF) in a similar role (Houart et al., 1988), although the evidence here is less robust.

What emerges from this mosaic of observations is not a single, unified mechanism, but rather a network of overlapping signals and tissues, each contributing to time- and context-dependent ways. Still, the central conceptual problem remains: *what defines the spatial logic of these signals?* Do gradients alone suffice to explain the precise, reproducible positioning of the head and tail, or must we invoke higher-order mechanisms, such as biomechanical constraints, global embryo geometry, or emergent systems-level properties of signalling networks?

The anterior–posterior axis problem thus illustrates the limits of reductionist models. While gradients and transcriptional codes are essential components, they may not constitute the whole story. Just as with Driesch's regulative embryos and Wolpert's questions on size determination, AP axis specification may ultimately require a new conceptual framework, one that integrates signal dynamics, tissue mechanics, and systems-level self-organisation into a coherent account of how embryos establish polarity and proportion.

6.3.5 Embryonic Competence

The concept of embryonic competence emerged from the classical organiser experiments of Spemann and Mangold in the 1920s. Still, it was Conrad Waddington in 1940 who gave the phenomenon its name and theoretical weight (Waddington, 1936). Competence refers to the ability of embryonic cells to respond to specific inductive signals that guide their differentiation into particular tissues or organs. Importantly, this responsiveness is neither static nor universal: it is restricted in both space and time. The gastrula-stage ectoderm is competent to form neural tissue under the influence of the organiser, whereas mesoderm and endoderm are not. Likewise, the ectoderm that is competent to become mesoderm at the blastula stage loses this potential by early gastrulation (Servetnick and Grainger, 1991).

Waddington's concept marked a conceptual break with preformationist notions and placed interaction at the centre of developmental theory. Development was no longer seen as the mere unfolding of pre-formed parts, but as a dynamic dialogue between inductive signals and the changing capacity of tissues to interpret them. This shift paved the way for modern molecular developmental biology. Yet despite its foundational role, competence itself has often been treated as a secondary consideration, overshadowed by the search for inductive molecules and signalling pathways. What makes competence so intriguing is that it is as critical for correct patterning as the inductive signals themselves. Induction without competence is meaningless. Competence determines the spatial boundaries of tissue responses and ensures proportional organ development. Several experimental observations underscore this. For instance, in *Xenopus*, only ~40% of animal pole cells differentiate into mesoderm when exposed to vegetal cells (Dale et al., 1985). Similarly, only the inner ectodermal layer can form brain tissue, even though the outer layer is equally exposed to the inductive signals (Asashima and Grunz, 1982). Moreover, dorsal and ventral animal pole cells exposed to identical doses of FGF generate distinct mesodermal derivatives (Sokol and Melton, 1991). These examples demonstrate that competence is not uniformly distributed across tissues but is regionally and temporally regulated.

Initially, Waddington suggested that competence might be explained simply by the presence or absence of receptors for inductive signals. Indeed, in *Xenopus*, FGFR1 expression on ectodermal cells is necessary for mesoderm induction by FGF signals from the endoderm (Kimelman and Kirschner 1987). However, the persistence of FGFR1 expression beyond the competence window indicates that receptor availability alone cannot account for the precise timing of competence (Friesel and Brown, 1992). Attention has therefore shifted towards chromatin accessibility as a determinant of competence. ATAC-seq and ChIP-based studies have shown that, following mesoderm induction, chromatin at dorsal gene loci becomes condensed, limiting transcriptional responsiveness (Esmaeili et al., 2020). This provides a plausible mechanism for loss of mesodermal competence. Yet, when neural crest competence was examined, the correlation broke down: chromatin accessibility remained open even after competence was lost (Esmaeili et al., 2020). Thus, while epigenetic regulation is certainly a factor, it does not fully explain the phenomenon.

More recently, non-molecular factors have entered the discussion. Physical forces such as hydrostatic pressure have been shown to regulate competence to neural crest induction in amphibian ectoderm (Alasaadi et al., 2024). This finding suggests that competence is not reducible to gene expression or signalling pathways alone, but is instead an emergent property of cells and tissues—arising from an interplay of biochemical, mechanical, and perhaps even bioelectrical cues.

The broader implication is that embryonic competence may resist a purely reductionist explanation. Rather than being controlled by a single molecular switch, competence may reflect the holistic state of a tissue: its receptor profile, chromatin landscape, mechanical properties, electrical potentials, and metabolic status all working in concert. Competence is, in this sense, a systems-level property, a readiness of the tissue to engage in a developmental dialogue.

The neglect of competence as a subject of study has left a major gap in our understanding of embryogenesis. If competence defines where and when inductive signals are effective, then the regulation of competence is just as fundamental to embryonic patterning as the signals themselves. Future research must therefore embrace integrative approaches that combine molecular genetics with biophysics, imaging, and systems biology to uncover the principles by which tissues gain and lose the ability to respond.

In conclusion, competence is not merely a passive receptivity to signals but a dynamic, actively regulated state that ensures the fidelity and robustness of development. Just as Driesch's regulative embryos revealed the adaptive capacities of early embryonic systems, the study of competence reveals the conditional nature of induction: the same signal can yield very different outcomes depending on the state of the receiving tissue. Understanding competence may therefore be the key to unlocking deeper principles of developmental robustness, regeneration, and the engineering of tissues and organs in vitro.

6.3.6 Ontogeny Recapitulates Phylogeny

Few ideas in the history of biology have been as influential and as controversial as the notion that ontogeny recapitulates phylogeny. Throughout the nineteenth century, recapitulation theory occupied a central place in embryology. Charles Darwin himself remarked that many aspects of embryonic development could only be understood in light of this idea. In its most extreme form, the theory proposed that the embryos of higher organisms pass through stages resembling the adult forms of lower ones (Northcutt, 1990). A mammalian embryo was thought to become "fishlike," then "amphibian-like," before reaching its mature form. The most prominent advocate was the German biologist Ernst Haeckel (1834–1919), who formulated the "biogenetic law": *ontogeny is a short and rapid recapitulation of phylogeny.* Haeckel's dramatic comparative drawings of embryos, many of which were later criticised or even shown to be inaccurate, cemented the idea in textbooks and popular science for decades. He attempted nothing less than to unify descriptive embryology under a single conceptual framework.

But by the early twentieth century, recapitulation in its strict form was rejected (Hall, 2003). Careful embryological studies revealed that embryos of different species never actually pass through stages resembling the adults of others. At best, they transiently display morphological features reminiscent of ancestral forms, and even these often appear in highly modified ways. Yet even though the strong version of recapitulation has been discredited, a more nuanced interpretation retains explanatory power. In some cases, developmental stages still reflect vestiges of deep evolutionary history (Werth, 2014). For example, mammalian embryos temporarily develop pharyngeal arches reminiscent of fish gills, though they later give rise to jaws, ears, and throat structures; human embryos form a tail-like structure, later resorbed or integrated into the coccyx; the development of the vertebrate heart passes through configurations resembling simpler ancestral circulatory patterns. These examples raise the important question of why some of these structures are conserved during development. The consensus current view is that while ontogeny does not literally replay phylogeny, developmental processes are constrained by historical scaffolding: new evolutionary features are layered upon pre-existing developmental programmes, rather than designed anew.

This raises an important paradox. Comparative genomics and developmental genetics demonstrate that evolution is efficient: genes, tissues, or structures without function are usually lost. Why, then, do embryos sometimes retain transient, seemingly obsolete structures? Why does development "waste effort" producing tails or gill slits that will later regress? These are important questions that remain largely open, or with speculative answers, in modern developmental biology. One of these speculative answers is that these ancestral vestiges persist not because they are functionless, but because they remain embedded in the architecture of developmental programmes. Evolution is conservative with respect to early development: changes at these stages can be catastrophic, disrupting the entire construction of the organism. As a result, ancestral features may be retained not for their function but as developmental constraints, scaffolds upon which later processes depend.

The history of the recapitulation idea illustrates both the perils of overextension and the insight of a partial truth. Haeckel's rigid biogenetic law is untenable, but the persistence of ancestral traces in development is undeniable. The persistence of ancestral structures in embryos forces us to think in terms of evolutionary constraints and developmental robustness. Evolution can and does discard unnecessary organs, tissues, and molecules, but some elements remain, probably because they are woven into the fabric of developmental architecture in ways that are not easily bypassed. Thus, recapitulation today is best understood not as a literal replay of evolution, but as a suggestion of the historical continuity of developmental programmes, shaped by both selection and constraint.

In this balanced sense, recapitulation should not be treated as a rigid law but as a lens through which to interrogate the interplay between development and evolution. Rather than asking *whether* ontogeny literally repeats phylogeny, the modern challenge is to ask *why certain ancestral traces persist while others are erased*. This leads to several open questions, for example, *what developmental mechanisms explain evolutionary constraints?* If embryos retain ancestral structures that no longer serve an obvious function, is this because these features are indispensable scaffolds for later development, or simply because they are difficult to rewire without catastrophic consequences? *Why do some vestiges disappear while others persist?* For example, tails regress in humans, but pharyngeal arches are conserved as precursors to essential craniofacial structures. What determines whether an ancestral feature is evolutionarily dispensable or becomes co-opted for a new role? *Is there a "developmental cost" to innovation?* Does the persistence of certain embryonic stages reflect a deep resistance to altering early developmental programmes, such that evolution preferentially modifies later stages where the risks of disruption are lower?

These unresolved questions highlight that recapitulation, in its modern sense, remains a fertile area of inquiry. The phenomenon forces us to confront the *historical contingency of development*: evolution does not act on a blank slate but on an inherited framework of embryonic processes. Understanding why some ancestral traces remain while others vanish will require integrating evolutionary theory, developmental genetics, and systems-level models of constraint and plasticity. Far from being a discarded relic of nineteenth-century thought, recapitulation continues to provoke some of the most fundamental questions at the intersection of development and evolution.

6.4 THE FUTURE OF DEVELOPMENTAL BIOLOGY

The field of developmental biology has undergone a remarkable transformation over the past half-century. Once grounded primarily in descriptive embryology and classical transplantation experiments, it now sits at the intersection of genetics, molecular biology, physics, computational science, and engineering. The progress has been extraordinary: the molecular basis of Spemann's organiser has been elucidated, gene regulatory networks underlying axis specification have been mapped, and many of the once "black box" processes of morphogenesis can now be dissected in real time. These achievements are too numerous to list in full here, but together they illustrate a discipline that has repeatedly reinvented itself through conceptual and technical revolutions. For an excellent summary of the technical and conceptual advances in developmental biology over the past fifty years, see Liberali and Schier (2024).

Much of this success has been driven by methodological innovation. CRISPR/Cas9 and related gene-editing technologies have expanded the scope of experimental embryology to almost any species, allowing comparative studies across a broader evolutionary range than ever before (Mali et al., 2013). Advances in live imaging and fluorescent reporters now enable the tracking of individual cells and their dynamic behaviours with subcellular resolution, revealing developmental processes that were once invisible (Nozumi et al., 2017). Optogenetic tools provide the unprecedented ability to manipulate gene activity in both space and time, offering causal tests of hypotheses that were previously out of reach (Krueger et al., 2019). Single-cell omics have allowed developmental biologists to reconstruct differentiation pathways with remarkable precision, creating atlases of cell states that map the transitions from pluripotency to specialised fates (Griffiths et al., 2018). Complementing these advances, organoid and stem-cell–based systems have provided experimental models that bridge the gap between reductionist assays and the full complexity of the embryo, enabling both mechanistic insight and biomedical applications.

These technologies will undoubtedly continue to yield deeper descriptions of developmental processes at the genetic, molecular, and cellular levels. Yet, it is important to pause and recognise that more detail does not always mean more understanding. Knowing the precise behaviour of thousands of genes or the trajectories of millions of cells does not in itself resolve the fundamental conceptual questions that developmental biology has wrestled with for more than a century.

Many of these questions, such as how embryos regulate after perturbation, how size is determined and maintained, how competence windows are controlled in time and space, and why some ancestral features persist while others are lost, require explanations at the level of systems and principles, not just parts. Molecular biology has been extremely successful at dissecting the individual genes and proteins involved in controlling specific mechanisms, but developmental biology now needs approaches that work at a larger scale: the scale of the embryo. The "entelechy" of Driesch, although conceptually flawed, emphasised the importance of understanding the embryo as a whole—for instance, how one half can regenerate to restore a complete organism. The same need for explanations at the tissue or embryonic scale applies to size control: restoring the size of an organ requires recognition of the organ as a whole.

A particularly promising area in this regard is biomechanics: the study of how physical forces, material properties, and tissue dynamics shape development (Goodwin and Nelson, 2021). Biological form cannot be explained by molecular signals alone; it also emerges from the integration of forces generated by cell adhesion, cytoskeletal rearrangements, extracellular matrix interactions, and hydrostatic pressures. Forces can be transmitted across tissues almost instantaneously—much faster than the diffusion of molecular gradients—and they scale with embryo size. Importantly, they are circularly integrated with genes and molecules. Mechanics is not simply an outcome of gene regulation but an active player in pattern formation and fate determination. Mechanical signals can control gene expression, influence competence, bias lineage choices, and orchestrate morphogenetic events. For example, physical forces can deform subcellular structures such as the nucleus, leading to epigenetic changes and the activation of specific differentiation programmes (McCreery et al., 2025). The embryo, then, should be understood not just as a genetic programme but as a self-organising physical system, where mechanics and molecules interact across scales. While genes and molecules act at subcellular and cellular scales, mechanical forces operate isotropically at the tissue scale, fulfilling many of the requirements raised by classical conceptual problems.

Embryonic competence is supposed to change across whole tissues instantaneously and isotropically. Recent work has shown that neural crest competence is regulated by hydrostatic pressure in the blastocoel cavity—an isotropic property that shifts instantly across the entire ectoderm as volume changes (Figure 6.4) (Alasaadi et al., 2024). Thus, mechanical forces such as hydrostatic pressure seem particularly well suited to regulate embryonic competence. Interestingly, Waddington, who

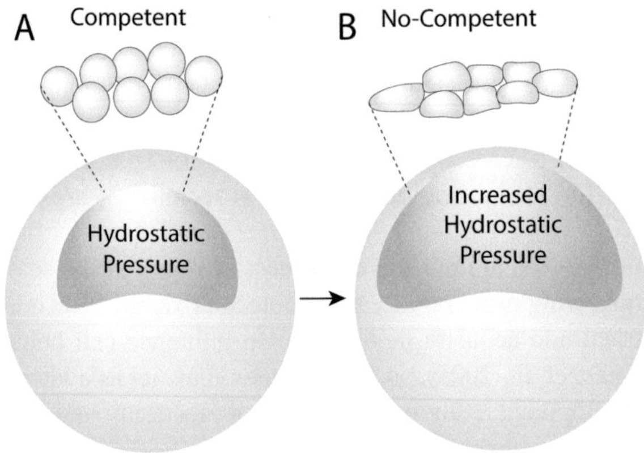

FIGURE 6.4 Embryonic competence is regulated by mechanics. (A) Competence for neural crest induction in amphibian embryos is regulated by the hydrostatic pressure of the blastocoel cavity (pink area). At competent stages, hydrostatic pressure in the blastocoel is low, meaning the adjacent ectoderm cells experience little compression. (B) At non-competent stages, hydrostatic pressure in the blastocoel increases, compressing the adjacent ectoderm cells. This compression affects Yap nuclear localisation, which in turn interferes with the Wnt signalling required for neural crest induction. The role of mechanical cues in modulating Wnt signalling corresponds to the definition of embryonic competence (Alasaadi et al., 2024).

proposed the concept 90 years ago, suggested that competence was more likely to be regulated by mechanical rather than chemical factors:

> It hardly seems probable that the function of these various neighbouring tissues is to exert some specific chemical influence on the processes which give rise to the competence. One can more easily suppose that the competence can be independently produced by the ectoderm, provided only that the mechanical conditions are favourable
>
> (Waddington, 1936)

Embryonic regulation has also been studied in avian embryos, where subdividing the epiblast disk not only redirects cell fates to form a complete, well-proportioned embryo at the original site but can even generate additional, fully formed embryos (Spratt and Haas, 1960). Recent work has revealed that actomyosin-dependent mechanical forces are central to embryonic self-organisation, shaping both tissues and gene expression to canalise early development robustly yet flexibly (Caldarelli et al., 2024). With the growing interest in developmental biomechanics, it is expected that other regulatory processes will be explained in the coming years.

The challenge for the coming decades will be to integrate scales: to understand how molecular instructions and mechanical forces cooperate to produce reliable outcomes despite variability and noise. This will demand new theoretical frameworks and computational models capable of explaining how local interactions among molecules and cells give rise to the robust global architecture of the embryo. It will also require new biomechanical techniques to measure and manipulate forces in vivo.

For instance, tissue stiffness is crucial for controlling cell behaviours such as migration and differentiation, yet most research has been conducted in vitro, with only a few in vivo studies. Atomic force microscopy (AFM) is commonly used to measure stiffness, but it is invasive and not suitable for all embryonic models. Emerging methods such as Brillouin Microscopy are better suited for complex embryos (Bilenca et al., 2024), and further innovations in physical techniques are likely in the near future.

But mechanical forces are not the only cues operating at the scale of the whole embryo and propagating more rapidly than molecular signals. Recent evidence indicates that electrical and acoustic fields can also influence cell behaviour (Ferreira et al., 2025; Kumeta et al., 2025), suggesting they may act as additional modulators of embryonic development. Although the study of electricity and sound in this context is still at an early stage compared with biomechanics, these cues deserve attention if we are to capture the full, integrative nature of embryonic systems.

The future of developmental biology will also require renewed attention to comparative and evolutionary developmental biology, as only by studying a wide diversity of organisms can we disentangle what is essential, what is contingent, and what is truly conserved across evolution.

The future of developmental biology is therefore not only technological but also conceptual. As powerful as new tools are, their value lies in addressing questions that remain open and pressing. How does an embryo measure its own size? How does it regulate its form after disturbance? What determines the windows of competence that constrain or permit responses to inductive signals? Why do some ancestral traits linger in development while others vanish? And at the deepest level: what are the organising principles that make embryonic development so robust, yet so evolvable?

In summary, developmental biology has never been in a stronger position to tackle these fundamental questions. The convergence of genetics, imaging, biophysics, and computational modelling promises not just more data but a more integrated understanding of how embryos build themselves. The field stands at an exciting frontier, where classical questions meet cutting-edge approaches. If history is a guide, the next 50 years will bring not only technical advances but also the conceptual breakthroughs that define what it means to explain development. The outlook is optimistic, not just for the accumulation of knowledge, but for a deeper synthesis of how living forms come into being.

ACKNOWLEDGEMENTS

Work in the laboratory of R.M. is supported by grants from the Medical Research Council (MR/S007792/1), Biotechnology and Biological Services Research Council (M008517; BB/T013044), and Wellcome Trust (102489/Z/13/Z).

REFERENCES

Alan, R., 1977. Conrad Hal Waddington. 8 November 1905–26 September 1975. *Biogr. Mem. Fellows R. Soc.* 23, 575–622.

Alasaadi, D.N., Alvizi, L., Hartmann, J., Stillman, N., Moghe, P., Hiiragi, T., Mayor, R., 2024. Competence for neural crest induction is controlled by hydrostatic pressure through *Yap*. *Nat. Cell Biol.* 26, 530–541.

Asashima, M., Grunz, H., 1982. Effects of inducers on inner and outer Gastrula Ectoderm layers of Xenopus Laevis. *Differentiation* 23, 206–212.

Beddington, R.S.P., 1994. Induction of a second neural axis by the mouse node. *Development* 120, 613–620.

Bilenca, A., Prevedel, R., Scarcelli, G., 2024. Current state of stimulated Brillouin scattering microscopy for the life sciences. *J. Phys. Photonics.* 6, 032001.

Caldarelli, P., Chamolly, A., Villedieu, A., Alegria-Prévot, O., Phan, C., Gros, J., 2024. Self-organized tissue mechanics underlie embryonic regulation. *Nature* 633, 887–894.

Callebaut, M., 2008. Historical evolution of preformistic versus neoformistic epigenetic thinking in embryology. *Belg. J. Zool.* 138, 20–35.

Capdevila, J., Izpisua Belmonte, J.C., 2001. Patterning mechanisms controlling vertebrate limb development. *Annu. Rev. Cell. Dev. Biol.* 17, 87–132.

Chanoine, C., 2022. Hans Driesch et la fin de toute vision préformationniste du développement. *Med. Sci. (Paris)* 38, 205–210.

Collins, P., Billett, F.S., 1995. The terminology of early development: history, concepts, and current usage. *Clin. Anat.* 8, 418–425.

Cosans, C.E., 2010. *History of classical anatomy*. In *eLS*. John Wiley & Sons, Ltd.

Dale, L., Smith, J.C., Slack, J.M., 1985. Mesoderm induction in Xenopus Laevis: a quantitative study using a cell lineage label and tissue-specific antibodies. *J. Embryol. Exp. Morphol.* 89, 289–312.

De Robertis, E.M., Driever, W., Mayor R., 2024. Celebrating the centennial of the most famous experiment in embryology: Hilde Mangold, Hans Spemann and the organizer. *Cells Dev.* 178, 203921.

De Robertis, E.M., Moriyama, Y., 2016. The Chordin morphogenetic pathway. Editor(s): Paul M. Wassarman. *Current Topics in Developmental Biology*, Academic Press, 116, 231–245.

Driesch, H., 1891. Entwicklungsmechanische Studien I-II. *Z. Wiss. Zool.* 53, 160–184.

Dye, F.D., 2002. *Dictionary of Developmental Biology and Embryology*, Wiley-Liss, New York.

Esmaeili, M., Blythe, S.A., Tobias, J.W., Zhang, K., Yang, J., Klein, P.S., 2020. Chromatin accessibility and histone acetylation in the regulation of competence in early development. *Dev. Biol.* 462, 20–35.

Ferreira, F., Moreira, S., Zhao, M. *et al.*, 2025. Stretch-induced endogenous electric fields drive directed collective cell migration in vivo. *Nat. Mater.* 24, 462–470.

Foley, A.C., Skromne, I.S., Stern, C.D., 2000. Reconciling different models of forebrain induction and patterning: a dual role for the hypoblast. *Development* 127, 3839–3854.

Friesel, R., Brown, S.A., 1992. Spatially restricted expression of fibroblast growth factor receptor-2 during Xenopus development. *Development* 116, 1051–1058.

Goodwin, K., Nelson, C.M., 2021. Mechanics of development. *Dev. Cell.* 56, 240–250.

Griffiths, J.A., Scialdone, A., Marioni, J.C., 2018. Using single-cell genomics to understand developmental processes and cell fate decisions. *Mol. Syst. Biol.* 14, e8046.

Hall, B.K., 2003. Evo-Devo: evolutionary developmental mechanisms. *Int. J. Dev. Biol.* 47, 491–495.

Hamburger, V., 1997. Wilhelm Roux: visionary with a blind spot. *J. Hist. Biol.* 30, 229–238.

Hemmati-Brivanlou, A., Kelly, O.G., Melton, D.A., 1994. Follistatin, an antagonist of activin, is expressed in the Spemann organizer and displays direct neuralizing activity. *Cell* 77, 283–295.

Hemmati-Brivanlou, A., Melton, D., 1997. Vertebrate neural induction. *Annu. Rev. Neurosci.* 20, 43–60.

Hisaki, T., Kawai, I., Sugiura, K., Naito, K., Kano, K., 2014. Regulation of embryonic size in early mouse development in vitro culture system. *Zygote* 22, 340–347.

Holtfreter, J., 1925. Defekt- und Transplantationsversuche an der Anlage von Leber und Pancreas jüngster Amphibienkeime. *Wilhelm Roux. Arch. Entwickl Mech. Org.* 105, 330–384.

Horder, T., 2010. History of Developmental Biology. In *Encyclopaedia of Life Sciences*, John Wiley & Sons, Ltd (Ed.).

Houart, C., Westerfield, M., Wilson, S.W., 1988. A small population of anterior cells patterns the forebrain during zebrafish gastrulation. *Nature* 391, 788–792.

Hubert, K.A., Wellik, D.M., 2023. Hox genes in development and beyond. *Development* 150, dev192476.

Kimelman, D., Kirschner, M., 1987. Synergistic induction of mesoderm by FGF and TGF-beta and the identification of an mRNA coding for FGF in the early Xenopus embryo. *Cell* 51, 869–877.

Knoetgen, H., Teichmann, U., Wittler, L., Viebahn, C., Kessel, M., 2000. Anterior neural induction by nodes from rabbits and mice. *Dev. Biol.* 225, 370–380.

Krueger, D., Izquierdo, E., Viswanathan, R., Hartmann, J., Pallares Cartes, C., De Renzis, S., 2019. Principles and applications of optogenetics in developmental biology. *Development* 146, dev175067.

Kumeta, M., Otani, M., Toyoda, M. *et al.*, 2025. Acoustic modulation of mechanosensitive genes and adipocyte differentiation. *Commun. Biol.* 8, 595.

Lewis, N.E., Rossant, J., 1982. Mechanism of size regulation in mouse embryo aggregates. *J. Embryol. Exp. Morph.* 72, 169–181.

Liberali, P., Schier, A.F., 2024. The evolution of developmental biology through conceptual and technological revolutions. *Cell* 187, 3461–3495.

Mali, P., Esvelt, K.M., Church G.M., 2013. Cas9 as a versatile tool for engineering biology. *Nat. Methods.* 10, 957–963.

Mangold, O., 1933. Uber die Induktionsfahighkeit der verschiedenen Bezirke der Neurula von Urodelen. *Naturwissenschaften* 21 761–766.

Marchant, L., Linker, C., Ruiz, P., Guerrero, N., Mayor, R., 1998. The inductive properties of mesoderm suggest that the neural crest cells are specified by a BMP gradient. *Dev. Biol.* 198, 319–329.

McCreery, K.P. et al., 2025. Mechano-osmotic signals control chromatin state and fate transitions in pluripotent stem cells. *Nat. Cell Biol.* 27, 1757–1770. https://doi.org/10.1038/s41556-025-01767-x

Nguyen, V.H., Schmid, B., Trout, J., Connors, S.A., Ekker, M., Mullins, M.C., 1998. Ventral and lateral regions of the zebrafish gastrula, including the neural crest progenitors, are established by a bmp2b/swirl pathway of genes. *Dev. Biol.* 199, 93–110.

Nieuwkoop, P.D., 1999. The neural induction process; its morphogenetic aspects. *Int. J. Dev. Biol.* 43, 615–623.

Nieuwkoop, P.D., Nigtevecht, G.V., 1954. Neural activation and transformation in explants of competent ectoderm under the influence of fragments of anterior notochord in Urodeles. *J. Embryol. Exp. Morph.* 2, 175–193.

Northcutt, R.G., 1990. Ontogeny and phylogeny: a re-evaluation of conceptual relationships and some applications. *Brain Behav. Evol.* 36, 116–140.

Nozumi, M., Nakatsu, F., Katoh, K., Igarashi, M., 2017. Coordinated movement of vesicles and Actin bundles during nerve growth revealed by Superresolution Microscopy. *Cell Rep.* 18, 2203–2216.

Orietti, L.C., Rosa, V.S., Antonica, F., Kyprianou, C., Mansfield, W., Marques-Souza, H., Shahbazi, M.N., Zernicka-Goetz, M., 2021. Embryo size regulates the timing and mechanism of pluripotent tissue morphogenesis. *Stem Cell Reports.* 16, 1182–1196.

Parada, C., Banavar, S.P., Khalilian, P., Rigaud, S., Michaut, A., Liu, Y., Joshy, D.M., Campàs, O., Gros, J., 2022. Mechanical feedback defines organizing centers to drive digit emergence. *Dev. Cell.* 57, 854–866.

Power, M.A., Tam, P.P., 1993. Onset of gastrulation, morphogenesis and somitogenesis in mouse embryos displaying compensatory growth. *Anat. Embryol. (Berl).* 187, 493–504.

Roselló-Díez, A., Torres, M., 2011. Regulative patterning in limb bud transplants is induced by distalizing activity of apical ectodermal ridge signals on host limb cells. *Dev. Dyn.* 240, 1203–1211.

Sander, K., 1997. "Mosaic work" *and* "assimilating effects" in embryogenesis: Wilhelm Roux's conclusions after disabling frog blastomeres. In: K. Sander (ed), *Landmarks in Developmental Biology 1883–1924*. Springer, Berlin, Heidelberg.

Sasai, Y., Lu, B., Steinbeisser, H., Geissert, D., Gont, L.K., De Robertis, E.M., 1994. Xenopus chordin: a novel dorsalizing factor activated by organizer-specific homeobox genes. *Cell* 79, 779–790.

Servetnick, M., Grainger, R.M., 1991. Changes in neural and lens competence in Xenopus ectoderm: evidence for an autonomous developmental timer. *Development* 112, 177–188.

Silverman, M.E., 2007. De Motu Cordis: the Lumleian Lecture of 1616: an imagined playlet concerning the discovery of the circulation of the blood by William Harvey. *J. R. Soc. Med.* 100, 199–204.

Smith, W.C., Harland, R.H., 1992. Expression cloning of noggin, a new dorsalizing factor localized to the Spemann organizer in Xenopus embryos. *Cell* 70, 829–840.

Snow, M.H.L., Tam, P.P.L., 1979. Is compensatory growth a complicating factor in mouse teratology? *Nature* 279, 555–557.

Sokol, S., Melton, D.A., 1991. Pre-existent pattern in Xenopus Animal Pole cells revealed by induction with activin. *Nature* 351, 409–411.

Spemann, H, Mangold, H., 2024. Induction of embryonic primordia by implantation of organizers from a different species. *Cells Dev.* 178, 203940.

Spratt, N.T., Haas, H., 1960. Integrative mechanisms in development of the early chick blastoderm. I. Regulative potentiality of separated parts. *J. Exp. Zool.* 145, 97–137.

Stern, C.D., 2001. Initial patterning of the central nervous system: How many organizers? *Nat. Rev. Neurosci.* 2, 92–98.

Summerbell, D., Lewis, J.H., Wolpert, L., 1973. Positional information in chick limb morphogenesis. *Nature* 244, 492–496.

Suzuki, H., Yaguchi, J., Tsuyuzaki, K. *et al.*, 2025. Unraveling the regulative development and molecular mechanisms of identical sea urchin twins. *Nat. Commun.* 16, 8005.

Thomas, P., Beddington, R.S.P., 1996. Anterior primitive endoderm may be responsible for patterning the anterior neural plate in the mouse embryo. *Curr. Biol.* 6, 1487–1496.

Varlet, I., Collignon, J., Robertson, E.J., 1997. Nodal expression in the primitive endoderm is required for specification of the anterior axis during mouse gastrulation. *Development* 124, 1033–1044.

Waddington, C.H., 1936. The origin of competence for lens formation in the Amphibia. *J. Exp. Biol.* 13, 86–91.

Werth, A.J., 2014. Vestiges of the natural history of development: historical holdovers reveal the dynamic interaction between ontogeny and phylogeny. *Evolution: Educ. Outr.* 7, 12.

Williamson, J., 2024. Celebrating Spemann-Mangold at 100: an interview with Eddy De Robertis. *Cells Dev.* 178, 203917.

Wolpert L., 2010. Arms and the man: the problem of symmetric growth. *PLoS Biol.* 8(9), e1000477.

Zhao, Z., Chen, X., Dowbaj, A.M. et al., 2022. Organoids. *Nat. Rev. Methods Primers* 2, 94.

Zimmerman, L.B., De Jesús-Escobar, J.M., Harland, R.M., 1996. The Spemann organizer signal noggin binds and inactivates bone morphogenetic protein 4. *Cell* 86, 599–606.

Section IV

Shedding New Light on Old Problems

7 Playing the Game of Regeneration

Elly M. Tanaka

The complexity of salamander limb regeneration has fascinated but also intimidated generations of biologists. The phenomenon attracts personalities with diverse motivations—those that seek pure understanding at the tissue, cell, or molecular level, those interested in evolution, and those that seek to cure human deficiencies. Limb regeneration is like a sports team—encompassing dynamic and complex interdependencies of individualized cellular players to score a goal. The process also displays adaptability to the type and location of amputation and is evidently a highly reactive system. Every fan can find their favorite aspect for obsession. Critical insights for unraveling this complexity have come from the explosive advances in genome engineering, genome sequencing, and imaging, which have brought the system into focus and poised it to divulge nuggets of insight to all those constituencies.

As for novice sports fans, the jargon and complexity of regeneration can seem confusing, contradictory, and overwhelming to a new acolyte or interested observer of regeneration. Keeping track of different players, teams that play with different strategies, and keeping track of scoring are all aspects of being a fan that take time to learn. Here, my aim is to personally accompany the reader through a tour of the cellular players of regeneration and patterning—and to synthesize several pieces of work that superficially may seem contradictory but that have provided fundamental insights into cellular plasticity.

7.1 RECREATING A BODY PART THAT WAS ALREADY BUILT ONCE

In a salamander, the amputation of an arm initiates the production of a stem and progenitor cell zone through which multiple tissues are rebuilt to achieve the same interdependencies and relative placement as before. Who is contributing those cells and how? Does a stem cell do everything, or do different cells have pre-defined roles? Like a complex play, different team members are doing each of these things. By tracking epidermis, dermis (mesodermal part of skin), muscle, and Schwann cells using GFP-expressing transgenically marked tissue, Kragl, Knapp, and colleagues showed that these tissue layers generally maintained their identities during regeneration, except for the dermis—a phenomenon that will be addressed in more detail below (Kragl et al., 2009). These results indicated that during limb regeneration, cells adhere to lineage boundaries that are present in the neurula stage of development. When specifically examining the muscle lineage in the axolotl, genetic fate mapping showed that muscle satellite cells are the reserve stem cells that migrate into

DOI: 10.1201/9781003560784-11

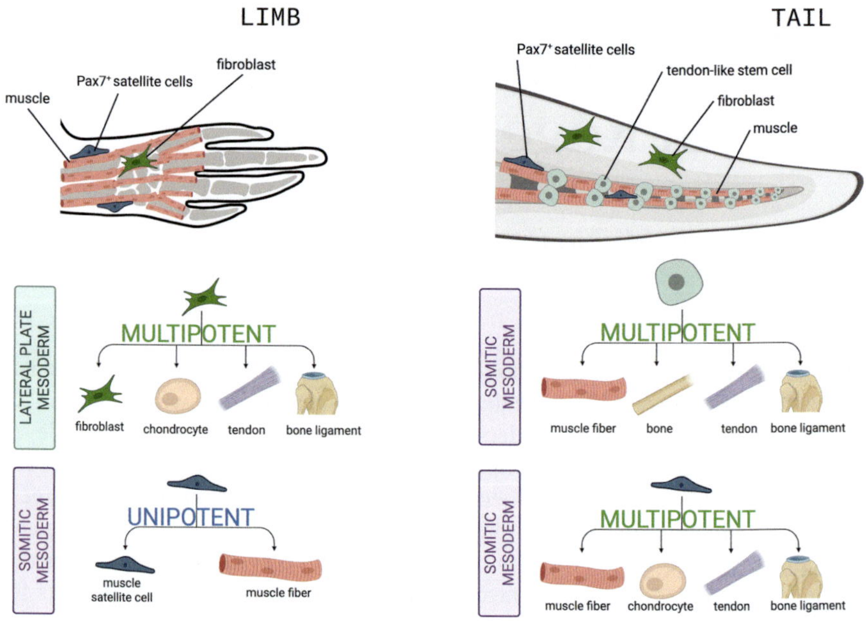

FIGURE 7.1 Mesodermal cell lineages and potency during limb regeneration and tail regeneration. Top: The mature limb and tail are depicted with mesodermal cell types that contribute substantially during axolotl limb or tail regeneration. During limb regeneration (left), PRRX1$^+$ fibroblasts dedifferentiate into a lateral plate-derived limb bud progenitor that produces bone, cartilage, tendons, ligaments, and fibroblasts. PAX7$^+$ satellite cells are unipotent muscle stem cells. During tail regeneration, tail muscle and connective tissue are formed during regeneration from an LNFG$^+$ stem cell that sits between muscle fibers and has tendon-like features. The PAX7$^+$ muscle satellite cells also show multipotency to form muscle and connective tissue. Blood vessels are not included in this depiction. The developmental origins of the cells are indicated to the left.

the blastema to regenerate the muscle (Figure 7.1), although in the newt, a different type of salamander, muscle fibers dedifferentiate to generate progenitor cells that regenerate muscle (Kragl et al., 2009; Sandoval-Guzman et al., 2014). This means that even muscle shows species-specific differences in whether stem cells move forward or muscle moves backward to generate the cell pool for regeneration.

While limb cells adhere to their developmental history, one lineage shows notable features and plays an important organizational role in regeneration. Limb bones, cartilage, ligaments, tendons, dermis and interstitial fibroblasts all arise during development from a multipotent progenitor in the lateral plate mesoderm and are collectively called connective tissue (Barresi and Gilbert, 2023). Remarkably, if during development somitic muscle progenitors are disabled in the embryo, the lateral plate mesoderm cells make a muscle-less but patterned arm skeleton with all ligaments and tendons (Chevallier and Kieny, 1982). This observation led to the conclusion that the lateral plate mesoderm cells and not muscle cells harbor the information for patterning the limb. During limb regeneration, each cell type responds to injury differently.

For example, cartilage cells can divide and make more cartilage, but they do not cross the amputation plane and do not contribute to other tissues (Currie et al., 2016; Steen, 1968). Periosteal cells that are typically associated with bone fracture repair migrate to cover the bone and extend the regenerating bone, but make little contribution to lower arm segments (Gerber et al., 2018). In contrast the dermal and interstitial fibroblasts play an extremely important role in regeneration (Figure 7.1). These cells, which in humans typically form scar tissue, dedifferentiate back to the limb lateral plate mesoderm cell type, which means that the fibroblast reacquires the multipotency to form bones, cartilage, ligaments, tendons, dermis and interstitial fibroblasts but not muscle (Gerber et al., 2018). So, on one hand, cells are adhering to their lineage history in the embryo, with the fibroblasts dedifferentiating and acquiring multipotency within the lateral plate mesoderm lineage.

How was it established that fibroblasts dedifferentiate rather than harbor a latent stem cell that does the work—especially since fibroblasts are an amorphous-looking cell type? Single-cell sequencing of FACS-isolated fibroblast lineage at time intervals less than one cell cycle could resolve that cells quantitatively transitioned from one transcriptional phenotype to another (Gerber et al., 2018), and furthermore, live imaging of digit tip regeneration showed the quantitative migration of cells into the blastema, with cells accumulating at the tip of the regenerate (Currie et al., 2016). The molecular phenotyping enabled by single-cell sequencing allowed for the conclusion that these fibroblasts achieve a transcriptional state closely resembling that of the progenitor cell in the developing limb bud (Gerber et al., 2018).

7.2 FORWARD OR REVERSE: PATTERNING THE REGENERATE

Production of different tissue types is not sufficient to regenerate a functional arm. The exquisite patterning of the skeleton with muscle, nerve, and skin is essential. In regeneration, this patterning process is adaptive to whether the amputation occurred in the upper arm, lower arm, or hand. How does the arm coordinate this complex process and adapt to the injury—in other words, how do the fibroblasts and muscle cells respond appropriately to the location of the amputation and build the right parts of the skeleton? The fibroblast cells that dedifferentiate into the multipotent progenitors constitute the master coordinators of regeneration, as they did during development. Inactivation of muscle stem cells yields regeneration of a perfectly patterned limb skeleton that is muscleless (Namenwirth, 1974). Several further experiments showed that the fibroblast cells contain intrinsic information about their location in the arm that guides the process of regeneration and ensures that the right pattern of bones is produced. GFP-expressing fibroblasts, muscle, or Schwann cells from the hand were transplanted to the upper arm prior to or during regeneration (Kragl et al., 2009; Nacu et al., 2013). Upon regeneration, the hand fibroblast-derived cells contributed only to the regenerated hand, while the hand muscle or Schwann cells contributed to all segments of the regenerated arm. This meant that the fibroblasts contain an intrinsic identity of where they are from, so that when one amputates an arm, the fibroblasts dedifferentiate and start the process of building a limb from the level at which they derive. Molecular profiling of modified histones in upper arm, lower arm, and hand fibroblasts showed that cells from the different levels have different set points in the chromatin modifications (Kawaguchi et al., 2024). Hand fibroblasts display repressive histone marks on homeodomain protein genes associated with upper and lower arm development, while these repressive marks are lacking on homeodomain protein genes associated with hand development. The upper arm fibroblasts show the converse pattern of histone marks. This pattern of histone marks corresponds to the upper arm, lower arm, and hand fibroblast cells having inherited different chromatin states during development that poise them to go forward from the point where they differentiated to build missing limb parts once they dedifferentiate into a multipotent (lateral plate mesoderm-type) limb bud cell. The other cell types, such as muscle stem cells and Schwann cells, apparently do not harbor this chromatin set point and can seed all segments of the regenerating limb to adapt to the frame established by the lateral plate mesoderm-derived cells (Kragl et al., 2009; Nacu et al., 2013).

So, while limb regeneration is a seemingly quixotic process where a beautifully patterned limb arises after it is lost, in essence, it is an ordered, almost conservative process where cells adhere to somitic, lateral plate, epidermal, and neuroectodermal lineages, and the master lateral plate mesoderm cells produce missing parts based on inheritance of information from development using developmental principles.

7.3 EVERYTHING GOES

Those seeking a clean, unifying principle to explain regeneration will be disappointed. All variants of mechanisms are observed depending on the body part, tissue, or even differences for the same tissue in different species. It is clear that nature has explored a very large space of possibilities to permit regeneration to succeed. The good news is that it means that the underlying molecular networks are highly adaptable and malleable, suggesting that biology is a playground where mixing and matching inspired by principles discovered in natural regeneration could lead to successful regeneration.

The studies of the limb set up expectations of similarity in regeneration of other tissues, but it has recently emerged that regeneration of another salamander appendage, the tail, occurs differently (Masselink et al., 2026). The tail contains a similar spectrum of mesodermal constituents to the limb, such as muscle, bone/cartilage (vertebrae), tendons, ligaments, dermal and interstitial fibroblasts, but tracing these various tissues during regeneration resulted in several surprises. Dermal fibroblasts in the tail did not contribute extensively to the regenerated vertebrae! In the tail, a stem cell with tendon-like features resides between muscle segments and upon tail amputation proliferates and produces progeny that form not only bony vertebrae, tendons, and ligaments, but also muscle—a diversity of cell types that is not seen in limb regeneration (Figure 7.1). Tracing of tail muscle satellite cells revealed that these cells are not unipotent, as in the limb, but also contribute to vertebrae, dermis, and presumably tendons and ligaments (Figure 7.1) (Wang et al., 2026)! Taken together, these tracing results in the tail show that the mesodermal lineage in the tail has quite different potencies and properties compared to the limb. During tail development, both muscle and connective tissue arise from somitic mesoderm, compared to the limb, where the connective tissue arises from the lateral plate. It is likely that the expanded potency of the tail mesodermal lineages is related to the history of a common progenitor in the somite. Chao and colleagues further investigated the molecular basis for the expanded potency of satellite cells in the tail versus the limb (Wang et al., 2026; Yi et al., 2025). The potency is an intrinsic property of the tail satellite cells, since the tail cells showed expanded potency even when transplanted into the limb. Molecular profiling of the tail satellite cells and functional analysis showed that tail muscle stem cells implement extended ERK activation and TGFB signaling to mediate the plasticity to form connective tissue (Wang et al., 2026; Yi et al., 2025). Differences in cell potential associated with different body regions are not exclusive to regeneration and have long been known for the neural crest, where the head neural crest has a cartilage forming capacity that is not observed in other body regions (Weston, 1962). Interestingly, TGFB signaling is associated with the expanded potential of the head neural crest (Rothstein et al., 2025).

Expanded potential may be a feature of neural stem cells during tail regeneration. Electroporation of a GFAP-promoter–driven GFP plasmid into the spinal cord lumen allowed tracking of neural stem cells after tail amputation. Contribution to muscle and cartilage was observed (Echeverri and Tanaka, 2002), although spinal cord labeling through an integrated GFP transgene or Sox2 genetic fate mapping did not apparently result in contribution to mesoderm (McHedlishvili et al., 2007; Wang et al., 2026). Molecular characterization of the spinal cord regeneration process has revealed that the mesoderm-associated gene Brachyury is upregulated in the regenerating spinal cord and is under negative regulation by miR-200a (Walker et al., 2022). Downregulation of miR-200a during spinal cord lesion led to upregulation of Brachyury and neural stem cell contribution to muscle using the GFAP:GFP plasmid electroporation tracing approach. It will be exciting and important to examine this phenomenon using genomically integrated transgenic markers.

The tail patterning process also proceeds by a conceptually different path when compared to the limb. During tail development, segmented muscles and vertebrae are templated by the formation of epithelial somites along the body axis (Barresi and Gilbert, 2023). This somite formation is influenced by the molecular oscillation of Notch signaling components in the presomitic mesoderm (Palmeirim et al., 1997). During tail regeneration, somites are not observed, leading to a hypothesis that somite-mediated segmentation of vertebrae is not relevant during regeneration. When a central oscillatory factor, *Hes7*, was mutated via CRISPR-mediated gene editing, developing animals showed defects in vertebral patterning (Masselink et al., 2026). Upon amputation of the CRISPANT tails, a perfectly segmented tail regenerated, demonstrating a fundamental difference in mechanism between development and regeneration. A broad survey of developmental regulators via spatial transcriptomics showed that many of the conserved embryonic regulators of segmentation were present in the embryo but absent in the regenerating tail. Examination of the regeneration process revealed the formation of a cartilage rod, which over time shows periodic bulging. Theoretical modeling combined with perturbation of collagen crosslinking has given rise to the model where the regenerating cartilage rod is under compression, which causes a periodic buckling along the cartilage rod, which templates the formation of vertebral segments. Interestingly, the frequency of vertebrae formation scales with rod diameter, providing a natural scaling mechanism in axolotls that regenerate at different ages and sizes.

7.4 PERSPECTIVES

The regeneration of a complex body structure such as the limb or tail shows the remarkable diversity and flexibility afforded by nature. We have learned very much on the cellular level, covered here, as well as on the molecular level, much of which has not been covered in this chapter. How cells execute regeneration clearly relates to their developmental history, which apparently leaves traces in the epigenome of adult cells so that they do not go back too far in lineage, and so that at least some cells have a trace of their position to template the new process. Whether a lack of regeneration represents a loss in these chromosomal memory processes, or whether the injury response is decoupled from the key genes that restart developmental pathways, is yet to be explored. The presence in the axolotl tail of competent progenitors that launch a non-developmental program to form vertebrae is intriguing. Whether such cell equivalents exist in other animals, and whether it is feasible to produce cartilage rods with the appropriate physical properties to undergo buckling instability are exciting prospects for the future.

This is a highly personal reflection on some of the astounding aspects of regeneration I have encountered over the years, and is by no means a comprehensive review of the field. Much of the literature cited itself cites foundational work in the field, and the reader is encouraged to delve into those sources to understand the significant precedents that led up to the work described.

REFERENCES

Barresi, M.J.F., Gilbert, S.F., 2023. *Developmental biology. International thirteenth edition*. New York, NY: Oxford University Press.

Chevallier, A., Kieny, M., 1982. On the role of the connective tissue in the patterning of the chick limb musculature. *Wilhelm Rouxs Arch Dev Biol* 191, 277–280.

Currie, J.D., Kawaguchi, A., Traspas, R.M., Schuez, M., Chara, O., Tanaka, E.M., 2016. Live imaging of axolotl digit regeneration reveals spatiotemporal choTablereography of diverse connective tissue progenitor pools. *Dev Cell* 39, 411–423.

Echeverri, K., Tanaka, E.M., 2002. Ectoderm to mesoderm lineage switching during axolotl tail regeneration. *Science* 298, 1993–1996.

Gerber, T., Murawala, P., Knapp, D., Masselink, W., Schuez, M., Hermann, S., Gac-Santel, M., Nowoshilow, S., Kageyama, J., Khattak, S., et al., 2018. Single-cell analysis uncovers convergence of cell identities during axolotl limb regeneration. *Science* 362, eaaq0681.

Kawaguchi, A., Wang, J., Knapp, D., Murawala, P., Nowoshilow, S., Masselink, W., Taniguchi-Sugiura, Y., Fei, J.-F., Tanaka, E.M., 2024. A chromatin code for limb segment identity in axolotl limb regeneration. *Dev Cell* 59, 2239–2253.e9.

Kragl, M., Knapp, D., Nacu, E., Khattak, S., Maden, M., Epperlein, H.H., Tanaka, E.M., 2009. Cells keep a memory of their tissue origin during axolotl limb regeneration. *Nature* 460, 60–65.

Masselink, W., Gerber, T., Singh Jamwal, V., Falcon, F., Deshayes, T., Grindle, R., Seaman, R., Röcklinger, M., Papadopoulos, S. C., Deneke, G., et al., 2026. Buckling instability underlies vertebral segmentation during axolotl tail regeneration.

McHedlishvili, L., Epperlein, H.H., Telzerow, A., Tanaka, E.M., 2007. A clonal analysis of neural progenitors during axolotl spinal cord regeneration reveals evidence for both spatially restricted and multipotent progenitors. *Development* 134, 2083–2093.

Nacu, E., Glausch, M., Le, H.Q., Damanik, F.F., Schuez, M., Knapp, D., Khattak, S., Richter, T., Tanaka, E.M., 2013. Connective tissue cells, but not muscle cells, are involved in establishing the proximo-distal outcome of limb regeneration in the axolotl. *Development* 140, 513–518.

Namenwirth, M., 1974. The inheritance of cell differentiation during limb regeneration in the axolotl. *Dev Biol* 41, 42–56.

Palmeirim, I., Henrique, D., Ish-Horowicz, D., Pourquié, O., 1997. Avian hairy gene expression identifies a molecular clock linked to vertebrate segmentation and somitogenesis. *Cell* 91, 639–648.

Rothstein, M., Azambuja, A.P., Kanno, T.Y., Breen, C., Simoes-Costa, M., 2025. TGF-β signaling controls neural crest developmental plasticity via SMAD2/3. *Dev Cell* 60, 1686–1701.e7.

Sandoval-Guzman, T., Wang, H., Khattak, S., Schuez, M., Roensch, K., Nacu, E., Tazaki, A., Joven, A., Tanaka, E.M., Simon, A., 2014. Fundamental differences in dedifferentiation and stem cell recruitment during skeletal muscle regeneration in two salamander species. *Cell Stem Cell* 14, 174–187.

Steen, T.P., 1968. Stability of chondrocyte differentiation and contribution of muscle to cartilage during limb regeneration in the axolotl (Siredon mexicanum). *J Exp Zool* 167, 49–78.

Walker, S.E., Sabin, K.Z., Gearhart, M.D., Yamamoto, K., Echeverri, K., 2022. Regulation of stem cell identity by miR-200a during spinal cord regeneration. *Development* 149, dev200033.

Wang, L., Song, L., Yi, C., Zhou, J., Yong, Z., Hu, Y., Pan, X., Qiao, N., Cai, H., Zhao, W., et al., 2026. Divergent stem cell mechanisms govern the primary body axis and appendage regeneration in the axolotl. *Sci Adv* 12, eadx5697.

Weston, J.A., 1962. The migration and localization of neural crest cells labelled with triated thymidine. *Am Zool* 2, 458.

Yi, C., Wang, L., Song, L., Fei, J.F., 2025. Phased ERK function on muscle stem cell plasticity in axolotl. Biorxiv

8 Reassessing Chordate Evolution

Did Tunicates Originate from a Vertebrate Ancestor?

Michael Levine and Nicholas Treen

The vertebrate body plan is one of the miracles of life on Earth. It is composed of a prominent head with centralized organs of special sense, a dorsal hollow neural tube, somites, and other distinguishing characteristics along an elongated trunk that distinguishes it from other animal groups, such as mollusks and arthropods (Holland and Chen, 2001). The basic vertebrate "tadpole" is the substrate for all higher animal life, including ourselves.

During the past 50 years, there has been enormous progress in identifying the genes and regulatory networks responsible for the specification of the different cell types and tissues comprising many or most vertebrate structures and organs, from sensory placodes to somites (Streit, 2018; Miao and Pourquié, 2024). However, there is one burning question that persists as a central mystery in animal evolution, namely, what is the origin of the vertebrate body plan?

The fossil record has nominated a few "proto-vertebrate" ancestors, such as *Haikouichthys* and *Myllokunmingia* (Shu et al., 1999; Shu et al., 2003). Both are about 520 million years old and display a head, muscular tail, and somites (or somite-like axial structures). However, even assuming that one or both of these fossils represent an ancient or ancestral chordate, the origins of the basic vertebrate body plan remain uncertain.

Our closest relatives, cephalochordates (e.g., amphioxus) and tunicates (aka urochordates), offer some clues (Satoh et al., 2014). Cephalochordates possess somites but essentially lack a head (although they have some head organs such as gill slits and photoreceptors). Tunicates have a rudimentary head as swimming tadpoles, but as adults, they lack just about every distinguishing feature of the vertebrate body plan. To understand how these features arose in evolution, we must consider other phyla related to the chordates, namely, echinoderms and hemichordates, even though they bear little semblance to the "crown and summit" of the animal kingdom, the vertebrata (Darwin, 1871).

To understand vertebrate origins, we must consider the last shared ancestor of both chordate and nonchordate Deuterostomes. Once again, the fossil record provides some candidates, such as Yunnanozoon from the lower Cambrian (Chen et al.,

 DOI: 10.1201/9781003560784-12

1999), but it is hard to relate these to the chordate body plan, despite the presence of certain morphological features such as gill slits.

There was one compelling effort to envision vertebrate origins in the context of morphological evolution: the Garstang hypothesis. Garstang was struck by the morphological similarities of certain echinoderms and hemichordates. In particular, the auricularian larvae of holothurians (sea cucumbers; echinoderm) and the tornarian larvae of enteropneusts (acorn worms; hemichordate) look quite similar, suggesting a key affinity ("ambulacraria") of these two otherwise very distinctive phyla. Garstang suggested that a morphological involution event along the aboral ectoderm of an ancestral ambulacrarian, akin to neurulation, spawned the origins of the basic chordate/vertebrate body plan (Garstang, 1928a).

Variations on the Garstang hypothesis have been suggested in more recent years (e.g., Bone, 1960; Romer, 1967); altogether, these represent the most serious efforts to understand vertebrate origins as a series of morphological events in evolution. Gene expression assays and other contemporary technologies are consistent with a key prediction of the hypothesis, namely, inversion of the dorsal-ventral axis in the transition from ambulacraria to chordates (Lowe et al., 2006). However, rather than being a sudden event as envisioned by Garstang, the "involution process" leading to axis inversion was more likely a gradual sequence of events during deuterostome evolution.

In summary, evo-devo studies of echinoderms and hemichordates are beginning to offer plausible scenarios for the origins of the vertebrate body plan. However, for almost as long as vertebrate origins have been considered, there has been a complicating factor: where do the tunicates fit into all of this? Do they represent a derived aberration? A living fossil of the authentic "proto-vertebrate" ancestor? Or something else?

Garstang himself suggested that vertebrates evolved by pedomorphosis from the larva of tunicate-like ancestors with filter-feeding adults (Garstang, 1928b), but based on recent molecular and developmental evidence, this scenario is now considered very unlikely (reviewed in Holland et al., 2015; Schlosser, 2021).

Efforts to answer these questions are being greatly enriched by analyzing the evolution of specialized cell types and tissues in vertebrates, such as cranial neural crest and placodes (Koontz et al., 2023). A variety of technologies have facilitated these efforts, including single-cell RNA-Seq assays, multicolor *in situ* hybridization assays, and comparative genome analyses. In this chapter, we focus on reviewing the most recent findings that clarify what we can learn about vertebrate evolution from our closest living invertebrate relatives, the tunicates.

As discussed above, there are three major groups of chordates: tunicates, cephalochordates, and vertebrates (Satoh et al., 2014). For many years, it was believed that the cephalochordates (e.g., amphioxus) were the sister group to the vertebrates. This was widely accepted since the overt morphologies of the two groups share striking similarities, most notably, sequential segmented somites along the rostral-caudal body axis (Yong et al., 2021). By contrast, tunicates look weird, particularly as adults. They lack key features of the vertebrate body plan. For example, a class of tunicates, the larvaceans, possess a head/tail body axis and a prominent notochord but lack segmented somites (Ferrández-Roldán et al., 2019). Other tunicates, such as ascidians, also exhibit vertebrate properties as larvae, including a slightly enlarged head, a dorsal hollow neural tube, and a prominent notochord (Kumano and Nishida, 2007). Nonetheless, adult cephalochordates look undeniably more vertebrate-like than any tunicate at any stage of their life cycles. Consequently, tunicates were relegated—quite reasonably—to a basal group within the chordates, possibly representing aspects of the earliest ancestral chordates (Garstang, 1928b and others).

The earliest molecular phylogenies, based on ribosomal DNA sequences, provided support for cephalochordates being the sister group to vertebrates, with tunicates being a monophyletic outgroup (Wada and Satoh, 1994; Cameron et al., 2000). All this changed with the publication of a transformative paper in 2006 (Delsuc et al., 2006). Comparative genome sequence analyses suggested that tunicates, such as the larvacean *Oikopleura*—and not cephalochordates—are the sister group of the vertebrates, despite outward appearances (Figure 8.1). Subsequent sequence comparisons strengthened this conclusion (Delsuc et al., 2008, 2018; Telford and Copley 2011; Pisani et al., 2015; Marlétaz et al., 2024). The result is so shocking that it is still not fully accepted in the evo-devo research community. Contemporary papers and even book chapters sometimes depict animal phylogenies with cephalochordates as the sister group of vertebrates.

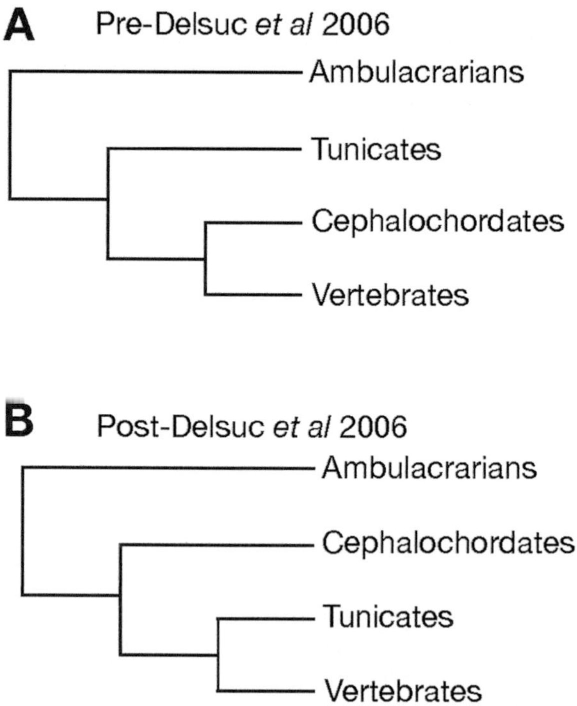

FIGURE 8.1 The position of tunicates and cephalochordates in deuterostome phylogeny. (A) Tunicates were placed as an outgroup of chordates based on morphological and early molecular evidence. (B) Whole-genome analysis resulted in the revision of the deuterostome phylogenetic tree with tunicates being placed as a sister-group to vertebrates.

In addition to comparative genomics, a variety of evo-devo studies suggest a kinship of tunicates and vertebrates. For example, adult ascidians (aka sea squirts) possess a beating heart that arises from embryonic anlagen evocative of those seen in vertebrates (Davidson, 2007, Figure 8.2A). It is specified by a variety of conserved regulatory determinants, including MesP and FoxF, which work together with FGF signaling to trigger the expression of genes required for the directed migration of heart progenitor cells (Satou et al., 2004; Davidson et al., 2006; Christiaen et al., 2008). Subsequent asymmetric cell divisions produce primary and secondary heart fields (Stolfi et al., 2010), similar to those seen in vertebrates. There is a striking link between the beating heart and "head muscles" involved in feeding (cardiopharyngeal

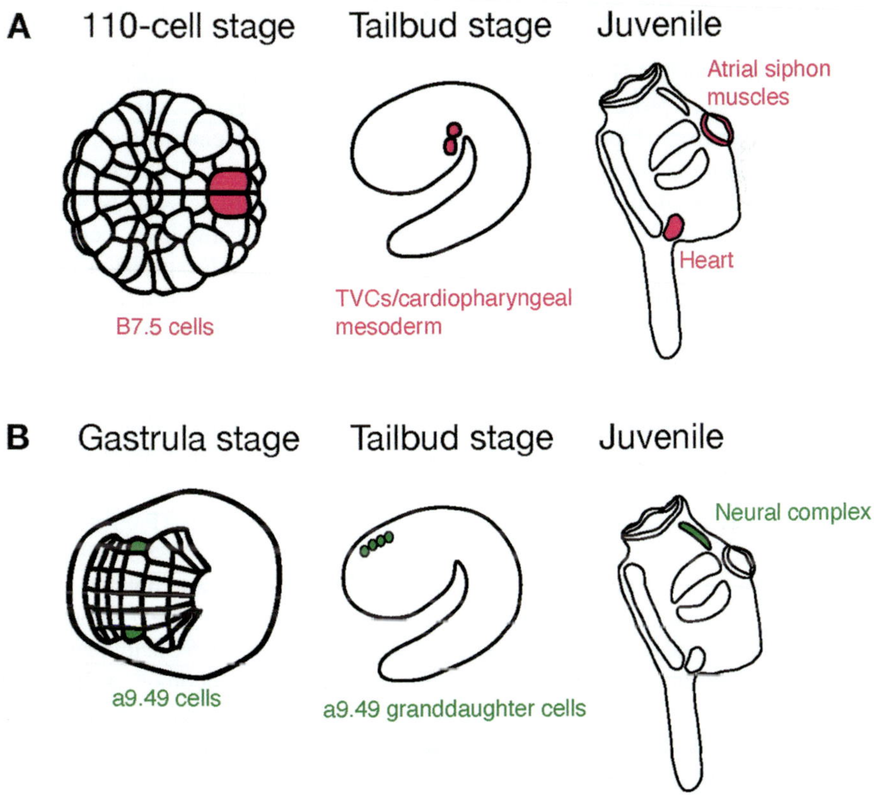

FIGURE 8.2 Examples of simplified vertebrate traits in *Ciona*. (A) The *Ciona* B7.5 cells at the 110-cell stage produce two pairs of migrating trunk ventral cells (TVCs) that form the *Ciona* cardiopharyngeal mesoderm. In the juvenile *Ciona*, the heart is formed from cardiopharyngeal mesoderm cells; these cells also make significant contributions to the muscles in the atrial (inflow) siphon. (B) The a9.49 cells at the gastrula stage have neural crest-like properties. The a9.49 cells will each divide to make four granddaughter cells in the sensory vesicle. These cells will produce the pigment cells of the tadpole and will be retained through metamorphosis to produce parts of the juvenile neural complex. As well as producing neural crest-like derivatives, the a9.49 lineage has the multipotency that is a defining property of neural crest.

mesoderm) (Diogo et al., 2015; Wang et al., 2019). Cephalochordates lack cardiopharyngeal mesoderm and a beating heart.

Over 40 years ago, Gans and Northcutt argued that vertebrates evolved a "new head", which possesses cell types and processes that are distinct from those seen in invertebrates (Gans and Northcutt, 1983). These include a cranium and other tissues that protect the enlarged brains of vertebrates. The new head also includes centralized organs of special sense for smell and hearing. These defining properties of the new head arise from cranial neural crest and placodes. There is evidence that both of these "vertebrate-specific" tissues are also present in ascidians.

The lateral plate ectoderm of neurulating *Ciona* embryos expresses a variety of determinants implicated in the specification of the pre-placodal ectoderm of vertebrates. These include Six1/2, Dmrt, FoxC, and Pax3/7 (Imai et al., 2004; Mazet et al., 2005). In head regions, the lineages expressing these determinants form palp mechanosensory cells and aATENs, which express GnRH and possess dual neurosecretory and chemosensory properties (Abitua et al., 2015). In tail regions, the lateral plate ectoderm gives rise to bipolar tail neurons (BTNs) (Imai and Meinertzhagen, 2007a), which were initially classified as being related to dorsal root ganglia, a derivative of trunk neural crest (Stolfi et al., 2015). However, more recent studies suggest that they probably correspond to placodal derivatives of cranial ganglia in vertebrates (Schlosser, 2021; Papadogiannis et al., 2022). Moreover, misexpression of pre-placodal determinants such as FoxC triggers the interconversion of different placodal cell types, such as the transformation of BTNs into palp mechanosensory cells and vice versa (Horie et al., 2018).

These observations are consistent with the possibility that the pre-placodal ectoderm of the last shared ancestor of tunicates and vertebrates contained an "expanded protoplacode", extending into trunk/tail regions (Horie et al., 2018; Papadogiannis et al., 2022). In vertebrates, this territory has been consolidated within head regions, with the notable exception of lateral line sensory cells in fishes (Piotrowski and Baker, 2014). The expanded proto-placode hypothesis raises the possibility that tail/trunk regions of ancestral proto-vertebrates lacked neural crest. By contrast, there is considerable evidence for a "rudimentary" cranial neural crest in *Ciona* tadpoles.

Swimming larval tadpoles contain a simple brain, the sensory vesicle, that relays external sensory signals to the motor ganglion, which controls locomotion (Imai and Meinertzhagen, 2007b; Akahoshi et al., 2021). The most important decision in the lifecycle of ascidians is where to attach and undergo metamorphosis. Once attached, it is game over for the population since adults are totally sedentary. The only opportunity for dispersal is that one day during the life cycle when larvae swim with great vigor to new locations for attachment and metamorphosis. The identification of salubrious environments for attachment depends on a variety of signals, including light and gravity (Kaijiwara and Yoshida, 1985; Tsuda et al., 2003; Jiang et al., 2005; Bostwick et al., 2020).

The detection of light and gravity depends on two pigmented cell types located within the lumen of the sensory vesicle, the ocellus and otolith, respectively. The progenitors of these cells arise from the dorsal midline of the neural tube in anterior regions, forming the sensory vesicle (Haupaix et al., 2014; Racioppi et al., 2014). The two a11.193 cells converge at the midline after closure of the trunk neural tube.

The cell occupying the more posterior position is exposed to a localized Wnt7 signal that triggers differentiation of the ocellus, while the anterior cell lacks this signal and follows an alternate fate, differentiation of the otolith (Abitua et al., 2012).

The Wnt7 regulatory network specifying the ocellus and otolith is strikingly similar to that controlling the differentiation of diverse pigmented cell types in the neural crest of zebrafish (Dorsky et al., 1998). This conservation, along with the expression of lateral plate determinants such as Pax3/7 and Msx, led to the proposal that the pigmented sensory cells of *Ciona* are homologous to neural crest-derived pigment cells in vertebrates, such as melanophores and iridophores (Abitua et al., 2012).

Such homology prompted the seemingly blasphemous suggestion that a forerunner of neural crest predated the evolution of the first vertebrates. Today, this seems reasonable since a variety of evo-devo studies suggest that true novelty is rare and often preceded by seemingly simpler rudiments, like the relationship of fish fins and tetrapod limbs (Stewart et al., 2019).

However, pigment cells are derivatives of neural crest and not neural crest per se (Le Douarin et al., 1992). A key property of neural crest is multipotency: the specification of multiple cell types, such as neurons and pigment cells (Motohashi and Kunisada, 2015), and, in the case of cranial neural crest, ectomesenchyme, which may be a truly novel cell type that forms the cranium and connective tissues in the jaws (Fabian and Crump, 2023). Are the progenitors of the pigmented sensory cells in *Ciona* multipotent? Recent lineage tracing experiments finally addressed this crucial question.

A comprehensive single-cell transcriptome atlas of staged *Ciona* embryos identified a variety of marker genes for the developing dorsal midline of tailbuds and larvae (Cao et al., 2019). Among these is Lmx1, a regulatory gene that is specifically expressed along the dorsal midline during closure of the neural tube. As seen in vertebrates, closure first occurs in trunk regions, pauses, and then resumes in the developing sensory vesicle (Hashimoto et al., 2015). The pigment precursors (a11.193) are the posterior-most cells of the zippering sensory vesicle. They arise from a mother cell, a10.97, that also produces the a11.194 cells, which are the next more anterior cells to converge along the midline during closure of the sensory vesicle (Haupaix et al., 2014; Racioppi et al., 2014). Single-nucleus lineage tracing was done to determine the fate of these sister cells of the otolith and ocellus (Todorov et al., 2024, Figure 8.2B).

The a10.97 lineage labels the otolith and ocellus in the sensory vesicle of swimming larval tadpoles. As expected, it also labels two "ependymal" cells derived from the pair of sister a11.194 cells. These function as neural progenitors to produce the juvenile and adult nervous system, including the cerebral ganglion, that forms during metamorphosis (Horie et al., 2011). In *Ciona*, there is a complete and dramatic transformation of the larval nervous system. It is far more complete than that seen for the metamorphosis of fish and frogs from tadpoles to adults, which involves adding new neurons to the pre-existing nervous system. In ascidians, the larval nervous system is essentially destroyed, except for quiescent ependymal cells, which proliferate during metamorphosis to produce the CNS of juveniles (Horie et al., 2011).

These studies suggest that multipotent neural crest cells preceded the advent of vertebrates. So far, only two documented cell types have been shown to arise from these cells: pigment cells and neural progenitor cells. It is easy to imagine that the co-option of different regulatory genes in this lineage allowed for the production of additional cell types in vertebrates. Most notably, co-option of the mesenchyme determinant Twist might be the key to the specification of ectomesenchymal cells arising from cranial neural crest (Abitua et al., 2012; Soldatov et al., 2019). Future lineage tracing experiments to show the fate of all neural cells past the 10th division will conclusively determine the pervasiveness of multipotency from the *Ciona* neural plate derivatives.

It is currently unclear whether other potential neural crest lineages arising from the dorsal midline of the sensory vesicle produce additional differentiated cell types in *Ciona*. We have emphasized the multipotency of the a10.97 lineage, but there are additional lineages located in more anterior regions of the neural tube that have not yet been traced. It is possible that these form ependymal cells as well as additional cell types. For example, derivatives of the a10.99 lineage appear to give rise to a pair of eminens neurons, which undergo long-range migration in the developing CNS (Lemaire et al., 2021). Cell migration is another key property of vertebrate neural crest, in addition to multipotency.

It is intuitive to interpret the cardiopharyngeal mesoderm and neural crest in *Ciona* as emergent properties in the last common ancestor of tunicates and vertebrates (Diogo et al., 2015). Unfortunately, there are no compelling fossils for this ancestor, so we have no idea what it looked like. We wish to put forward an alternative possibility: the fallen fish hypothesis.

Evolution is most commonly seen as a forward progression that forever improves ancestral tissues, organs, and morphologies. This way of thinking is pervasive, reaching a pinnacle in Darwin's discussion of the evolution of organs of "extreme perfection" such as eyes. However, we know of many cases of retrograde evolution. The loss of limbs in snakes (Leal and Cohn, 2016), eyes in cavefish (Krishnan and Rohner, 2017), and teeth in birds (Louchart and Viriot, 2011) represent a few well-studied examples. Rather than being emergent, the rudimentary cardiopharyngeal mesoderm and neural crest seen in ascidians are perhaps degenerate.

We suggest that the last shared ancestor of vertebrates and tunicates was an ancient cartilaginous "fish" akin to *Myllokunmingia* (Shu et al., 1999). We propose that the earliest tunicate ancestor evolved from a *Myllokunmingia*-like organism. The key change to evolve tunicates was that a larval form underwent a mode of metamorphosis in which many organs that are expanded in adult vertebrates, such as the sense organs, a motile tail, and much of the nervous system, degenerate. However, some other organs, such as the branchial basket, become highly elaborate. The end result is to form a sessile adult rather than a motile fish. What led to such a radical change in the body plan?

A key event was the acquisition of cellulose synthase (CesA) by horizontal gene transfer (Nakashima et al., 2004; Sasakura et al., 2016). By acquiring this gene, a specific population of the mobile ancestor could become sessile since CesA is required for the synthesis of the protective tunic that gives this group of chordates its name. Mutations in CesA block metamorphosis, leading to a "swimming juvenile" phenotype (Sasakura et al., 2005). We propose that a specific population of tunicate ancestors acquiring CesA underwent a "shadow" mode of evolution to form an array of sessile and motile tunicates that include doliolids, salps, larvaceans, and ascidians.

There is considerable diversification of the adult head regions of different tunicates. A particularly dramatic example is seen for a deep ocean ascidian called *Megalodicopia*. It is a predatory tunicate that is evocative of a venus fly trap. *Megalodicopia* embryos and larvae appear similar to the highly conserved forms seen in all solitary urodele ascidians (Havenhand et al., 2006). The oral siphon has been expanded to form a jaw-like feeding structure that encloses and digests small prey. It is possible that one or both of the two key vertebrate-style lineages in ascidians, neural crest and cardiopharyngeal mesoderm, have been expanded to help produce this enlarged feeding structure.

An implication of this degenerate mode of tunicate evolution is that the tunicate-vertebrate ancestor, an organism with a vertebrate-like body plan, had not undergone the genome-wide duplication events seen for modern vertebrates (Dehal and Boore, 2005). This raises the possibility that the first vertebrates evolved before the genome-wide duplications.

The tunicate genomes are highly degenerate and compact (Seo et al., 2001; Dehal et al., 2002). They lack coherent Hox gene clusters as seen in amphioxus (Garcia-Fernández and Holland, 1994; Ikuta et al., 2004). Nonetheless, they did not reduce duplicated ancestral genomes. They clearly possess rearranged versions of single-copy chordate genomes (Satou et al., 2019). If tunicates are indeed fallen fish, then the evolution of the most striking innovations of vertebrates did not depend on genome-wide duplication events as generally believed.

The poor fossil record of tunicates forces us to be speculative about the emergence of tunicates and the nature of the tunicate-vertebrate ancestor. The major groups ofchordate fossils known (*Yunnanozoons*, *Haikouella*, *Metaspriggina*, and *Myllokunmingia*) all have comparable features to extant cephalochordates and vertebrates, as well as larvae of tunicates, but very little direct similarity to the adult forms of tunicates.

Was the tunicate-vertebrate ancestor more like a tunicate or more like a vertebrate? It is compelling to suggest that vertebrates formed from a filter-feeding creature that lost an ascidian-like adult body plan and adapted the larval body for prolonged swimming and predation (i.e., pedomorphosis as suggested by Garstang). Although this is plausible, we struggle to understand how an ascidian embryo could be modified to form anything resembling vertebrate embryos. The ascidians *Ciona*, *Phallusia*, and *Halocynthia* are not closely related. *Ciona* and *Phallusia*, both *Phlebobranch* ascidians, diverged around 274 MYA. *Phlebobranch* and *stolidobranch* ascidians diverged around 390 MYA (Delsuc et al., 2018).

After hundreds of millions of years, it is reasonable to assume the embryos of different ascidian species would be vastly different, but at least at the early embryonic stages the embryos and patterns of gene expression between *Ciona*, *Phallusia*, and *Halocynthia* are so similar, it is notable when small differences are seen, for example, at the 64-cell stage the embryos of the three species are so similar, an expert might struggle to distinguish between them and gene expression patterns for regulatory genes typically show minimal differences (Figure 8.3A).

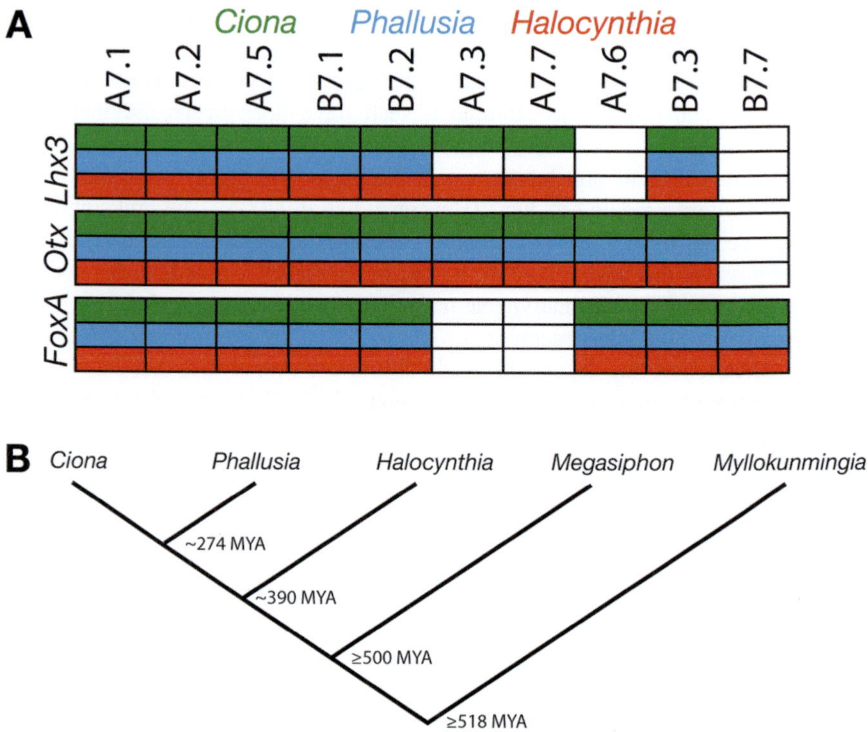

FIGURE 8.3 Ascidians have ancient and largely unchanged embryos. (A) Expression of three regulatory genes (Lhx3, Otx, and FoxA) in 64-cell stage ascidian embryos. Each box represents a single cell. The box is filled if the RNA has been detected in that cell. Expression patterns are qualitatively identical, with the exception of *Phallusia* Lhx3, which has lost expression in the A7.3 and A7.7 cells. Data taken from Madgwick et al., 2019, Kobayashi et al., 2010, Wada et al., 1996, Kumano et al. 2006. (B) Phylogeny of living tunicates as well as extinct ancestors. Atrial and oral siphons can be seen in all living ascidians as well as in the *Megasiphon* fossil. The *Myllokunmingia* fossil does not have tunicate-like siphons and has a vertebrate-like form.

The lack of differences seen between modern ascidian embryos suggests to us that they cannot tolerate changes and are therefore poorly suited for generating the vast diversity of vertebrate embryos. The reliance on early cell fate specification as well as fixed cell lineages may have resulted in an embryo that poorly tolerates even minor changes. We find the scenario in which an animal with a vertebrate-style embryo gave rise to tunicates to be at least as plausible as that of an animal with a tunicate-like embryo adapting to produce vertebrate embryos (see Holland et al., 2015 and Schlosser, 2021 for extensive considerations of this and views on possible scenarios for where tunicates fit into theories of vertebrate origins).

Ascidian embryo similarities would not be a problem if early tunicates were different from their modern forms. However, this is probably not true. A recently discovered fossil of an adult ascidian is over 500 million years old (Nanglu et al., 2023).

This fossil is extremely well preserved, and it is remarkable how similar it is to modern ascidians (Figure 8.3B). The fossil has clear oral and atrial siphons, suggesting an unchanged craniopharyngeal lineage (as well as other cell lineages such as the A7.6 mesenchyme) compared to modern ascidians. The existence of morphologically modern ascidians at a time when the earliest vertebrates are emerging should be given some serious thought. We are also struck by the ability of other classes of tunicates to simplify and degenerate. The larvacean *Oikopleura* reduced its pool of beating cardiomyocytes to just six cells by eliminating a number of key cardiogenic determinants seen in *Ciona* (Ferrández-Roldán et al., 2021).

To have a complete understanding of the chordate evolution and how vertebrates emerged, we must better understand how tunicates fit into these scenarios, specifically how tunicates emerged. A particularly illuminating fossil would be the tunicate-vertebrate ancestor, but if the fallen fish hypothesis is correct, then we may already have this in hand.

Future studies will provide detailed analyses of metamorphosis in tunicates. Our understanding of gene expression at the RNA-level in *Ciona* embryos is essentially complete, but we know extremely little about the juvenile and adult stages. Recent advances in single-cell and spatial transcriptomics on the *Styela* endostyle (Jiang et al., 2024), as well as similar studies showing the contribution of neural crest to the embryonic lamprey endostyle (Stundl et al., 2025), represent examples of the next frontier for this research. Similar investigations on the gill slits, siphons, heart, stomach, and adult nervous systems could provide comparative insights that would help us understand how these align with their vertebrate counterparts. Advances in comparative genomics should allow us to formally test instances of retrograde evolution, as proposed in this chapter.

REFERENCES

Abitua, P.B., Gainous, T.B., Kaczmarczyk, A.N., Winchell, C.J., Hudson, C., Kamata, K., Nakagawa, M., Tsuda, M., Kusakabe, T.G., Levine, M., 2015. The pre-vertebrate origins of neurogenic placodes. *Nature* 524, 462–465.

Abitua, P.B., Wagner, E., Navarrete, I.A., Levine, M., 2012. Identification of a rudimentary neural crest in a non-vertebrate chordate. *Nature* 492, 104–107.

Akahoshi, T., Utsumi, M.K., Oonuma, K., Murakami, M., Horie, T., Kusakabe, T., Oka, K., Hotta, K., 2021. A single motor neuron determines the rhythm of early motor behavior in Ciona. *Sci. Adv.* 7, eabl6053.

Bone, Q., 1960. The origin of Chordates. *Zool. J. Linn. Soc.* 252, 252–269.

Bostwick, M., Smith, E.L., Borba, C., Newman-Smith, E., Guleria, I., Kourakis, M.J., Smith, W.C., 2020. Antagonistic inhibitory circuits integrate visual and gravitactic behaviors. *Curr. Biol.* 30, 600–609.e2.

Cameron, C.B., Garey, J.R., Swalla, B.J., 2000. Evolution of the chordate body plan: new insights from phylogenetic analyses of deuterostome phyla. *Proc. Natl. Acad. Sci. U.S.A.* 97, 4469–4474.

Cao, C., Lemaire, L.A., Wang, W., Yoon, P.H., Choi, Y.A., Parsons, L.R., Matese, J.C., Wang, W., Levine, M., Chen, K., 2019. Comprehensive single-cell transcriptome lineages of a proto-vertebrate. *Nature* 571, 349–354.

Chen, J.Y., Huang, D.Y., Li, C.W., 1999. An early Cambrian craniate-like chordate. *Nature* 402, 518–522.

Christiaen, L., Davidson, B., Kawashima, T., Powell, W., Nolla, H., Vranizan, K., Levine, M., 2008. The transcription/migration interface in heart precursors of Ciona intestinalis. *Science* 320, 1349–1352.

Darwin, C. 1871. *The descent of man, and selection in relation to sex*. Vol 1.

Davidson, B., 2007. Ciona intestinalis as a model for cardiac development. *Semin. Cell Dev. Biol.* 18, 16–26.

Davidson, B., Shi, W., Beh, J., Christiaen, L., Levine, M., 2006. FGF signaling delineates the cardiac progenitor field in the simple chordate, Ciona intestinalis. *Genes Dev.* 20, 2728–2738.

Dehal, P., Boore, J.L., 2005. Two rounds of whole genome duplication in the ancestral vertebrate. *PLoS Biol.* 3, e314.

Dehal, P., Satou, Y., Campbell, R.K., Chapman, J., Degnan, B., De Tomaso, A., et al. 2002. The draft genome of Ciona intestinalis: insights into chordate and vertebrate origins. *Science* 298, 2157–2167.

Delsuc, F., Brinkmann, H., Chourrout, D., Philippe, H., 2006. Tunicates and not cephalochordates are the closest living relatives of vertebrates. *Nature* 439, 965–968.

Delsuc, F., Philippe, H., Tsagkogeorga, G., Simion, P., Tilak, M.K., Turon, X., López-Legentil, S., Piette, J., Lemaire, P., Douzery, E.J.P., 2018. A phylogenomic framework and timescale for comparative studies of tunicates. *BMC Biol.* 16, 39.

Delsuc, F., Tsagkogeorga, G., Lartillot, N., Philippe, H., 2008. Additional molecular support for the new chordate phylogeny. *Genesis* 46, 592–604.

Diogo, R., Kelly, R.G., Christiaen, L., Levine, M., Ziermann, J.M., Molnar, J.L., Noden, D.M., Tzahor, E., 2015. A new heart for a new head in vertebrate cardiopharyngeal evolution. *Nature* 520, 466–473.

Dorsky, R.I., Moon, R.T., Raible, D.W., 1998. Control of neural crest cell fate by the Wnt signalling pathway. *Nature* 396, 370–373.

Fabian, P., Crump, J.G., 2023. Reassessing the embryonic origin and potential of craniofacial ectomesenchyme. *Semin. Cell Dev. Biol.* 138, 45–53.

Ferrández-Roldán, A., Fabregà-Torrus, M., Sánchez-Serna, G., Duran-Bello, E., Joaquín-Lluís, M., Bujosa, P., Plana-Carmona, M., Garcia-Fernàndez, J., Albalat R., Cañestro, C., 2021, Cardiopharyngeal deconstruction and ancestral tunicate sessility. *Nature* 599, 431–435.

Ferrández-Roldán, A., Martí-Solans, J., Cañestro, C., Albalat, R., 2019. Oikopleura dioica: an emergent chordate model to study the impact of gene loss on the evolution of the mechanisms of development. *Results Probl. Cell Differ.* 68, 63–105.

Gans, C., Northcutt, R.G., 1983. Neural crest and the origin of vertebrates: a new head. *Science* 220, 268–273.

Garcia-Fernández, J., Holland, P.W.H., 1994. Archetypal organization of the amphioxus Hox gene cluster. *Nature* 370, 563–566.

Garstang, W., 1928a. The origin and evolution of larval forms. *Rep. Br. Assoc. Adv. Sci.* 1928, 77–98.

Garstang, W., 1928b. The morphology of the Tunicata, and its bearings on the phylogeny of the Chordata. *Q. J. Microsc. Sci.* 72, 51–187.

Hashimoto, H., Robin, F.B., Sherrard, K.M., Munro, E.M., 2015. Sequential contraction and exchange of apical junctions drives zippering and neural tube closure in a simple chordate. *Dev. Cell* 30, 241–255.

Haupaix, N., Abitua, P.B., Sirour, C., Yasuo, H., Levine, M., Hudson, C., 2014. Ephrin mediated restriction of ERK1/2 activity delimits the number of pigment cells in the Ciona *CNS. Dev. Biol.* 394, 170–180.

Havenhand, J.N., Matsumoto, G.I., Seidel, E., 2006. Megalodicopia hians in the Monterey submarine canyon: distribution, larval development, and culture. *Deep Sea Res. Part I Oceanogr. Res. Pap.* 53, 215–222.

Holland, N.D., Chen, J., 2001. Origin and early evolution of the vertebrates: new insights from advances in molecular biology, anatomy, and palaeontology. *Bioessays* 23, 142–151.

Holland, N.D., Holland, L.Z., Holland P.W., 2015. Scenarios for the making of vertebrates. *Nature* 520, 450–455.

Horie, R., Hazbun, A., Chen, K., Cao, C., Levine, M., Horie, T., 2018. Shared evolutionary origin of vertebrate neural crest and cranial placodes. *Nature* 560, 228–232.

Horie, T., Shinki, R., Ogura, Y., Kusakabe, T.G., Satoh, N., Sasakura, Y., 2011. Ependymal cells of chordate larvae are stem-like cells that form the adult nervous system. *Nature* 469, 525–528.

Ikuta, T., Yoshida, N., Satoh, N., Saiga, H., 2004. Ciona intestinalis Hox gene cluster: its dispersed structure and residual colinear expression in development. *Proc. Natl. Acad. Sci. U.S.A.* 101, 15118–15123.

Imai, J.H., Meinertzhagen, I.A., 2007a. Neurons of the ascidian larval nervous system in Ciona intestinalis: II. Peripheral nervous system. *J. Comp. Neurol.* 501, 335–352.

Imai, J.H., Meinertzhagen, I.A., 2007b. Neurons of the ascidian larval nervous system in Ciona intestinalis: I. Central nervous system. *J. Comp. Neurol.* 501, 316–334.

Imai, K.S., Satou, Y., Imai, H., Satoh, N., 2004. Gene expression profiles of transcription factors and signaling molecules in the ascidian embryo: towards a comprehensive understanding of gene networks. *Development* 131, 4047–4058.

Jiang, A., Han, K., Wei, J., Su, X., Wang, R., Zhang, W., Liu, X., Qiao, J., Liu, P., Liu, Q., Zhang, J., Zhang, N., Ge, Y., Zhuang, Y., Yu, H., Wang, S., Chen, K., Lu, W., Xu, X., Yang, H., Fan, G., Dong, B., 2024. Spatially resolved single-cell atlas of ascidian endostyle provides insight into the origin of vertebrate pharyngeal organs. *Sci. Adv.* 10, eadi9035.

Jiang, D., Tresser, J.W., Horie, T., Tsuda, M., Smith, W.C., 2005. Pigmentation in the sensory organs of the ascidian larva is essential for normal behavior. *J. Exp. Biol.* 208, 433–438.

Kaijiwara, S., Yoshida, M., 1985. Changes in behavior and ocellar structure during the larval life of solitary ascidians. *Biol. Bull.* 169, 565–577.

Kobayashi, M., Takatori, N., Nakajima, Y., Kumano, G., Nishida, H., Saiga, H., 2010. Spatial and temporal expression of two transcriptional isoforms of Lhx3, a LIM class homeobox gene, during embryogenesis of two phylogenetically remote ascidians, Halocynthia roretzi and Ciona intestinalis. *Gene Expr. Patterns* 10, 98–104.

Koontz, A., Urrutia, H.A., Bronner, M.E., 2023. Making a head: neural crest and ectodermal placodes in cranial sensory development. *Semin. Cell Dev. Biol.* 138, 15–27.

Krishnan, J., Rohner, N., 2017. Cavefish and the basis for eye loss. *Phil. Trans. R. Soc. B* 372, 20150487.

Kumano, G., Nishida, H., 2007. Ascidian embryonic development: an emerging model system for the study of cell fate specification in chordates. *Dev. Dyn.* 236, 1732–1747.

Kumano, G., Nishida, H., & Satoh, N. (2006). Overlapping expression of FoxA and Zic confers responsiveness to FGF signaling to specify notochord in ascidian embryos. *Dev. Biol.* 300(2), 770–784.

Le Douarin, N.M., Dupin, E., Baroffio, A., Dulac, C., 1992. New insights into the development of neural crest derivatives. *Int. Rev. Cytol.* 138, 269–314.

Leal, F., Cohn, M.J., 2016. Loss and re-emergence of legs in snakes by modular evolution of Sonic hedgehog and HOXD enhancers. *Curr. Biol.* 26, 2966–2973.

Lemaire, L.A., Cao, C., Yoon, P.H., Long, J., Levine, M., 2021. The hypothalamus predates the origin of vertebrates. *Sci. Adv.* 7, eabh2734.

Louchart, A., Viriot, L., 2011. From snout to beak: the loss of teeth in birds. *Trends Ecol. Evol.* 26, 663–673.

Lowe, C.J., Terasaki, M., Wu, M., Freeman Jr, R.M., Runft, L., Kwan, K., Haigo, S., Aronowicz, J., Lander, E., Gruber, C., Smith, M., Kirschner, M., Gerhart, M., 2006. Dorsoventral patterning in hemichordates: insights into early chordate evolution. *PLoS Biol* 4, e921.

Madgwick, A., Magri, M.S., Dantec, C., Gailly, D., Fiuza, U.-M., Guignard, L., Hettinger, S., Gomez-Skarmeta, J.L., Lemaire, P., 2019. Evolution of embryonic cis-regulatory landscapes between divergent Phallusia and Ciona ascidians. *Dev. Biol.* 448, 71–87.

Marlétaz, F., Timoshevskaya, N., Timoshevskiy, V.A., Parey, E., Simakov, O., Gavriouchkina, D., Suzuki, M., Kubokawa, K., Brenner, S., Smith, J.J., Rokhsar, D.S., 2024. The hagfish genome and the evolution of vertebrates. *Nature* 627, 811–820.

Mazet, F., Hutt, J.A., Milloz, J., Millard, J., Graham, A., Shimeld, S.M., 2005. Molecular evidence from Ciona intestinalis for the evolutionary origin of vertebrate sensory placodes. *Dev. Biol.* 282, 494–508.

Miao, Y., Pourquié, O., 2024. Cellular and molecular control of vertebrate somitogenesis. *Nat. Rev. Mol. Cell Biol.* 25, 517–533.

Motohashi, T., Kunisada, T., 2015. Extended multipotency of neural crest cells and neural crest-derived cells. *Curr. Top. Dev. Biol.* 111, 69–95.

Nakashima, K., Yamada, L., Satou, Y., Azuma, J.I., Satoh, N., 2004. The evolutionary origin of animal cellulose synthase. *Dev. Genes Evol.* 214, 81–88.

Nanglu, K., Lerosey-Aubril, R., Weaver, J.C., Ortega-Hernández, J., 2023. A mid Cambrian tunicate and the deep origin of the ascidiacean body plan. *Nat. Commun.* 14, 3832.

Papadogiannis, V., Pennati, A., Parker, H.J., Rothbächer, U., Patthey, C., Bronner, M.E., Shimeld, S.M., 2022. Hmx gene conservation identifies the origin of vertebrate cranial ganglia. *Nature* 605, 701–705.

Piotrowski, T., Baker, C.V.H., 2014. The development of lateral line placodes: taking a broader view. *Dev. Biol.* 389, 68–81.

Pisani, D., Pett, W., Dohrmann, M., Feuda, R., Rota-Stabelli, O., Philippe, H., Lartillot, N., Wörheide, G., 2015. Genomic data do not support comb jellies as the sister group to all other animals. *Proc. Natl. Acad. Sci. U.S.A.* 112(50), 15402–15407.

Racioppi, C., Kamal, A.K., Razy-Krajka, F., Gambardella, G., Zanetti, L., di Bernardo, D., Sanges, R., Christiaen, L.A., Ristoratore, F., 2014. Fibroblast growth factor signalling controls nervous system patterning and pigment cell formation in Ciona intestinalis. *Nat. Commun.* 5, 4830.

Romer, A.S., 1967 Major steps in vertebrate evolution. *Science* 158 1629–1637.

Sasakura, Y., Nakashima, K., Awazu, S., Matsuoka, T., Nakayama, A., Azuma, J.I., Satoh, N., 2005. Transposon-mediated insertional mutagenesis revealed the functions of animal cellulose synthase in the ascidian Ciona intestinalis. *Proc. Natl. Acad. Sci. U.S.A.* 102, 15134–15139.

Sasakura, Y., Ogura, Y., Treen, N., Yokomori, R., Park, S.J., Nakai, K., Saiga, H., Sakuma, T., Yamamoto, T., Fujiwara, S., Yoshida, K., 2016. Transcriptional regulation of a horizontally transferred gene from bacterium to chordate. *Proc. R. Soc. B* 283, 20161712.

Satoh, N., Rokhsar, D., Nishikawa, T., 2014. Chordate evolution and the three-phylum system. *Proc. R. Soc. B* 281, 20141729.

Satou, Y., Imai, K.S., Satoh, N., 2004. The ascidian Mesp gene specifies heart precursor cells. *Development* 131, 2533–2541.

Satou, Y., Nakamura, R., Yu, D., Yoshida, R.E., Hamada, M., Fujie, M., Hisata, K., Takeda, H., Satoh, N., 2019. A nearly complete genome of Ciona intestinalis type A (C. robusta) reveals the contribution of inversion to chromosomal evolution in the genus Ciona. *Genome Biol. Evol.* 11, 3144–3157.

Schlosser, G., 2021. *Evolutionary origin of sensory and neurosecretory cell types*. Vertebrate cranial placodes, Vol. 2. CRC Press.

Seo, H.C., Kube, M., Edvardsen, R.B., Jensen, M.F., Beck, A., Spriet, E., Gorsky, G., Thompson, E.M., Lehrach, H., Reinhardt, R., Chourrout, D., 2001. Miniature genome in the marine chordate Oikopleura dioica. *Science* 294, 2506.

Shu, D.G., Conway Morris, S., Han, J. Zhang, Z.F., Yasui, K., Janvier, P., Chen, L., Zhang, X.L., Liu, J.N., Li, Y., Liu, H.Q., 2003. Head and backbone of the Early Cambrian vertebrate Haikouichthys. *Nature* 421, 526–529.

Shu, D.G., Luo, H.L., Conway Morris, S., Zhang, X.L., Hu, S.X., Chen, L., Han, J., Zhu, M., Li, Y., Chen, L.Z., 1999. Lower Cambrian vertebrates from south China. *Nature* 402, 42–46.

Soldatov, R., Kaucka, M., Kastriti, M.E., Petersen, J., Chontorotzea, T., Englmaier, L., Akkuratova, N., Yang, Y., Häring, M., Dyachuk, V., Bock, C., Farlik, M., Piacentino, M.L., Boismoreau, F., Hilscher, M.M., Yokota, C., Qian, X., Nilsson, M., Bronner, M.E., Croci, L., Hsiao, W.Y., Guertin, D.A., Brunet, J.-F., Consalez, G.G., Ernfors, P., Fried, K., Kharchenko, P.V., Adameyko, I., 2019. Spatiotemporal structure of cell fate decisions in murine neural crest. *Science* 364, eaas9536.

Stewart, T.A., Lemberg, J.B., Taft, N.K., Shubin, N.H., 2019. Fin ray patterns at the fintolimb transition. *Proc. Natl. Acad. Sci. U.S.A.* 117, 1612–1620.

Stolfi, A., Gainous, T.B., Young, J.J., Mori, A., Levine, M., Christiaen, L., 2010. Early chordate origins of the vertebrate second heart field. *Science* 329, 565–568.

Stolfi, A., Ryan, K., Meinertzhagen, I.A., Christiaen, L., 2015. Migratory neuronal progenitors arise from the neural plate borders in tunicates. *Nature* 527, 371– 374.

Streit, A., 2018. Specification of sensory placode progenitors: signals and transcription factor networks. *Int. J. Dev. Biol.* 62, 195–205.

Stundl, J., Desingu Rajan, A.R., Urrutia, H.A., Leyhr, J., Stundlova, J., Solovieva, T., Haitina, T., Sanchez, S., Musilova, Z., Martik, M.L., Bronner, M.E., 2025. Acquisition of neural crest promoted thyroid evolution from chordate endostyle. *Sci. Adv.* 11, eadv2657.

Telford, M.J., Copley, R.R., 2011. Improving animal phylogenies with genomic data. *Trends Genet.* 27, 186–195.

Todorov, L.G., Oonuma, K., Kusakabe, T., Levine, M.S., Lemaire, L.A., 2024. Neural crest lineage in the protovertebrate model Ciona. *Nature* 635, 912–916.

Tsuda, M., Sakurai, D., Goda, M., 2003. Direct evidence for the role of pigment cells in the brain of ascidian larvae by laser ablation. *J. Exp. Biol.* 206, 1409–1417.

Wada, H., Satoh, N., 1994. Details of the evolutionary history from invertebrates to vertebrates, as deduced from the sequences of 18S rDNA. *Proc. Natl. Acad. Sci. U.S.A.* 91, 1801–1804.

Wada, S., Katsuyama, Y., Sato, Y., Itoh, C., Saiga, H., 1996. Hroth, an orthodenticle related homeobox gene of the ascidian Halocynthia roretzi: its expression and putative roles in axis formation during embryogenesis. *Mech. Dev.* 60, 59–71.

Wang, W., Niu, X., Stuart, T., Jullian, E., Mauck, W.M. 3rd, Kelly, R.G., Satija, R., Christiaen, L., 2019. A single-cell transcriptional roadmap for cardiopharyngeal fate diversification. *Nat. Cell Biol.* 21, 674–686.

Yong, L.W., Lu, T.M., Tung, C.H., Chiou, R.J., Li, K.L., Yu, J.K., 2021. Somite compartments in Amphioxus and its implications on the evolution of the vertebrate skeletal tissues. *Front. Cell Dev. Biol.* 9, 607057.

Index

Pages in *italics* refer to figures.

A

B

C

D

E

F

M

N

O

P

For Product Safety Concerns and Information please contact our
EU representative GPSR@taylorandfrancis.com Taylor & Francis
Verlag GmbH, Kaufingerstraße 24, 80331 München, Germany